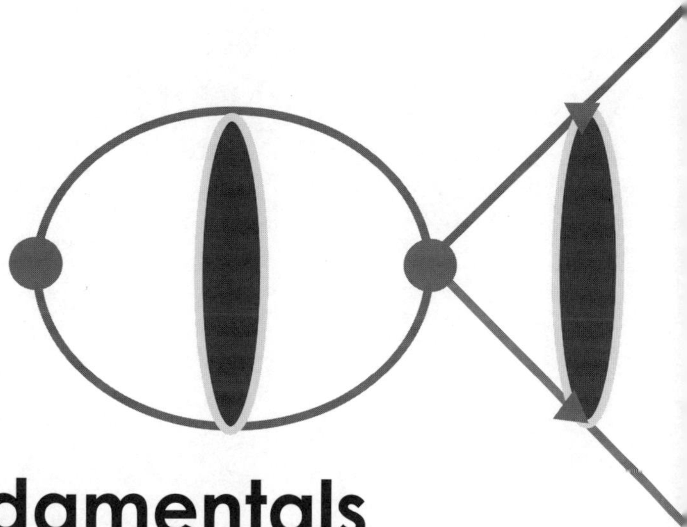

# Fundamentals in Hadronic Atom Theory

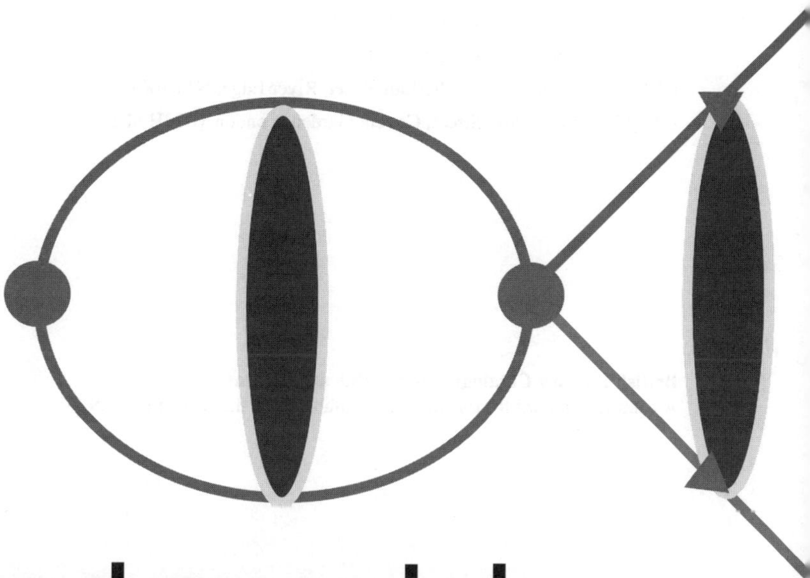

# Fundamentals in Hadronic Atom Theory

**A. Deloff**

*Soltan Institute for Nuclear Studies, Warsaw, Poland*

World Scientific
New Jersey • London • Singapore • Hong Kong

*Published by*

World Scientific Publishing Co. Pte. Ltd.
5 Toh Tuck Link, Singapore 596224
*USA office:* Suite 202, 1060 Main Street, River Edge, NJ 07661
*UK office:* 57 Shelton Street, Covent Garden, London WC2H 9HE

**British Library Cataloguing-in-Publication Data**
A catalogue record for this book is available from the British Library.

**FUNDAMENTALS IN HADRONIC ATOM THEORY**



ISBN 981-238-371-9

Printed in Singapore by World Scientific Printers (S) Pte Ltd

# Preface

The purpose of this book is to describe some of the major advances in our understanding of the fundamental properties of hadronic atoms and this text is an outgrowth of our personal research in the field over the years. The book has two parts: the first part is devoted to methods and the second part contains applications and a brief survey of the experimental situation. It has been assumed that the reader has a good knowledge of quantum mechanics and has completed a first course on quantum field theory.

Since our primary intention was to present in a unified and integrated manner the foundations of the macroscopic approach, we have stayed in principle within the conventional picture of hadronic atoms based on a two-body model Hamiltonian in which all strong interaction effects have been simulated by an absorptive effective potential representing the complicated interaction between the hadron and the nucleus. Thus, strictly speaking we present here a quantum mechanical theory of the hadronic hydrogen as in this case the two-body model happens to be exact. It is understood that for heavier elements all many-body complications have been deferred as they are supposed to be accounted for in the derivation of the effective potential which in general can be expected to be an immensely complex object. In this book, however, the problem of a microscopic derivation of the effective potential has not been our primary concern. Instead, we adopt a pragmatic philosophy that one has to explore first a simple macroscopic prototype, especially that the phenomenological approach has had considerable successes in reproducing the hadronic atoms data, using effective potentials of the simplest possible form. Nevertheless, it should be kept in mind that some important many-body aspects of the problem might be missed in result of this simplification.

The introduction of a complex nuclear potential into the description of a hadronic atom has two observable effects: a shift in the energy of the otherwise hydrogen-like level position, and a broadening of the atomic level arising from the fact that the system can now decay through the absorptive strong interaction. When the measured level energies and widths in hadronic atoms are identified with the complex eigenvalues of the the model Hamiltonian the calculated spectrum could be directly confronted with experiment. However, since the underlying Hamiltonian is a non-selfadjoined operator the calculation of the spectrum is a non-trivial task, and, in particular, the usual quantum mechanical rules such as orthogonality, normalization and completeness are no longer valid. Since this point has received little attention in the literature we demonstrate explicitly how the standard quantum mechanical scheme could be extended in order to incorporate the above rules in a consistent way. Having established the calculational scheme, we can use the powerful analytic methods developed long time ago for studying the singularities of the S-matrix in potential scattering. These methods are directly applicable to the problem at issue since the task of finding bound states corresponding to hadronic atoms is equivalent to locating the poles of the appropriate S-matrix in the upper part of the complex momentum plane (k-plane). The presence of the Coulomb potential is here a major complication since the origin of the k-plane becomes a highly singular point where the S-matrix has both a logarithmic branch point, which makes necessary introducing a branch cut, and an essential singularity. These singularities appear as a consequence of the infinite range of the Coulomb potential and of crucial importance for understanding the hadronic atoms phenomena. Also this point does not seem to have received enough appreciation in the literature.

In order to get further insight into the fundamental properties of hadronic atoms, we consider a quantum mechanical two-potential problem in which all complications have been stripped down to essentials. So, we have deliberately disregarded many possible extensions which might complicate the problem in an uninteresting way assuming that the absorptive nuclear potential is local, central, energy independent and has a short range, i.e comparable with the nuclear sizes. Although at first sight this might have appeared as a severe limitation, yet we argue that this model contains the basic ingredients necessary to capture much of the important physics, and, in particular it provides means to examine certain phenomena of considerable theoretical interest arising as a result of mixing between the atomic and the nuclear effects. Also, as we shall see, many of the extensions we choose to leave out could be easily re-introduced once the more fundamental questions have been resolved. On the formal side, when the nuclear potential is supplemented with the hadron-nucleus Coulomb interaction, one has to cope with a quantum mechanical bound state problem which is sufficiently complex to exhibit a highly non-trivial dynamical structure. Consequently, our second objective was to assemble all the necessary mathematical tools required to take on this complicated problem.

The actual nuclear potentials are known to be quite strong so that in general one has to consider situations when they might be capable of supporting bound states of their own. In addition to the atomic spectrum one would then expect to have nuclear bound states but since binding results from a collaborative action of two potentials we prefer to call them *quasi-nuclear* bound states: they become identical with the usual nuclear bound states when the Coulomb interaction is turned-off. In contrast with the atomic states, these quasi-nuclear bound states can be expected to have much bigger binding, commeasurable with the energy scale typical for strong interactions. On the other hand, it is well known that the atomic level shifts could also be quite large and it is far from obvious how it might be possible to distinguish the atomic states from the quasi-nuclear states. In this book we show that this could be done by investigating pole trajectories of the associated S-matrix. We believe that a clear-cut ramification of the spectrum of a two-potential Hamiltonian is a new development of considerable importance. To amplify this point, suppose that we have located a number of bound states in our two-potential problem, i.e we have located a number of poles of the corresponding S-matrix in the k-plane. When the depth of the nuclear potential is varied these poles will be moving in the k-plane along certain trajectories and in the limit of vanishing depth the poles corresponding to atomic states should resume unperturbed Coulomb positions on the positive half of the imaginary axis. If this does not happen, we have to do with a quasi-nuclear bound state. In the latter case, when the nuclear attraction falls below certain minimal value, the bound state pole travels towards the negative real axis and finally crosses to the lower half of the k-plane. When this happens, the pole does not represent a bound state anymore. Such behavior is typical for a bound state pole generated by a short-ranged potential as has been well documented in the literature. Although the quasi-nuclear bound states appear to have similar features as the nuclear bound states but the presence of Coulomb potential introduces one important difference. The central point here is a remarkable saturation theorem, which to the best of our knowledge has not been presented in the literature. This theorem states that given a *real* nuclear interaction plus attractive Coulomb potential, then in such a two-potential problem we can have nothing else but the atomic spectrum and the bound states are in one-to-one correspondence with the unperturbed Coulomb levels, no matter how strong is the nuclear attraction. In this case the quasi-nuclear bound states simply do not exist. The point is that for real potentials all bound state poles must necessarily reside on the positive imaginary axis but the poles generated by the nuclear interaction are not allowed to enter the upper half of the k-plane. When the nuclear attraction increases the poles in the lower half of the k-plane approach to the origin, exactly as in the Coulomb-free case, but in the two-potential case passing through the origin is impossible without hitting the Coulomb singularities. Since the trajectory never enters the positive imaginary axis it is interesting to know what does the trajectory do when the attraction is increased arbitrarily. It turns out that by increasing the nuclear attraction the trajectory is pushed away from the origin and through the Coulomb cut enters another Riemann sheet. This picture may change providing that the absorption exceeds critical level. Only then the bound state pole, not confined then to the imaginary axis, is given the chance to evade the singular point at the origin. To generate a quasi-nuclear bound state, both the nuclear attraction and the absorption must exceed certain minimal values. Formal considerations have been illustrated by numerous explicit examples showing the behavior of pole trajectories when both the nuclear attraction and the absorption are varied. Pole trajectories for local potentials have been studied in several papers but the addition of the Coulomb potential introduces new singularities changing qualitatively the behavior of the pole trajectories and this case so far does not seem to have been presented in the literature. The quasi-nuclear bound states might well exist in nature and one such example is the antiprotonic-hydrogen where the possible existence of bound antinucleon-nucleon or baryonium states has been advocated by a number of authors using both nuclear and quark models. The strongly bound baryonium states are expected to have large binding energies so that they are markedly different from the bound states corresponding to atomic levels of antiprotonic hydrogen.

When the attraction of the nuclear potential increases the atomic spectrum becomes strongly distorted and this gives rise to many interesting effects. For real nuclear potentials the key role plays an utmost important no-crossing theorem which states that all bound states poles of the S-matrix are necessarily simple, so that two pole trajectories must never cross in the upper half of the k-plane. The atomic binding energies when plotted as functions of the nuclear potential strength, in order to avoid crossing rearrange themselves in a complicated way displaying a collective step-like behavior. This phenomenon which is known as Zeldovich effect is shown here to be a consequence of the saturation and the no-crossing theorems. Since the formation of a quasi-nuclear level is prohibited, when the nuclear attraction is sufficiently large, the atomic levels behave the same way as nuclear level would. For complex nuclear potentials the no-crossing theorem is no longer valid and crossing of two pole trajectories in the upper part of the k-plane becomes possible which has profound consequences. At the crossing point the S-matrix has a second order pole and the corresponding two eigenfunctions become identical. Although we have than one function less but we show that the completeness relation still holds if the set of eigenfunctions is supplemented with the derivative with respect to $k$ of the eigenfunction corresponding to the crossing point eigenvalue. The other important point is that the eigenmomentum is a non-analytic function of the potential strength, and, in particular, has a branch point singularity at the critical value of the strength which gives rise to crossing. As a result, in the vicinity of this critical strength both the binding energy and the level width show a rapid variation, a phenomenon observed a long time ago by Krell and known as Krell oscillations.

The two-potential model considered in this book is relevant also in many other problems, in particular in the analysis of the low energy scattering data. Because the standard Coulomb wave functions have complicated analytic properties in the complex k-plane the calculations of the low-energy scattering parameters have been cumbersome, often involving unjustified approximations. In consequence, many simple results have been obscured by technical aspects and this important practical problem has been avoided in the textbooks devoted to scattering theory. The introduction of a different set of Coulomb wave functions which are analytic in $k^2$ turns fuzziness into clarity. In particular, these functions may be expanded in powers of $k^2$, which allows to calculate very easily the scattering lengths and effective ranges for all angular momenta. We believe that an up-to-date presentation of this very old topic will be useful in practical applications.

Of course, a book on hadronic atoms would have been incomplete without reference to the modern developments based on Chiral Perturbation Theory. Admittedly, this promising approach is not in the mainstream of our exposition but we have appended a Chiral Theory Primer just to give the reader some feeling of what has been going in this area. This is a rather fashionable trend these days with a large outflux of papers which deserve a separate survey.

It is our experience, that a heavy load of formalism is much easier digested when a rapid access to a calculational stage has been appreciated. We would like to believe, that this text will be useful for a practitioner in the field looking for a complete box of tools so we have presented a variety of different methods which could be immediately used to solve the complex eigenvalue problem, testing numerically in each case their performance. In addition to that, we have attempted to make the presentation as self-contained as possible by covering the technical details and mathematical techniques which are necessary to solve the problems discussed in the text but with which the reader might be unfamiliar.

As mentioned above, in the second part of this book we make contact with experiment. We thought that it would be convenient to have the full hadronic atoms data base in one place. Therefore, we have taken the trouble of tabulating the experimentally available strong interaction level shifts and widths data. In our survey of the experimental material we start with the simplest two-body systems like the two-meson atomic bound states and the hadronic hydrogen atoms. Then, we turn to the much more complicated three-body system of the hadronic deuterium. In this case a microscopic treatment involves a Faddeev approach. Finally, we consider all the remaining heavier atomic species for which the microscopic approach becomes prohibitively difficult and therefore the analysis must be based on the optical, or effective potential concept.

Although we have drawn freely from many different sources, we make no pretense for completeness. Therefore, it is not unlikely that some important subjects might have been omitted; after all, the selection of topics thought to be fundamental obviously must be biased by personal prejudices of the author.

The organization of this book is as follows. We begin with a quick overview of the subject in the next chapter. Then, in Chapter 2, we argue that with complex potentials the standard quantum mechanical scalar product has to be modified by taking as the "bra" states the time reversed eigenfunctions. They replace the usual complex conjugate eigenfunctions. This extension allows to sustain orthogonality of the eigenfunctions but the norm integral becomes a complex number which can take any value and in particular may vanish. Chapter 3 contains a modern survey of Coulomb wave functions whose arguments have been extended into the complex domain. In Chapter 4 we discuss the Coulomb resolvent and the different scattering operators. Chapter 5 summarizes the basic results of the formalism describing two-potential scattering that are needed in the remaining parts of this work. A thorough discussion of the effective range approximation is presented in Chapter 6. We give explicit expressions for the scattering lengths and effective ranges for arbitrary angular momenta. Also in Chapter 6 we formulate three important theorems which control the motion of the poles of the S-matrix. We investigate the behavior of the pole trajectories both analytically and numerically. The atomic spectrum is considered in Chapter 7. We demonstrate that the general theorems formulated in Chapter 6 provide explanation of both, the Zeldovich and the Krell effect. Formal considerations are illustrated by numerous examples. We conclude this chapter discussing various forms of the small shift approximations which generalize the famous Deser–Trueman formula and we examine their accuracy. Chapter 8 addresses the role of Gamow states and shows how to normalize these states. These results are subsequently used to prove that even in the case of crossing of two pole trajectories in the upper half of the $k$ plane the solutions of the wave equation still obey a generalized completeness relation. In Chapter 9 we present an *ab initio* derivation of the total X-ray transition rate taking into account the nuclear interaction. We obtain the "line shape" and provide an estimate of the corrections due to the nuclear interaction. Chapter 10 is devoted to computational methods of solving the complex eigenvalue problem. We discuss in detail the matching method, the variational techniques, Fredholm theory and the momentum space approach. In each case we provide numerical examples testing the performance of the method. The general formalism developed in this book has been also illustrated in Chapter 11 by a number of simple models where analytic solutions could be obtained. In Chapter 12, concluding Part I, we give a pedestrian introduction to the Chiral Perturbation Theory.

In Part II theory is confronted with experiment. In Chapter 13 we discuss the two-meson atomic systems. The hadronic hydrogen atoms are analyzed in Chapter 14. The three-body techniques based on the Faddeev equation are presented in Chapter 15 and applied for analyzing the hadronic deuterium atoms. The optical potential concept is introduced in Chapter 16. We outline a formal derivation of the optical potential and discuss various semi-phenomenological forms that have been applied to fit the heavier ($A>4$) hadronic atoms data.

The author wishes to thank Professor R.H. Dalitz for his engagement in this project when it was in a formative stage, as well as for perceptive criticism and for making incisive comments on the early manuscript. This work has been supported, in part, by a KBN grant 5P03B04521.

# Contents

## Part II: Comparison with Experiment

# PART I
# Theoretical Background

# Chapter 1

# Hadronic Atoms – An Overview

When a negatively charged particle (different from electron) stopped in matter is captured into an atomic orbit we have to do with an exotic atom. In this book we shall be specifically interested in hadronic atoms, i.e. atomic systems in which one of the electrons has been replaced by a hadron. Nevertheless, most of the results presented below will remain valid for arbitrary exotic atoms. Since the captured particle besides the Coulomb force brings in also some other types of interactions with atomic nuclei, the investigation of exotic atoms provides a unique opportunity to study that interaction close to threshold. It is therefore not surprising that studies of hadronic atoms have had impact on a whole range of disciplines as diverse as physical chemistry, solid state physics, nuclear and elementary particle physics.

Because the hadrons are much more massive as compared with electrons, by a factor of roughly $10^3$, according to simple Bohr theory the Coulomb binding energies will go up by that factor while the radii of the classical atomic orbits will be decreased by the same amount. Thus, hadronic atoms could be envisaged as very small and tightly bound objects. As a result of scaling, the energies of the corresponding radiative transitions furnish the range appropriate for the X-rays. Typically, the transition energies from hadronic atoms span the interval between a few keV and a few MeV. The radii of the atomic orbits are proportional to $n^2$ where $n$ is the principal quantum number and for low orbital momenta become so small as to be comparable in size with the nuclear radii. Consequently, except perhaps for the very lightest elements, the hadron will be captured by the nucleus before ever reaching atomic levels with the lowest orbital momenta. Nuclear capture initiates a nuclear reaction with numerous exit channels and the hadron is likely to disappear altogether. By contrast with ordinary atoms a hadronic atom cannot be associated with a single stationary state. It should be rather envisaged as a complicated process which takes about $10^{-12}$ sec and in such a short time most of the hadrons can be regarded as stable particles. Yet, this time is also long enough to justify applying time independent formalism for describing processes caused by strong interactions with typical time scale of the order $10^{-18}$ sec.

The hadronic atom history from inception until the very end can be conveniently broken into three phases which we shall now describe briefly.

*(i) Capture phase.* A negatively charged particle passing through matter first looses energy by the ionization processes. Next, it will be slowed down further as a result of inelastic collisions with electrons whose velocities lie in the same range until the ultimate capture into a high atomic orbit can take place. The hadron then cascades down towards the nucleus visiting a series of its own atomic orbits and its energy is released by stripping the outer electrons (Auger effect), or by electric dipole radiative transitions (X-ray emission). The Auger effect wins this competition when there is a large overlap between electron and hadron wave functions, i.e. for large $n$ and small binding energies, so that it is most important in the upper part of the cascade. In quantitative terms the theory of the capture process is still incomplete and many questions remain unanswered.

*(ii) X-ray transitions.* When the hadron finds itself in orbits inside the classical radius of the electron K-shell there is no electron screening and as long as the nuclear interaction is not felt yet the problem at hand is essentially equivalent to that of a hydrogen-like atom. Although the energy levels of a hadronic atom cannot be determined directly but the energy differences are available from measurements of the X-ray transition energies. In the main sequence of X-ray transitions the Coulomb force is the dominant ingredient of the hadron-nucleus interaction so that the theoretical description of these transitions is highly accurate. Consequently, from the X-ray measurements it was possible to infer the values of some static quantities characterizing the orbiting hadron (such as mass, magnetic moment, *etc.*) just to make sure that they are in good agreement with values obtained in other experiments on a different way.

*(iii) Coulomb-nuclear interference.* Finally, the hadron is deposited in orbits so close to the nucleus that the nuclear interaction is no longer negligible and the latter is capable of shifting the atomic level from the position predicted by purely electro magnetic interactions. In addition to that the nuclear absorption reduces the lifetime of the last atomic state so that the X-ray line will be broadened. By conducting precise measurements of the

X-ray spectrum the energy shifts and the level widths associated with the last level can be directly measured. The nuclear capture affects also the one-before-last level but the effects are so small that it has not been possible to observe them directly. Nevertheless, since the radiative transition competes with the nuclear capture this causes a reduction of intensity feeding the last level what can be detected. Knowing the yield of the last X-ray transition one can often deduce the branching ratio of the transition probabilities of the two competing processes from which the width of the one-before-last level can be determined. The data for various hadronic atoms on the measured energy shifts and widths of their levels are now available for a whole range of nuclear species.

To be more precise, the energy shift $\epsilon$ is defined as a difference between the measured energy (i.e. the full energy) and the corresponding energy resulting from pure QED without strong interaction. With this sign convention positive (negative) $\epsilon$ would correspond to attractive (repulsive) forces providing that there was no absorption present. It should be stressed here that the strong interaction effects enter in a highly non-linear manner and in general cannot be evaluated perturbatively. Therefore, a negative shift may be also obtained with attractive potential if only the absorptive component of the potential is large enough. In some cases the complicated interplay between the real and imaginary component of the nuclear potential causes effects going against our intuition, e.g. an increase of absorption does not necessarily have to lead to an increase of the width and it is perfectly possible that the width might actually go down.

The atomic cascade together with some of the X-ray transitions are illustrated schematically on the Grotrian diagram displayed in Fig. 1 where we consider the simplest hadronic-hydrogen case. The atomic states are identified by the principal quantum number and the angular momentum $\ell$ with the usual notation where the letters $s, p, d, f, ...$ stand for the values of $\ell$ equal, respectively, $0, 1, 2, 3, ...$ . The different X-rays are identified by the n-value of the final state and all transitions feeding a given state are grouped into series where the letters $K, L, M, ...$ are used to denote $n = 1, 2, 3, ..$, and, similarly as optical spectroscopy, the $K, L$ and $M$ series have been called, respectively as Lyman , Balmer and Paschen series. The different transitions in a given series are labeled by the decrement of the principal quantum number $\Delta n$. It has been customary to use Greek subscripts for $\Delta n$ with $\alpha, \beta, \gamma, ...$ denoting $\Delta n = 1, 2, 3, ...$, respectively. So, in this notation $K_\alpha, K_\beta, K_\gamma$ lines correspond to $2p \to 1s, 3p \to 1s, 4p \to 1s$ transitions and $L_\alpha, L_\beta, L_\gamma$ lines to $3d \to 2p, 4d \to 2p, 5d \to 2p$ transitions *etc.* In Fig. 1

we show only the dominant electric dipole X-ray transitions in which the selection rule $|\Delta\ell| = 1$ implies that the angular momentum decreases by one between the initial and final state.

It should be noted that in the particular case of hadronic-hydrogen there is an additional complication due to Stark mixing of levels. The hadronic-hydrogen is an electrically neutral object of a very small size as compared with the hydrogen atom (e.g. kaonic-hydrogen is about thousand times smaller) and therefore can easily penetrate inside the comparatively large hydrogen atoms where it is exposed to a strong electric field of the proton. This leads to Stark mixing between states of different angular momentum increasing the probability of absorption from states which have low angular momenta but high $n$ values de-populating thereby the number of hadrons reaching the lowest orbits by radiative transitions. As a result of Stark mixing the hadronic-hydrogen X-ray yields are quite low, especially for liquid hydrogen targets where the density is high. As shown in Fig. 1, the nuclear interaction affects only the lowest levels. The $1s$ level is shifted by $\epsilon_{1s}$ with respect to the unperturbed Coulomb position, and, in addition to that, will also be broadened due to the absorptive interaction. Accordingly, in Fig. 1 the $1s$ level has been smeared and acquired a 'shape' characterized by the width $\Gamma_{1s}$. The shifts and widths of higher levels are usually so small that they cannot be observed directly with the currently available detectors. Nevertheless, the width of the $2p$ state can be determined indirectly from the intensity balance between the total yield of Balmer X-ray transitions feeding the $2p$ level and the yield of the $K_\alpha$ transition. We obtain then the following branching ratio

$$\frac{\Gamma_{2p\to 1s}}{\Gamma_{2p\to 1s} + \Gamma_{2p}} = \frac{\text{Yield}(K_\alpha)}{\sum_j \text{Yield}(L_j)},$$

where $\Gamma_{2p\to 1s}$ denotes the electro-magnetic $2p \to 1s$ total transition rate and $\Gamma_{2p}$ is the absorptive width of the $2p$ level. When the yields are known from experiment and $\Gamma_{2p\to 1s}$ is supplied from QED the width of the $2p$ level can be immediately deduced from the above formula.

Although the list of negatively charged hadrons which might be attached to nuclei is quite long, but so far the X-rays have been detected from $\pi^-$, $K^-$, $\Sigma^-$ and $\bar{p}$ atoms. A compilation of hadronic atom data can be found in [Batty (1982)] and in its recent update [Batty *et al.* (1997)] and these two reviews contain extensive references to the original experimental papers (cf. also [Horvath (1984)]). To measure X-rays from hadronic atoms a suitable low-momentum hadron beam is required. Then, a counter and

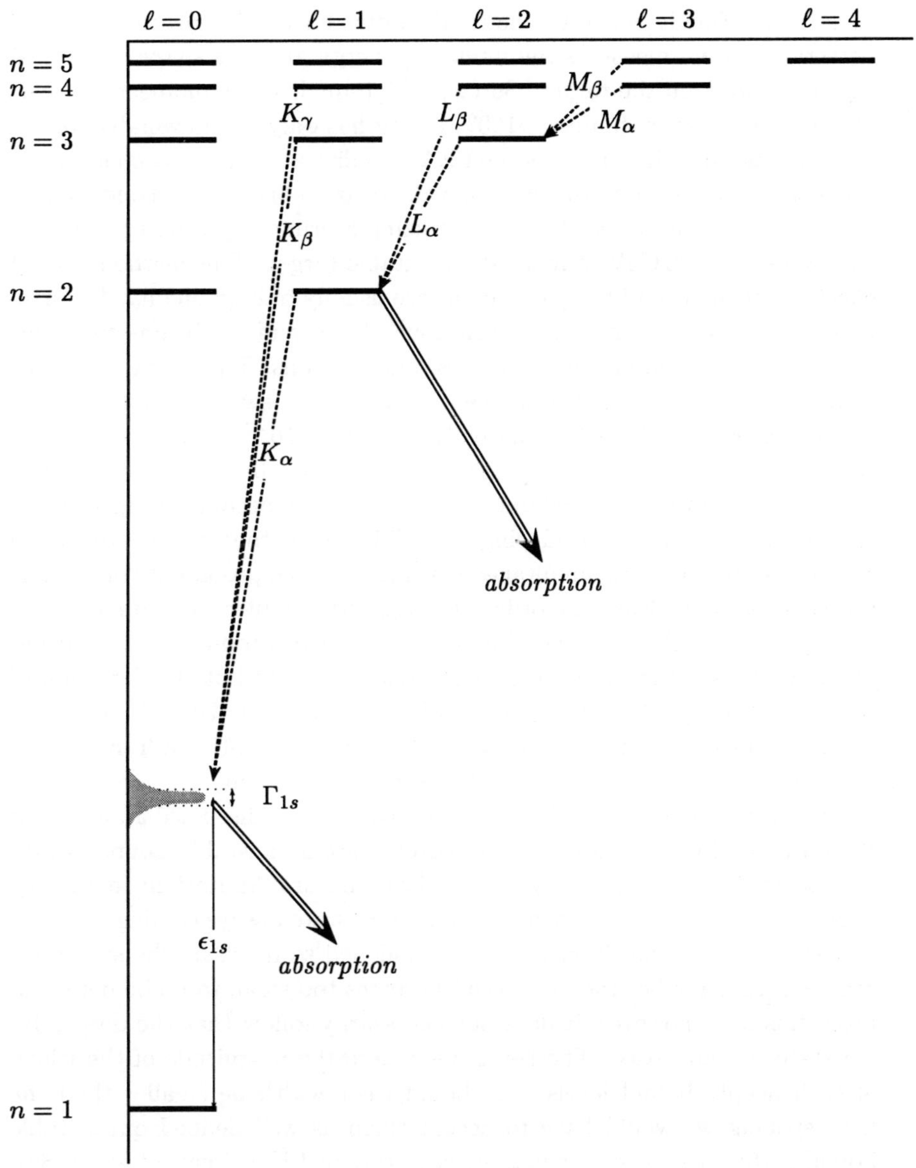


Fig. 1.1 Level scheme for a hadronic hydrogen showing some of the X-ray transitions.

a degrader system must be introduced to produce and identify particles stopping in the target selected as the host nucleus for the hadronic atom to be formed. Finally, when the hadronic atoms have been formed, an X-ray detection system measures the intensity, energy and line shape of the i X-rays emitted from the target. So far, a high-quality low-momentum beam of very small size with almost 100% purity has only been available for the antiprotons since the advent of the LEAR facility at CERN. By contrast, all other hadronic atom experiments have had to be based on secondary particle beams which are produced by utilizing a primary proton beam in the energy range 1–30 GeV to irradiate a suitable target. This method is most effective for pions and the pion beams are usually of high quality. Unfortunately, for kaons the intensities were much lower and the beams contained large contamination of other particles, mostly pions. The situation is even worse for the $\Sigma^-$ as the low-momentum hyperon beams are not available due to their very short lifetime of the order of $10^{-10}$ *sec.* Nevertheless, the $\Sigma^-$ atoms can be formed with the kaon beam because some fraction of the $K^-$ is always knocked out when these kaons become engaged in the reaction $K^-p \rightarrow \pi^+\Sigma^-$ producing $\Sigma^-$. This means that kaonic atoms and $\Sigma^-$ atoms are produced simultaneously although the probability of forming the latter is by at least one order of magnitude smaller. There have been attempts to use the secondary kaon interaction to produce the $\Xi^-$ atoms *via* a double strangeness exchange process $(K^-, K^+)$ but the detection of the X-rays from $\Xi^-$ atoms will be difficult (cf. [Gal (2001)]). One can speculate also about the $\Omega^-$ atoms and even more exotic hadronic atoms involving the anti charm and beauty degrees of freedom.

It is interesting to note that the radiative cascade discussed above is not the only possibility to investigate hadronic atoms unstable bound states. As a matter of fact, the X-ray cascade has a natural limitation: in a heavy nucleus it never descends down to the ground state always coming to a halt earlier at some higher level which is broadened so much by the strong interaction that further radiative yield becomes too small to be be detected. From this fact, however, it does not necessarily follow that the deeper lying states do not exist. The key issue is here the magnitude of the width of such deeply bound levels and should their width be smaller than the level spacing we would have to accept them as well defined quasi-stable bound states. Detailed calculations have revealed [Friedman *et al.* (1985); Toki *et al.* (1989)] that in the case of pionic atoms owing to favorable circumstances the widths of the deeply-bound states are indeed so narrow that they meet the above criteria which allow them to be regarded as quasi-stable states even for the $1s$ state in $^{208}$Pb. The reduction of the size of the width has been attributed to the repulsive component in the pion-nucleus potential pushing the pionic wave function outwards into the nuclear periphery where the absorptive potential is relatively weak. It has been suggested that to produce and observe such states pion transfer reactions such as $(n,p)$, $(d,{}^2\mathrm{He})$, $(n,d)$ and $(d,{}^3\mathrm{He})$ could be used in which pions, preferably with low-momentum, would be implanted directly into the nuclear interior. The theoretical predictions have been confirmed experimentally by the first successful observation of a $2p$ level in pionic $^{207}\mathrm{Pb}$ (cf. [Yamazaki *et al.* (1996)]). Undoubtedly, the discovery of these deeply-bound states is extremely important since they might provide information about the behavior of real pions in the nuclear environment.

# Chapter 2

# Extended Quantum Mechanical Framework

We shall be concerned with the problem in which a negatively charged hadron has been captured into an atomic orbit of a nucleus carrying a charge equal to $Z$ proton charges and we wish to concentrate our attention on the very last stage of a hadronic atom history when strong interaction effects become important. On a microscopic level we are immediately faced with a many-body and multi-channel situation. Since in its full complexity the problem appears to be prohibitively difficult a macroscopic approach has been proposed instead, based on a working hypothesis that a complex effective potential of a short range stipulates the underlying dynamics. This point of view has been also adopted in this book and little attempt is made at taking on the overriding issue of the validation of the macroscopic model. The derivation of the nuclear potential from a microscopic theory has been a long standing problem of paramount importance and will be considered in Part II of this book. However, for the time being, we choose to leave out this problem, regarding the nuclear potential as if it was supplied either from phenomenology, or derived from first principles. It is rather encouraging that on purely phenomenological level, this model has been very successful in a semi-quantitative description of a large bulk of hadronic atom data.

Hereafter, we shall be interested in the general properties of hadronic atoms but it is not our intention to fit a specific set of hadronic atom data (this will be done in Part II). Instead, we wish to pursue further the macroscopic approach by exploring all its consequences and for that purpose we consider a generic hadronic atom situation assuming a spherically symmetric non-relativistic Hamiltonian in which the point-like Coulomb potential has been supplemented by a short range nuclear potential of an unspecified shape. This potential is assumed to have a complex depth to simulate absorptive effects. As already mentioned in the Preface, the adopted model Hamiltonian has been deliberately stripped down to essentials but nonetheless we believe that it contains the most important physical ingredients. Indeed, in a hadronic atom the dominant role of the Coulomb force is indisputable and any compromise here leads to misconception. As far as nuclear interaction is concerned, we think that a minimal model at least must be able to simulate two basic qualitative features: firstly, that this interaction is of a short range, and, secondly, that it has an absorptive component. These features are apparent in all experiments and will certainly have to be accounted for at any higher level of sophistication. Since in the present study we make a number of simplifying assumptions we shall now discuss them in order.

*(i) Non-relativistic wave equation.* It is well known that in atomic physics relativistic effects play an important role and since the relativistic correction to the binding energy is proportional to the mass of the bound particle they appear to be equally important in the hadronic atom case. It has been argued that the hadronic atom problem requires a relativistic treatment and it has been a common practise to use for both spin zero and spin one-half hadrons the same Klein-Gordon equation as appropriate for the Coulomb problem fiddling an extra term containing the nuclear potential. Strictly speaking, this procedure is unjustified and ambiguous, but, in fact, resorting to Klein-Gordon equation has purely instrumental character, providing a convenient way of introducing relativistic corrections. The point is that in the actual fitting of hadronic atoms data the quantities of interest are the *differences* of energies calculated with and without the nuclear interaction in which the lions share of the relativistic corrections drops out as a result of cancellation. Therefore, the results are insensitive to the form of the adopted wave equation. It should be perhaps reminded here that the leading relativistic corrections to the Coulomb binding energy are of the order $\alpha^4 Z^4$ (where $\alpha$ is the fine structure constant) and require large $Z$ and low $\ell$ to become important. However, in contrast with ordinary atoms, in a heavy nucleus the absorption stops the X-ray cascade and states with low orbital momenta which are significantly affected by relativistic effects are not observed. For the last level the relativistic correction contributes typically no more that one or two percent of the total Coulomb energy and might be comparable in size with the shift caused by the strong interaction. The important point here is that the lowest order relativistic correction to the Coulomb binding energy can be very reliably calculated using the first order perturbation formula. Indeed, for the Dirac and the Klein-Gordon equations the correction of the order $\alpha^4 Z^4$ produced pertur-

batively is identical with the same order correction obtained by expanding the exact energy in powers of $\alpha^2 Z^2$. We think therefore, that it is theoretically much more satisfactory to apply non-relativistic quantum mechanics to the hadronic atom problem, calculating if need arises the relativistic corrections perturbatively. Thus, in our general discussion it appears fully justified to ignore all complications connected with relativity. This is the attitude adopted in this book and we make no attempt to go beyond the Schrödinger equation. With this choice there is less interpretational problems to be resolved and there is no danger that various artifacts connected with a particular relativistic equation could be misinterpreted as general properties of hadronic atoms. Furthermore, we did not want to get involved into a somewhat academic discussion whether the nuclear potential should be regarded as a Lorentz scalar or a fourth component of a vector.

*(ii) Electromagnetic corrections.* For simplicity, we suppress also all electro-magnetic corrections in our Hamiltonian but this does not mean that they are negligible. To be more precise, the most important here is the correction due to the extended charge distribution of the nucleus which modifies the Coulomb potential at small separations. The second correction is due to vacuum polarization and is also quite significant since hadronic orbits are located close to the nucleus where the hadrons are influenced by the strong electric field of the nucleus. Therefore, opposite to the atomic electron case, it is the vacuum polarization that is more important than the self-energy among the two lowest order radiative corrections. The vacuum polarization can be included by introducing a short ranged modification of the Coulomb potential (Uehling potential ). The corresponding expressions have been collected by [Fullerton *et al.*(1976)]. The common feature of the above electromagnetic corrections is their short-ranged nature, and as such, they can be lumped together with the nuclear potential into a single effective potential. Therefore, these corrections will never appear explicitly in this work.

*(iii) Local effective potential.* There is a number of good reasons to believe that, in general, the nuclear potential derived from a microscopic theory might be non-local, so that in coordinate representation the corresponding wave equation in the hadron-nucleus centre of mass frame, takes the form

$$\left(\nabla^2 + \frac{2\mu\alpha Z}{r} + k^2\right)\Psi(\mathbf{r}) = 2\mu \int \langle \mathbf{r}\,|V|\,\mathbf{r}'\rangle\, \psi(\mathbf{r}')\, \mathrm{d}^3 r', \tag{2.1}$$

where $\mu$ denotes the hadron-nucleus reduced mass (we also put hereafter

$\hbar = c = 1$). By definition, the potential $\langle \mathbf{r} \,|V|\, \mathbf{r}' \rangle$ is said to be non-local if it is non-diagonal in coordinate representation. When the potential is rotationally invariant, we can use the partial wave expansion

$$\langle \mathbf{r} \,|V|\, \mathbf{r}' \rangle = \frac{4\pi}{r\, r'} \sum_{\ell,m} V_\ell(r, r')\, Y_{\ell m}(\hat{\mathbf{r}})\, Y^*_{\ell m}(\hat{\mathbf{r}}'),$$

in which infinite number of kernels $V_\ell(r, r')$, in general different in each partial wave, has to be provided in order to determine completely a non-local potential. The potential is said to be local if all these kernels are diagonal and independent of $\ell$, i.e. when we have

$$V_\ell(r, r') = V(r)\, \delta(r - r').$$

In momentum representation the local potential is a function of the momentum transfer only, i.e. $\langle \mathbf{k}|V|\mathbf{k}' \rangle = \tilde{V}(|\mathbf{k} - \mathbf{k}'|)$ (where $\tilde{V}$ is the Fourier transform of $V$) and does not depend upon the individual momenta $\mathbf{k}$ or $\mathbf{k}'$. The smallest departure from locality takes place if in the potential finite powers of these momenta are are admitted which leads to the so called velocity dependent potentials. When the momenta $\mathbf{k}$ are regarded as operators in the coordinate space ($\mathbf{k} \to \mathrm{i}\nabla$), the kernels $V_\ell(r, r')$ are $\ell$ dependent and may involve differentiation operators. For this particular class of potentials, we can write

$$V_\ell(r, r') = \delta(r - r')\, \mathcal{V}_\ell(r, \frac{\mathrm{d}}{\mathrm{d}r}),$$

where $\mathcal{V}_\ell$ is a polynomial in the derivatives. From the formal point of view, the prospect to work with non-local potentials does not need to be a source of great concern as powerful numerical techniques have been developed to deal with such problems (cf. Chap. 11). On the other hand, in a semi-phenomenological approach our feel for the underlying physics is much better when local potential are used and over the years the tendency has been to avoid as much as possible non-local potentials. A notable exception is here the pionic atom case where the necessity to account for the strong p-wave $\pi N$ interaction has led to a non-local pion-nucleus potential [Kisslinger (1955)]. For pionic atoms, it has been customary to use the following ansatz [Krell *et al.* (1969)] in the wave equation

$$\Big[\nabla^2 + \frac{2\mu\alpha Z}{r} + k^2\Big]\Psi(\mathbf{r}) = \Big[q(\mathbf{r}) - \nabla\alpha_1(\mathbf{r})\nabla\Big]\Psi(\mathbf{r}), \tag{2.2}$$

where $q(\mathbf{r})$ and $\alpha_1(\mathbf{r})$ denote, respectively, the local and non-local part of the pion-nucleus potential. The non-locality of the potential enters through the gradient operators which are supposed to act on the pion wave function $\Psi(\mathbf{r})$. The advantageous feature of the above ansatz is that by means of the substitution

$$\Psi(\mathbf{r}) = \left[1 + \alpha_1(\mathbf{r})\right]^{-1/2} \Phi(\mathbf{r})$$

equation (2.2) can be cast to a form formally resembling the Schrödinger equation with a local potential $U(\mathbf{r})$, *viz*

$$\left[\nabla^2 + \frac{2\mu\alpha Z}{r} + k^2\right]\Phi(\mathbf{r}) = 2\mu U(\mathbf{r})\Phi(\mathbf{r}),$$

where $U(\mathbf{r})$ can be expressed in terms of $q(\mathbf{r})$ and $\alpha_1(\mathbf{r})$

$$2\mu U(\mathbf{r}) = \frac{1}{1+\alpha_1(\mathbf{r})}\left\{q(\mathbf{r}) + \alpha_1(\mathbf{r})k^2 + \tfrac{1}{2}\nabla^2\alpha_1(\mathbf{r}) - \frac{1}{1+\alpha_1(\mathbf{r})}\left[\tfrac{1}{2}\nabla\alpha_1(\mathbf{r})\right]^2\right\}.$$

In view of the above result, is seems that going beyond the local potential case at the present stage would be premature because in phenomenology either the hadron-nucleus potential has been adopted in a local form from the onset, or the potential is velocity dependent and non-local but in such case one may study an equivalent problem in which the potential is local.

*(iv) Spherical symmetry.* Our simplifying assumption that the problem has spherical symmetry cannot be always justified because the shapes of many nuclear species are known to be non-spherical. A typical example would be a permanently deformed nucleus whose density contains a static quadrupole component inducing additional electric quadrupole interaction. When such non-central term is added to the central potential, neither the nuclear spin $I$ nor the hadron's orbital momentum $\ell$ are good quantum numbers and the conserved quantity is the total angular momentum $\mathbf{F} = \mathbf{I} + \mathbf{L}$. In consequence, this leads to a hyperfine splitting of the unperturbed state $\ell$ (defined for the central part of the interaction) where the individual members of the multiplet are distinguished by the values of $F = \ell + I, \ell + I - 1, ..., |\ell - I|$. Actually, the resulting pattern of levels may be very complicated because the quadrupole term in the density induces also a non-central component in the nuclear potential which gives rise to additional shifts of the levels within the hyperfine multiplet. A detailed study of these phenomena, however, is outside the scope of this book. Since the hyperfine corrections to the energy are usually obtained by using perturbative methods, it will be sufficient to confine our attention to central

interactions.

*(v) Single channel approach.* Perhaps the least justified simplification in the model Hamiltonian is its single channel form. It is true that we endeavour to make up for it by explicitly introducing an absorptive component but the effective potential is still energy independent and in particular the information about the location of the inelastic thresholds has never been passed on to the Hamiltonian. Unfortunately, the many-channel extension requires introducing potential matrices involving large number of parameters such as threshold energies, strengths, ranges, *etc.* As a result the multichannel macroscopic Hamiltonian rapidly becomes over-parametrized and unless these parameters are supplied from a microscopic theory the problem turns out to be virtually inextricable. Some examples of multi-channel problems will be given in Part II.

Having explained the nature and the motivation of our assumptions we are ready now to put the hadronic atom problem on a formal basis. We assume that the underlying dynamics can be described by a standard two-body Schrödinger equation in the center of mass frame as appropriate for two spinless particles without any other internal degrees of freedom (cf. eq. (2.1)). The two potentials present in this equation are: the point-like Coulomb potential and the complex, short-ranged nuclear potential $V(r)$, where the latter is assumed to be spherically symmetric and local, i.e. in (2.1) we put $\langle \mathbf{r} | V | \mathbf{r}' \rangle = \delta(\mathbf{r} - \mathbf{r}')\, V(r)$. In consequence, the angular part of the wave function can be separated out in the usual way for each angular momentum $\ell$ and we are left with the following equation for the radial part $u_\ell(s,k,r)$

$$\left[\frac{\mathrm{d}^2}{\mathrm{d}r^2} - \frac{\ell(\ell+1)}{r^2} + \frac{2\mu\alpha Z}{r} + sv(r) + k^2\right] u_\ell(s,k,r) = 0. \tag{2.3}$$

In eq. (2.3) the strong interaction term $2\mu\, V(r)$ has been rewritten in a more convenient form as $-s\, v(r)$, where $v(r)$ is

$$v(r) = \frac{V(r)}{\displaystyle\int_0^\infty V(r) r\, \mathrm{d}r},$$

and for simplicity is assumed to be a real function. The nuclear potential in (2.3) has been scaled in such a way that its depth is specified by a complex dimensionless parameter $s$, defined as

$$s = -2\mu \int_0^\infty V(r) r\, \mathrm{d}r.$$

With the minus sign in front of the above integral and $v(r)$ positive the case $Im\, s > 0$ corresponds to an absorptive interaction. The essential feature of $v(r)$ is that it must have a short range, is less singular at $r = 0$ than $1/r$ and falls off faster than an exponential function for $r \to \infty$. The last assumption means that using phenomenological potentials with exponential fall off we should introduce a cut-off at some large value of $r$. It should be noted that this restriction does not result in a loss of generality because the additional singularities which we leave out by truncating the tail of the potential are located in the k-plane far away from the region of interest. In practical terms the cut-off can hardly be felt as it is impossible to tell the difference between a potential truncated at some large enough radius and one that has a smooth exponential fall off. Actually, the assumption that the real and imaginary part of the short range potential have the same shape is never needed except that it simplifies bookkeeping. As has been explained above, in our formulation we have deliberately disregarded many possible extensions which might complicate the problem in an uninteresting way. Nevertheless, we think that eq. (2.3) pertains to a fairly large class of realistic problems encountered in phenomenological description of the hadronic atoms data.

We shall be primarily interested in obtaining solutions of eq. (2.3) regular at $r = 0$ and for $r \to \infty$ satisfying the so called Siegert [Siegert (1939)] condition which admits only outgoing wave distorted by the Coulomb field. The boundary conditions are

$$\lim_{r \to 0} u_\ell(s, k, r) = r^{\ell+1}; \tag{2.4}$$

$$\lim_{r \to \infty} u_\ell(s, k, r) = \mathrm{e}^{\mathrm{i}(\rho - \eta \log 2\mathrm{i}\rho)}, \tag{2.5}$$

where $\rho = kr$ and $\eta = -\mu\alpha Z/k$. By solving eq. (2.3) with the boundary conditions (2.4)–(2.5) we obtain the eigenfunctions and the corresponding eigenvalues at which such solutions exist will be denoted as $k_n$ where $n = 1, 2, ...$ enumerates the eigenstates. In general, the resulting eigenvalues are complex numbers and they will correspond to bound states if $Im\, k_n \geq 0$, and, respectively, to Gamow states [Gamow (1928)] if $Im\, k_n < 0$. To find the location of the bound states in the complex k-plane let us consider eigenstates with $Im\, k_n > 0$ setting $k = k_n$ in eq. (2.3). Taking complex conjugate of eq. (2.3) with $k = k_{n'}$ we obtain second equation and multiplying (2.3) by $u_\ell^*$ and the conjugate equation by $u_\ell$, after subtracting and integrating over $r$, we have

$$(k_{n'}^{2*} - k_n^2) \int_0^\infty u_\ell(s^*, k_{n'}^*, r)^* \, u_\ell(s, k_n, r) \, dr = \\ = 2\,\mathrm{i}\,\mathrm{Im}\, s \int_0^\infty u_\ell(s^*, k_{n'}^*, r)^* \, v(r) \, u_\ell(s, k_n, r) \, dr. \tag{2.6}$$

When $n' = n$, (2.6) takes the form

$$2\,\mathrm{Re}\, k_n \, \mathrm{Im}\, k_n \int_0^\infty |u_\ell(s, k_n, r)|^2 \, dr = \\ = -\mathrm{Im}\, s \int_0^\infty |u_\ell(s, k_n, r)|^2 \, v(r) \, dr, \tag{2.7}$$

which shows that for absorptive potentials bound states can be located in the II quadrant ($Re\, k_n \leq 0$) whereas for emissive potentials, respectively, in the I quadrant ($Re\, k_n \geq 0$). When the potential is real then of course only the imaginary axis is available for the bound states.

We conclude that the eigenvalues of eq. (2.3) can be located all over the complex plane except for the I quadrant for absorptive potentials, and, respectively, except II quadrant for emissive potentials.

The energy $E = k^2/2\mu$ in general will also be a complex number whose real and imaginary part is identified, respectively, with the binding energy $B$ and the width $\Gamma$ of the level. For absorptive forces, we write

$$E = B - \mathrm{i}\tfrac{1}{2}\Gamma = \frac{(\mathrm{Re}\, k)^2 - (\mathrm{Im}\, k)^2 + \mathrm{i}\, 2\,\mathrm{Re}\, k \, \mathrm{Im}\, k}{2\mu}, \tag{2.8}$$

so that for $k$ in the second quadrant $\Gamma$ is necessarily a positive quantity. The binding energy $B$, however, to be negative requires that the condition $|Re\, k| < |Im\, k|$ is satisfied and in the k-plane such states are located above the bisector to the second quadrant. The states placed below this bisector, i.e. the states with positive $B$ have been sometimes called bound states embedded in the continuum (cf. [Gal *et al.* (1981)]). The situation is summarized in Fig. 2 which shows the location of different eigenvalues belonging to the absorptive Hamiltonian. The energy and the momentum are of course equivalent and either of them could be used as the basic independent variable. Nevertheless, a simple change of variable from $E$ to $k$ introduces a sign ambiguity when the square root of $k^2$ is taken. The usual convention is to choose as "physical" that one of the two square roots which yields $Im\, k > 0$ and if $Im\, k = 0$ then $Re\, k > 0$ becomes the signature. All other values of $k$ are referred to as "unphysical". The complex energy plane exhibits a two Riemann sheet structure where the physical sheet lies at the top of the unphysical sheet and the two sheets are joined in a criss-cross way by a square root branch cut running along the positive real axis. The inclusion of the Coulomb potential introduces a logarithmic branch point at $E = 0$ which calls for another branch cut in the energy plane making the multi-sheet structure very complicated. Much of these complications could be avoided by using $k$ as the independent variable and therefore the latter variable has been given preference in this work.

Returning to formula (2.6) and setting $n' \neq n$ we realize that the right hand side of (2.6) is in general not bound to vanish, so that two different eigenfunctions are not necessarily orthogonal. It could be easily shown that also the familiar Feynman–Hellmann theorem [Hellmann (1937); Feynman (1939)] does not hold any more. Indeed, if the underlying Hamiltonian $H_\ell$ is a function of a parameter $\xi$ then the derivative of the eigenenergy $E_n = k_n^2/2\mu$ with respect to that parameter, can be written as

$$\frac{\mathrm{d}E_n}{\mathrm{d}\xi} = \frac{\int_0^\infty |u_\ell(s,k_n,r)|^2 \left[\frac{\mathrm{d}H_\ell}{\mathrm{d}\xi} - \mathrm{Im}\, s\, v(r)\right] \mathrm{d}r}{\int_0^\infty |u_\ell(s,k_n,r)|^2 \, \mathrm{d}r} \tag{2.9}$$

but owing to the presence of the extra term in (2.9) proportional to Im s, the celebrated theorem is violated.

It is quite clear that the above difficulties result from the non-hermicity of the Hamiltonian and it is often said that if the Hamiltonian is non-Hermitean then time reversal invariance does not hold. On the other hand, the resulting S-matrix is symmetric even when the potential is complex and this suggests that the standard concept of invariance under time reversal ought to be extended. Naively speaking time reversal invariance means that we should be able to undo things, so when we are dealing with an absorptive potential we need also its emissive counterpart and both should be considered on equal footing. In classical physics if we have an accelerating force the inverse process involves slowing down force and invariance with respect to time reversal means only that the form of the dynamical equations is preserved, however if the force is a function of velocity it will be changed when the direction of velocity is reversed. In quantum mechanics, similarly as in classical physics, with every solution of dynamical equations we can associate another solution which corresponds to $t \to -t$.

In the problem under investigation we are studying the radial part of the wave function in a time independent approach. Under time reversal i.e. when $t \to -t$ both the wave function and the Hamiltonian need to

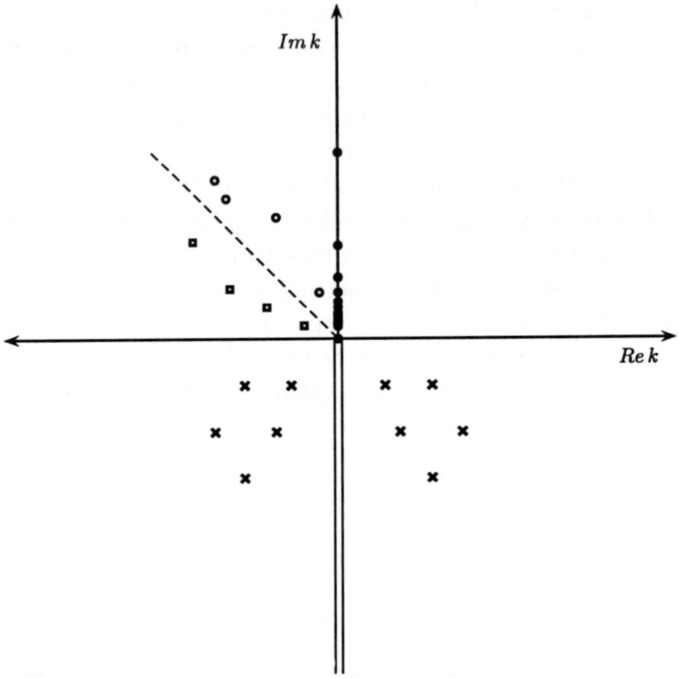


Fig. 2.1 Location of the various types of eigenvalues belonging to the absorptive Hamiltonian in the complex k-plane. They can appear anywhere except the first quadrant. The open circles and open squares correspond to bound states with, respectively, negative and positive binding energy, crosses denote Gamow states and full circles purely Coulombic eigenstates. Double line symbolizes the Coulomb branch cut.

be transformed. We consider a slightly more general situation when the Hamiltonian could be also momentum dependent and we introduce time reversal operator $T$ with the properties

$$T\, u_\ell(s,k,r) = u_\ell(s^*,-k^*,r)^*; \tag{2.10}$$

$$T\, H_\ell(s,k)\, T^{-1} = H_\ell(s^*,-k^*)^* \tag{2.11}$$

and T-invariance means that the above transformation does not change the Hamiltonian, i.e.

$$H_\ell(s^*,-k^*)^* = H_\ell(s,k). \tag{2.12}$$

When this condition holds, we also find that

$$u_\ell(s^*,-k^*,r)^* = u_\ell(s,k,r) \tag{2.13}$$

what follows from the fact that $Tu_\ell$ satisfies the same equation as $u_\ell$ and the same boundary conditions (2.4) and (2.5) which are manifestly T-invariant. Of course, T-invariance of the wave function implies T-invariance of the S-matrix. It is interesting to note that in the complex k-plane T-invariance can be viewed as reflection symmetry with respect to the imaginary axis. As a consequence of T-invariance for real potentials resonances occur in pairs so that the imaginary axis is then the axis of symmetry.

The orthogonality of two different eigenfunctions can now be restored by modifying the definition of a scalar product in function space. By taking as the "bra" state not the complex conjugate eigenfunction but the time reversed eigenfunction and using T-invariance (2.13), we obtain

$$\left(k_n^2 - k_{n'}^2\right) \int_0^\infty u_\ell(s, k_n, r)\, u_\ell(s, k_{n'}, r)\, \mathrm{d}r = 0, \tag{2.14}$$

which differs from the standard orthogonality condition only in that the eigenfunctions have been now extended into the complex domain. Actually, this is not a new idea and a similar procedure has been proposed by [Garcia-Calderon *et al.* (1976)], in a different context. It is easy to check that using the modified scalar product, we can also bring back the Feynman–Hellmann theorem to its usual form. According to the new rule, the norm integral reads

$$N^2 = \int_0^\infty u_\ell(s, k_n, r)^2\, \mathrm{d}r, \tag{2.15}$$

and with complex eigenfunctions the normalization (2.15) is not restricted to positive values any more, and, in fact, $N^2$ is itself a complex number. Thus, in particular, $N^2$ may well vanish and, as we shall see, this is not an uncommon occurrence. Clearly, under such circumstances the wave function cannot be normalized but since normalized wave functions enter the Green's function and the completeness relation this case requires closer examination. We shall return to this problem later on in Chapter 9.

We conclude this chapter by deriving a very general expression for the width which proves quie useful in various applications. Taking the imaginary part of (2.8) and invoking (2.7), we obtain

$$\tfrac{1}{2}\Gamma = -\frac{\displaystyle\int_0^\infty |u_\ell(s, k_n, r)|^2\, \mathrm{Im}\, V(r)\, \mathrm{d}r}{\displaystyle\int_0^\infty |u_\ell(s, k_n, r)|^2\, \mathrm{d}r}, \tag{2.16}$$

and it is worth noting that the above formula involves no approximations. It is apparent from (2.16) that the width is given as the overlap of the absorptive part of the nuclear potential and density of the radial probability to find the bound hadron at the position r.

## Chapter 3

# Coulomb Wave Functions

In this chapter we are going to give a resume of the pure Coulomb wave functions. Although there is a vast literature devoted to Coulomb wave functions and the subject has been reviewed in several places ( [Hull *et al.* (1959); Curtis (1964); Abramowitz *et al.* (1965); Seaton (2002)]) most of the papers deal with the standard $F_\ell(\eta,\rho)$ and $G_\ell(\eta,\rho)$ functions defined e.g. in [Abramowitz *et al.* (1965)]. These functions, however, have complicated analytic properties in the complex k-plane and are less useful for studying bound state problems. Therefore, we give preference to two other sets of Coulomb functions: the so called Jost solutions $(f_\ell^-, f_\ell^+)$ and the regular and irregular functions analytic in energy $(\phi_\ell, \theta_\ell)$ which define, respectively, the S-matrix and the K-matrix. Since the latter functions are perhaps less familiar and also because we wish to make this book self-contained we provide a utilitarian compendium of their properties. Apart from that our purpose was to present standard topics from a different perspective and include a number of new developments.

The Coulomb wave functions satisfy eq. (2.3) in which the nuclear potential has been set equal to zero ($s = 0$). In this chapter it will be convenient for us to replace the orbital quantum number $\ell$ by a real continuous variable $\lambda$ where $\lambda = \ell + \frac{1}{2}$ is the "physical" value of $\lambda$. The Coulomb wave functions will be sought as solutions of the extended wave equation

$$\left\{\frac{\mathrm{d}^2}{\mathrm{d}r^2} + k^2 + \frac{2\mu\alpha Z}{r} - \frac{\lambda^2 - 1/4}{r^2}\right\} u(\lambda, k, r) = 0. \tag{3.1}$$

The Coulomb wave functions $u(\lambda, k, r)$ fall into two categories: they may be specified by the boundary conditions which are either imposed at r=0, or at infinity (r$\rightarrow \infty$). The solution which for small r behaves as $r^{\lambda+\frac{1}{2}}$ with $\lambda > -\frac{1}{2}$ is called a regular solution $\varphi(\lambda, k, r)$. Since k enters equation (3.1) quadratically and the boundary condition at r=0 does not depend upon k, we have $\varphi(\lambda, -k, r) = \varphi(\lambda, k, r)$. However, $\varphi(-\lambda, k, r)$ is a new solution. In general, the parameter $\lambda$ determines the behavior close to the origin while k controls the asymptotic behavior at infinity. The solution $\varphi(-\lambda, k, r)$ with $\frac{1}{2} - \lambda < 0$ exhibits a $r^{\frac{1}{2} - \lambda}$ singularity at the origin and therefore is referred to as irregular solution. It has to be observed that the boundary condition at r=0 specifies uniquely only the regular solution. By contrast, the irregular solution is not unique because the boundary condition at r=0 is insensitive when an arbitrary admixture of the the regular solution is added to the irregular solution.

The second class of solutions is specified through the boundary conditions at infinity. Such solution were the wave function at infinity behaves as $e^{-ikr}$ was first studied by Jost and is known as Jost solution $f(\lambda, k, r)$. In this case the boundary condition depends upon k but does not depend upon $\lambda$ so that $f(\lambda, k, r) = f(-\lambda, k, r)$ while $f(\lambda, -k, r)$ is a new solution. To obtain the explicit forms of both, the regular, and, the Jost solutions, we put in (3.1) $u(\lambda, k, r) = z^{b/2}\, e^{-z/2}\, w(a, b, z)$ where $a = \lambda + \frac{1}{2} - i\eta, b = 2\lambda + 1$ and $z = 2ikr$. With this transformation, the just introduced wave function $w(a, b, z)$ satisfies the Kummer equation (cf. [Abramowitz *et al.* (1965)])

$$z\frac{d^2w}{dz^2} + (b - z)\frac{dw}{dz} - a\,w = 0,$$

whose solutions are well known. Using the notation of [Abramowitz *et al.* (1965)], these are the confluent hypergeometric function M(a,b,z) and the Kummer function U(a,b,z) (also sometimes called Tricomi function).

The two linearly independent Jost solutions $f(\lambda, \pm k, r)$ are defined by the boundary condition at infinity (cf. [de Alfaro *et al.* (1965); Newton (1982)]):

$$\lim_{r \to \infty} f(\lambda, \pm k, r) e^{\pm i\,[\rho - \eta \log(2\rho)]} = 1. \tag{3.2}$$

The Jost solutions can be written as [Bertero *et al.* (1967)]

$$f(\lambda, \pm k, r) = e^{\epsilon\eta\pi/2}(\pm 2i\,k\,r)^{\lambda + \frac{1}{2}} e^{\mp i\,k\,r}\, U(\lambda + \tfrac{1}{2} \mp i\eta, 2\lambda + 1, \pm 2ik\,r) \tag{3.3}$$

$$|\arg(\pm ik)| < \pi,$$

where $\epsilon = sgn(Re\ k)$. The Jost solutions (3.3) are entire and even functions of $\lambda$ for fixed k and r (r>0). They are analytic functions of k, except for an essential singularity and a branch point, both at k=0, for fixed $\lambda$ and

r. The restriction on $\arg(\pm ik)$ defines the principal branch of the U function and introduces a cut along the positive imaginary k-axis for $f(\lambda, k, r)$ and, respectively, along the negative imaginary k-axis for $f(\lambda, -k, r)$. The Jost solution may be also expressed in terms of the Whittaker function $W_{\pm i\eta,\lambda}(z)$ defined in [Abramowitz *et al.* (1965)]

$$W_{\pm i\eta,\lambda}(z) = e^{-\frac{1}{2}z} z^{\frac{1}{2}+\lambda} U(\tfrac{1}{2}+\lambda \mp i\eta, 1+2\lambda, z).$$

In the following we shall need the Jost solutions only for physical value of $\lambda$ and these functions are defined by the limit

$$f_\ell^{\pm}(k, r) \equiv \lim_{\lambda\to\ell+\frac{1}{2}} f(\lambda, \mp k, r)$$

and the explicit form of $f^{\pm}$, is

$$f_\ell^{\pm}(k, r) = e^{\epsilon\eta\pi/2}(\mp 2i\,k\,r)^{\ell+1} e^{\mp i\,k\,r} U(\ell + 1 \pm i\eta, 2\ell+2, \mp 2ik\,r). \tag{3.4}$$

For integer b=n+1; n=0,1,2,.. the function U(a,b,z) has a convenient series representation (cf. [Abramowitz *et al.* (1965)])

$$U(a, n+1, z) = \frac{(-1)^{n+1}}{n!\,\Gamma(a-n)}\{M(a, n+1, z)\ln(z) +$$

$$+\sum_{j=0}^{\infty} \frac{\Gamma(a+j)n!z^j}{\Gamma(a)(n+j)!j!}[\psi(a+j) - \psi(j+1) - \psi(n+j+1)]\}$$

$$+\frac{(n-1)!}{\Gamma(a)}\sum_{i=0}^{n-1}\frac{\Gamma(a-n+i)\Gamma(1-n)\,z^{i-n}}{\Gamma(a-n)\Gamma(1-n+i)\,i!}, \tag{3.5}$$

where $\psi(z)$ denotes the digamma function and for n=0 the last term in (3.5) is absent. This formula exposes the nature of the singularities present in the Jost solution. To make the Jost solutions single valued, requires introducing appropriate cuts in the k-plane: along the positive imaginary axis for $f_\ell^-(k, r)$, and, respectively, along the negative imaginary axis for $f_\ell^+(k, r)$.

It is apparent from (3.4) and (3.5) that the Jost solutions for $r \to 0$ have $r^{-\ell}$ singularity and the value of $r^\ell f_\ell^{\pm}(k, r)$ at the origin defines the Jost function $F_\ell^{\pm}(k)$ which can be explicitly evaluated. We have

$$F_\ell^{\pm}(k) \equiv (2\ell+1)\lim_{r\to 0} r^\ell f_\ell^{\pm}(k, r) = \frac{e^{\epsilon\eta\pi/2}\Gamma(2\ell+2)}{(\mp 2ik)^\ell\Gamma(\ell+1\pm i\eta)}, \tag{3.6}$$

and $F_\ell^\pm(k)$ is analytic everywhere except at $k = 0$ where the Jost function has essential singularity. When the strength of the Coulomb potential is reduced to zero, one has

$$\lim_{\alpha Z\to 0} F_\ell^\pm(k) = \mathrm{e}^{\pm \mathrm{i}\frac{1}{2}\pi}\frac{(2\ell+1)!!}{k^\ell}.$$

The solutions (3.3) are time reversal invariant, i.e. we have

$$f_\ell^\pm(-k^*, r)^* = f_\ell^\pm(k, r). \tag{3.7}$$

Another quantity of interest is the Wronskian $w[f, g]$ where $f$ and $g$ are arbitrary solutions of the wave equation defined as $w[f, g] = fg' - gf'$ where prime denotes differentiation with respect to $r$. It follows immediately from the wave equation (3.1) that the Wronskian of any two solutions of (3.3) is independent of $r$. Therefore, to compute the Wronskian $w[f_\ell^-, f_\ell^+]$, it is sufficient to use using the asymptotic form (cf. (3.2)) and the answer is

$$w\left[f_\ell^-(k, r), f_\ell^+(k, r)\right] = 2\mathrm{i}k. \tag{3.8}$$

In the limit of vanishing Coulomb strength, these solutions go over into the appropriate spherical Hankel functions, *viz.*

$$\lim_{\alpha Z\to 0} f_\ell^\pm(k, r) = \mathrm{e}^{\pm \mathrm{i}\pi\ell/2}\rho h_\ell^\pm(\rho); \tag{3.9}$$

where [Messiah (1969)] definitions of spherical Bessel functions have been adopted everywhere throughout this book.

We turn now to the second class of Coulomb solutions, specified by the boundary condition at r=0. Accordingly, we define the regular solution $\varphi(\lambda, k, r)$ by the condition

$$\lim_{r\to 0} r^{\mp\lambda-\frac{1}{2}}\varphi(\pm\lambda, k, r) = \frac{(-2\eta k)^{\frac{1}{2}\pm\lambda}}{\Gamma(1+2\lambda)}, \tag{3.10}$$

where on the right hand side we have introduced certain factor which will prove useful later on. The regular solution will be given in terms of the confluent hypergeometric function (cf. [Bertero *et al.* (1967)])

$$\varphi(\pm\lambda, k, r) = \frac{(-2\eta kr)^{\frac{1}{2}\pm\lambda}}{\Gamma(1\pm 2\lambda)}\mathrm{e}^{-\mathrm{i}kr}M(\pm\lambda + \tfrac{1}{2} - \mathrm{i}\eta, 1 + \pm 2\lambda, 2\mathrm{i}kr), \tag{3.11}$$

so that $\varphi(\pm\lambda, k, r)$ is an entire and even function of k (note that $2\eta$k does not depend on k) for fixed $\lambda$ and r (r>0). Owing to the gamma function present in (3.11), the simple poles at the points $\lambda = -\frac{1}{2}$m (m=0,1,2..) appearing in the M function are cancelled and the regular solution is analytic function of $\lambda$ for fixed k and r (r>0). The Wronskian of the two regular solutions with opposite $\lambda$ can be readily calculated by going to the small r limit. A simple calculation, gives

$$w[\varphi(\lambda,k,r),\varphi(-\lambda,k,r)] = \frac{2\eta k}{\pi}\sin 2\pi\lambda, \tag{3.12}$$

and, as evident from (3.12), this Wronskian vanishes whenever $2\lambda$ assumes integer values which happens precisely for physical $\lambda$. Consequently, for physical $\lambda$ these two solutions become linearly dependent and the boundary condition at r=0 allows to obtain only the regular wave function. This difficulty is specific to the Coulomb problem, and, e.g. in the free case, in result of the substitution $\ell \to -\ell-1$ in $(kr)j_\ell(kr)$ we obtain the irregular solution $(kr)n_\ell(kr)$ which is entire and even function of k ($n_\ell$ denotes here the spherical Neuman function). The impossibility to obtain the irregular solution analytic in $k^2$ from (3.11) for the Coulomb case of course does not preclude yet its very existence, and, as will become apparent in the following, it merely means that one has to resort to other methods. Actually, knowing the Jost solutions (3.3) any other solution can be generated by taking an appropriate superposition of the two Jost solutions.

For physical $\lambda$, it will be convenient for us to define the regular solution as

$$\phi_\ell(k,\, r) = r^{\ell+1}\mathrm{e}^{\pm ikr}M(\ell+1\pm \mathrm{i}\eta, 2\ell+2, \mp 2\mathrm{i}\rho), \tag{3.13}$$

and in the limit $\lambda \to \ell+\frac{1}{2}$ the functions $\varphi(\pm\lambda,k,r)$ will be both proportional to $\phi_\ell(k,\, r)$

$$\lim_{\lambda\to\ell+\frac{1}{2}} \varphi(\lambda,k,r) = \frac{(-2\eta k)^{\ell+1}}{(2\ell+1)!}\phi_\ell(k,\, r); \tag{3.14a}$$

$$\lim_{\lambda\to\ell+\frac{1}{2}} \varphi(-\lambda,k,r) = -A_\ell(\eta^2)\frac{(-2\eta k)^{\ell+1}}{(2\ell+1)!}\phi_\ell(k,\, r), \tag{3.14b}$$

where $A_\ell(\eta^2)$ is a polynomial in $k^2$

$$A_\ell(\eta^2) = \prod_{m=0}^{\ell}(1+\frac{m^2}{\eta^2}). \tag{3.15}$$

The regular solution (3.13) is manifestly time reversal invariant and is analytic function of $k^2$ in any finite part of the complex k-plane. When the

Coulomb potential is switched off, we obtain

$$\lim_{\alpha Z\to 0}\phi_\ell(k,\,r) = r^{\ell+1}\mathrm{e}^{\pm\mathrm{i}\rho}{}_1F_1(\ell+1,2\ell+2;\mp 2\mathrm{i}\rho) = \frac{(2\ell+1)!!}{k^{\ell+1}}\rho j_\ell(\rho). \quad (3.16)$$

The Wronskian of the regular solution and the Jost solution may be easily obtained by taking the small r limit

$$w\left[f_\ell^{\pm}(k,\,r),\phi_\ell(k,\,r)\right] = F_\ell^{\pm}(k),$$

from which immediately follows the formula expressing the regular solution in terms of the Jost solutions

$$\phi_\ell(k,\,r) = \frac{1}{2\mathrm{i}k}\left[F_\ell^-(k)f_\ell^+(k,\,r) - F_\ell^+(k)f_\ell^-(k,\,r)\right]. \quad (3.17)$$

As mentioned above, by taking a linear combination of the two Jost solutions (3.3) any other solution can be obtained and in particular it is possible to construct an irregular companion to the regular solution (3.13) which will also be an entire function of $k^2$. Introducing the irregular wave $\theta_\ell(k,\,r)$ analytic in $\mathrm{k}^2$, we can write quite generally

$$\theta_\ell(k,\,r) = -\frac{1}{2\mathrm{i}k}\left[g_\ell^-(k)F_\ell^-(k)f_\ell^+(k,\,r) - g_\ell^+(k)F_\ell^+(k)f_\ell^-(k,\,r)\right], \quad (3.18)$$

where following [Lambert (1969)] the arbitrary normalization of $\theta_\ell(k,\,r)$ has been fixed by imposing the Wronskian condition

$$w\left[\theta_\ell(k,\,r),\phi_\ell(k,\,r)\right] = 1. \quad (3.19)$$

The hitherto unknown functions $g_\ell^{\pm}(k)$ have to be adjusted in such a way as to make $\theta_\ell(k,\,r)$ entire function of $k^2$. Inserting (3.18) and (3.17) in (3.19) and using (3.8), we are led to the constraint

$$g_\ell^+(k) - g_\ell^-(k) = \frac{2\mathrm{i}k}{F_\ell^+(k)\,F_\ell^-(k)}. \quad (3.20)$$

This formula may be also cast to the form

$$g_\ell^+(k) - g_\ell^-(k) = 2\mathrm{i}k^{2\ell+1}C_\ell^2(\epsilon\eta) \quad (3.21)$$

where

$$C_\ell(\eta) = \frac{2^\ell}{(2\ell+1)!}\exp\tfrac{1}{2}[\log\Gamma(\ell+1+\mathrm{i}\eta) + \log\Gamma(\ell+1-\mathrm{i}\eta) - \pi\eta]. \quad (3.22)$$

The functions $g_\ell^{\pm}(k)$ may be established by taking advantage of the fact that for non-physical values of $\lambda$ the irregular solution is known explicitly and given in terms of the confluent hypergeometric function. In this case, it is also possible to represent the Jost solution as a linear combination of the regular and irregular solutions. Indeed, using the formula expressing U(a,b,z) in terms of M(a,b,z), taken from [Abramowitz *et al.* (1965)]

$$U(a,b,z) = \frac{\Gamma(1-b)}{\Gamma(a-b-1}M(a,b,z) + \frac{\Gamma(1-b)}{\Gamma(a}z^{1-b}M(a-b+1,2-b,z),$$

from (3.3) we obtain the desired relation

$$f(\lambda,k,r) = \frac{\pi e^{\epsilon\eta\pi/2}(i\eta)^{\lambda-\frac{1}{2}}}{\Gamma(\lambda+\frac{1}{2}-i\eta)}\,\frac{\varphi(-\lambda,k,r)+a(k,\lambda)\,\varphi(\lambda,k,r)}{\sin 2\pi\lambda}, \quad (3.23)$$

where

$$a(k,\lambda) = -(i\eta)^{-2\lambda}\frac{\Gamma(\frac{1}{2}+\lambda-i\eta)}{\Gamma(\frac{1}{2}-\lambda-i\eta)}. \quad (3.24)$$

with $a(k,\ell+\frac{1}{2}) = A_\ell(\eta^2)$. We wish now to go to the physical limit in (3.23), in which case the numerator in the second fraction vanishes (cf. (3.14)), but so does $\sin 2\pi\lambda$ in the denominator resulting in an indeterminate 0/0 form. However, by applying de l'Hospital theorem, one has

$$-\frac{2\Gamma(\ell+1-i\eta)}{e^{\epsilon\eta\pi/2}(i\eta)^\ell}f_\ell^-(k,r) = -\frac{(-2\eta k)^{\ell+1}}{(2\ell+1)!}\phi_\ell(k,r)\left.\frac{\partial a(k,\lambda)}{\partial\lambda}\right|_{\lambda=\ell+\frac{1}{2}} +$$

$$+\lim_{\lambda\to\ell+\frac{1}{2}}\frac{\partial}{\partial\lambda}\left[\varphi(-\lambda,k,r)+A_\ell(\eta^2)\,\varphi(\lambda,k,r)\right], \quad (3.25)$$

where

$$\left.\frac{\partial a(k,\lambda)}{\partial\lambda}\right|_{\lambda=\ell+\frac{1}{2}} = A_\ell(\eta^2)\,\left[\psi(\ell+1-i\eta)+\psi(-\ell-i\eta)-2\ln(i\eta)\right]. \quad (3.26)$$

The last term on the right-hand-side in (3.25) is by construction analytic function of $k^2$ and without loss of generality, may be written in the form

$$\lim_{\lambda\to\ell+\frac{1}{2}}\frac{\partial}{\partial\lambda}\left[\varphi(-\lambda,k,r)+A_\ell(\eta^2)\,\varphi(\lambda,k,r)\right] =$$

$$= -\frac{N_\ell}{2ik}\left\{[g_\ell^-(k)+P(k^2)]F_\ell^-(k)f_\ell^+(k,r) - \right.$$

$$\left. - [g_\ell^+(k)+P(k^2)]F_\ell^+(k)f_\ell^-(k,r)\right\}, \quad (3.27)$$

where $N_\ell$ denotes a normalization constant and $P(k^2)$ is an arbitrary polynomial. Inserting (3.27) and (3.17) in (3.25), we arrive at an equation of the form $Xf^+ + Yf^- = 0$ which has to be satisfied for all r. This implies that the appropriate coefficients X and Y are bound to vanish independently. The two resulting constraints are then used to determine the normalization constant $N_\ell$ and $g_\ell^-(k)$. A simple calculation, yields

$$g_\ell^-(k) = \frac{(2\eta k)^{2\ell+1}}{2[(2\ell+1)!]^2} \left.\frac{\partial a(k,\lambda)}{\partial\lambda}\right|_{\lambda=\ell+\frac{1}{2}} - P(k^2). \tag{3.28}$$

To facilitate a judicious choice of the polynomial P(k$^2$), we cast formula (3.26) to the form

$$\left.\frac{\partial a(k,\lambda)}{\partial\lambda}\right|_{\lambda=\ell+\frac{1}{2}} = A_\ell(\eta^2)\ [\psi(1-\mathrm{i}\eta) + \psi(-\mathrm{i}\eta) - 2\ \ln(\mathrm{i}\eta)] + \\ + A_\ell(\eta^2) \sum_{m=1}^{\ell} \frac{1}{m^2+\eta^2}, \tag{3.29}$$

where the last term on the right-hand-side is evidently a polynomial in k$^2$. As a matter fact, this polynomial is completely redundant and therefore will be removed by adopting P(k$^2$) in such a form that this polynomial is cancelled in (3.28). This is accomplished by taking P(k$^2$) in the form

$$P(k^2) = \frac{(2\eta k)^{2\ell+1}}{2[(2\ell+1)!]^2} A_\ell(\eta^2) \sum_{m=1}^{\ell} \frac{1}{m^2+\eta^2},$$

and the ultimate formula for $g_\ell^\pm(k)$, reads

$$g_\ell^\pm(k) = \frac{(2\eta k)^{2\ell+1}}{[(2\ell+1)!]^2} A_\ell(\eta^2) \left\{\psi(\pm\mathrm{i}\eta) \pm \frac{1}{2\mathrm{i}\eta} - \log\left[\mp\mathrm{i}\eta\, sgn(\alpha Z)\right]\right\}. \tag{3.30}$$

This form of $g_\ell^\pm(k)$ used in (3.18) yields the irregular Coulomb wave function $\theta_\ell(k, r)$ identical with that derived by [Lambert (1969)]. Formula (3.30) may be also written in an equivalent form

$$g_\ell^\pm(k) = \frac{(2\eta k)^{2\ell+1}}{[(2\ell+1)!]^2} A_\ell(\eta^2) \left[h(\eta) \pm \frac{\mathrm{i}}{2\eta} C_0^2(\epsilon\eta)\right], \tag{3.31}$$

where

$$h(\eta) = \tfrac{1}{2}\left[\psi(\mathrm{i}\eta) + \psi(-\mathrm{i}\eta)\right] - \tfrac{1}{2}\log\eta^2; \tag{3.32}$$

and it is evident that both $h(\eta)$ and $C_0^2(\epsilon\eta)$ are even functions of $\eta$.

It is worth noting that both $g_\ell^\pm(k)$ and $\theta_\ell(k,\, r)$ are time reversal invariant

$$g_\ell^\pm(-k^*)^* = g_\ell^\pm(k), \tag{3.33}$$

$$\theta_\ell(-k^*, r)^* = \theta_\ell(k,\, r). \tag{3.34}$$

Expression (3.30) exposes the locations of the singularities present in $g_\ell^\pm(k)$: at $k = 0$ there is a logarithmic branch point and also an essential singularity, the function $g_\ell^+(k)$ for the attractive case has poles at $k = k_n$, whereas $g_\ell^-(k)$ has poles at $k = -k_n$. To make $g_\ell^\pm(k)$ a single valued function we introduce a cut in the complex k-plane along the negative imaginary axis for $g_\ell^+(k)$ and, respectively, along positive imaginary axis for $g_\ell^-(k)$. It should be stressed that all these singularities originate entirely on account of the Coulomb potential and therefore they will go away when the Coulomb field is switched off. In the latter case $g_\ell^\pm(k)$ becomes just a centrifugal barrier penetration factor

$$\lim_{\alpha Z\to 0} g_\ell^\pm(k) = \frac{\pm \mathrm{i} k^{2\ell+1}}{[(2\ell+1)!!]^2}, \tag{3.35}$$

and the function (3.18) reduces to the Coulomb-free solution

$$\lim_{\alpha Z\to 0} \theta_\ell(k,\, r) = \frac{k^\ell}{(2\ell+1)!!}\rho n_\ell(\rho). \tag{3.36}$$

Very important is the zero-energy (or large $|\eta|$) limit of $g_\ell^\pm(k)$. For $|\eta| \to \infty$ one has the following asymptotic expansions

$$g_\ell^\pm(k) = \frac{|2\mu\alpha Z|^{2\ell+1}}{[(2\ell+1)!]^2}\left\{\frac{1}{12\eta^2} + \frac{1}{12\eta^4}(\alpha_\ell + \frac{1}{10}) + \cdots\right\}; \tag{3.37}$$

for repulsion $\alpha Z < 0$, where

$$\alpha_\ell = \tfrac{1}{6}\,\ell(\ell+1)(2\ell+1); \tag{3.38}$$

and, respectively,

$$g_\ell^\pm(k) = -\frac{(2\mu\alpha Z)^{2\ell+1}}{[(2\ell+1)!]^2}\left\{\pm \mathrm{i}\pi + \frac{1}{\eta^2}(\pm \mathrm{i}\pi\alpha_\ell + \frac{1}{12}) + \right.$$
$$\left. + \frac{1}{12\eta^4}(\pm \mathrm{i}\pi\beta_\ell + a_\ell + \frac{1}{10}) + \cdots\right\}; \tag{3.39}$$

for attraction $\alpha Z > 0$, where

$$\beta_\ell = \frac{\ell(\ell^2-1)(4\ell^2-1)(5\ell+6)}{30}. \tag{3.40}$$

As we shall see later on, the fact that for Coulomb attraction the zero-energy limit of $g_\ell^\pm(k)$ yields a constant imaginary value will have profound consequences.

Since $\phi_\ell(k,\, r)$ and $\theta_\ell(k,\, r)$ are entire functions of $k^2$ it should be possible to expand them around $k = 0$ in a power series in $k^2$

$$\phi_\ell(k,\, r) = \phi_\ell(0,\, r) + k^2\,\dot{\phi}_\ell(0,\, r) + \tfrac{1}{2}k^4\,\ddot{\phi}_\ell(0,\, r) + \cdots\,; \tag{3.41}$$

$$\theta_\ell(k,\, r) = \theta_\ell(0,\, r) + k^2\,\dot{\theta}_\ell(0,\, r) + \tfrac{1}{2}k^4\,\ddot{\theta}_\ell(0,\, r) + \cdots\,; \tag{3.42}$$

where a dot hereafter denotes differentiation with respect to $k^2$ and, indeed, such Taylor expansions have been derived by many authors (references may be traced down from [Humblet (1984); Humblet (1985)]). Of particular interest are the first two coefficients in the Taylor expansion and for attractive Coulomb potential, one has

$$\phi_\ell(0,\, r) = \frac{(2\ell+1)!}{(2\mu\alpha Z)^{\ell+1}}(\tfrac{1}{2}x)J_{2\ell+1}(x); \tag{3.43}$$

$$\begin{aligned}\dot{\phi}_\ell(0,\, r) = {} & \frac{(2\ell+1)!}{(2\mu\alpha Z)^{\ell+3}} \times \\ & \times \Big[-(\ell+1)(\tfrac{1}{2}x)^3\, J_{2\ell+3}(x) + \tfrac{1}{3}(\tfrac{1}{2}x)^4\, J_{2\ell+4}(x)\Big]; \end{aligned} \tag{3.44}$$

$$\theta_\ell(0,\, r) = -\pi\frac{(2\mu\alpha Z)^\ell}{(2\ell+1)!}(\tfrac{1}{2}x)\,Y_{2\ell+1}(x); \tag{3.45}$$

$$\begin{aligned}\dot{\theta}_\ell(0,\, r) = {} & \frac{(2\mu\alpha Z)^{\ell-2}}{(2\ell+1)!} \times \\ & \times \Big[-\tfrac{2}{3}\ell(\ell+1)(2\ell+1)\pi(\tfrac{1}{2}x)\,Y_{2\ell+1}(x) + \\ & +(\ell+1)\pi(\tfrac{1}{2}x)^3\,Y_{2\ell+3}(x) - \tfrac{1}{3}\pi(\tfrac{1}{2}x)^4\,Y_{2\ell+4}(x) + \tfrac{1}{3}(\tfrac{1}{2}x)\,J_{2\ell+1}(x)\Big]; \end{aligned} \tag{3.46}$$

with $x = 2\sqrt{|2\mu\alpha Zr|}$, where $J_\nu(x)$ and $Y_\nu(x)$ are, respectively, Bessel and Neumann functions of order $\nu$. The corresponding expansions for repulsive Coulomb potential involve modified Bessel and Neumann functions, i.e.

$I_\nu(x)$ and $K_\nu(x)$, respectively, and the expansion coefficients, are

$$\phi_\ell(0,\,r) = \frac{(2\ell+1)!}{|2\mu\alpha Z|^{\ell+1}}\left(\tfrac{1}{2}x\right) I_{2\ell+1}(x); \tag{3.47}$$

$$\dot{\phi}_\ell(0,\,r) = -\frac{(2\ell+1)!}{|2\mu\alpha Z|^{\ell+3}} \times \\ \times\left[(\ell+1)\left(\tfrac{1}{2}x\right)^3 I_{2\ell+3}(x) + \tfrac{1}{3}\left(\tfrac{1}{2}x\right)^4 I_{2\ell+4}(x)\right]; \tag{3.48}$$

$$\theta_\ell(0,\,r) = 2\frac{|2\mu\alpha Z|^{\ell}}{(2\ell+1)!}\left(\tfrac{1}{2}x\right) K_{2\ell+1}(x); \tag{3.49}$$

$$\dot{\theta}_\ell(0,\,r) = \frac{|2\mu\alpha Z|^{\ell-2}}{(2\ell+1)!}\left[\tfrac{4}{3}\ell(\ell+1)(2\ell+1)\left(\tfrac{1}{2}x\right) K_{2\ell+1}(x) - \right. \\ \left. - 2(\ell+1)\left(\tfrac{1}{2}x\right)^3 K_{2\ell+3}(x) + \tfrac{2}{3}\left(\tfrac{1}{2}x\right)^4 K_{2\ell+4}(x) - \tfrac{1}{3}\left(\tfrac{1}{2}x\right) I_{2\ell+1}(x)\right]. \tag{3.50}$$

Explicit formulae expressing the functions $\phi_\ell(k,\,r)$ and $\theta_\ell(k,\,r)$ as Bessel function expansions have been derived by [Humblet (1985)] and they may be converted into power series in $\eta^{-2}$. Most importantly, the resulting power series representations are absolutely and uniformly convergent in any finite part of the complex $k^2$ plane. For attractive Coulomb field, one has

$$\phi_\ell(k,\,r) = -\frac{(2\ell+1)!}{(2\mu\alpha Z)^{\ell+1}}\sum_{n=0}^{\infty}(-\eta^{-2})^n \times \\ \times \sum_{j=0}^{n}\beta_{nj}\left(-\tfrac{1}{2}x\right)^{2n+1+j} J_{2\ell+1+2n+j}(x); \tag{3.51a}$$

$$\theta_\ell(k,\,r) = \frac{(2\mu\alpha Z)^{\ell}}{(2\ell+1)!}A_\ell(\eta^2)\sum_{n=0}^{\infty}(-\eta^{-2})^n \times \\ \times\left[\pi\sum_{j=0}^{n}\beta_{nj}\left(-\tfrac{1}{2}x\right)^{2n+1+j} Y_{2\ell+1+2n+j}(x) - \right. \\ \left. -\sum_{s=1}^{n}(-1)^s\frac{|B_{2s}|}{2s}\sum_{j=0}^{n-s}\beta_{n-s\,j}\left(-\tfrac{1}{2}x\right)^{2n-2s+1+j} J_{2\ell+1+2n-2s+j}(x)\right], \tag{3.51b}$$

where $B_{2j}$ are the Bernoulli numbers. The numerical coefficients $\beta_{nj}$ vanish for j>n and also for negative values of n or j. They are obtained from the recurrence relation

$$4(2n+j)\,\beta_{nj} = (2\ell+2n+j)\,\beta_{n-1\,j} + \beta_{n-1\,j-1}$$

with the starting value $\beta_{00} = 1$. For a repulsive Coulomb potential, we have

$$\phi_\ell(k,\, r) = \frac{(2\ell+1)!}{|2\mu\alpha Z|^{\ell+1}} \sum_{n=0}^{\infty} (-\eta^{-2})^n \times$$
$$\times \sum_{j=0}^{n} \beta_{nj}\, (\tfrac{1}{2}x)^{2n+1+j}\, I_{2\ell+1+2n+j}(x); \qquad (3.52a)$$

$$\theta_\ell(k,\, r) = -\frac{|2\mu\alpha Z|^{\ell}}{(2\ell+1)!} A_\ell(\eta^2) \sum_{n=0}^{\infty} (-\eta^{-2})^n \times$$
$$\times \Big[ 2 \sum_{j=0}^{n} \beta_{nj}\, (-\tfrac{1}{2}x)^{2n+1+j}\, K_{2\ell+1+2n+j}(x) +$$
$$+ \sum_{s=1}^{n} (-1)^s \frac{|B_{2s}|}{2s} \sum_{j=0}^{n-s} \beta_{n-s\, j}\, (\tfrac{1}{2}x)^{2n-2s+1+j}\, I_{2\ell+1+2n-2s+j}(x) \Big]. \qquad (3.52b)$$

and it is apparent from (3.51) and (3.52) that $\phi_\ell(k,\, r)$ and $\theta_\ell(k,\, r)$ are both entire functions of $k^2$. Since the x dependent expansion coefficients in (3.51) and (3.52) are real, $\phi_\ell(k,\, r)$ and $\theta_\ell(k,\, r)$ are real functions of the complex variable $k^2$. It has to be noted, however, that the function $\theta_\ell(k,\, r)$, given by (3.51b) or (3.52b), is proportional to $A_\ell(\eta^2)$ and therefore is bound to vanish identically for all r whenever $A_\ell$ is equal to zero. This happens for $\ell > 0$ and unphysical momenta $k = \pm \mathrm{i}\, \mu\alpha Z/n$ with n=1,2,..,$\ell$ and at these isolated values of k the function $\theta_\ell(k,\, r)$ is ill-defined.

Expressing $f_\ell^{\pm}(k,\, r)/F^{\pm}(k)$ in terms of $\phi_\ell(k,\, r)$ and $\theta_\ell(k,\, r)$ we can obtain the $k \to 0$ limit. For attraction, we have

$$\lim_{k\to 0} \frac{f_\ell^{\pm}(k,\, r)}{F_\ell^{\pm}(k)} = -\pi \frac{(2\mu\alpha Z)^\ell}{(2\ell+1)!} (\tfrac{1}{2}x) \left[ Y_{2\ell+1}(x) \pm \mathrm{i} J_{2\ell+1}(x) \right], \qquad (3.53)$$

whereas for repulsion, we obtain

$$\lim_{k\to 0} \frac{f_\ell^{\pm}(k,\, r)}{F_\ell^{\pm}(k)} = 2 \frac{|2\mu\alpha Z|^\ell}{(2\ell+1)!} (\tfrac{1}{2}x) K_{2\ell+1}(x). \qquad (3.54)$$

Finally, we come to consider the standard Coulomb wave functions $F_\ell(\eta, \rho)$ and $G_\ell(\eta, \rho)$, defined through the boundary conditions at infinity. For $|\rho| \to \infty$ and Re(k)>0, the latter conditions are

$$F_\ell(\eta, \rho) = \sin[\rho - \eta \log 2\rho - \tfrac{1}{2}\pi\ell + \sigma_\ell(\eta)]; \qquad (3.55a)$$
$$G_\ell(\eta, \rho) = \cos[\rho - \eta \log 2\rho - \tfrac{1}{2}\pi\ell + \sigma_\ell(\eta)]; \qquad (3.55b)$$

where $\sigma_\ell(\eta)$ denotes the Coulomb phase shift with $\eta$ not necessarily restricted to real values

$$\sigma_\ell(\eta) = \frac{\log\Gamma(\ell+1+\mathrm{i}\eta) - \log\Gamma(\ell+1-\mathrm{i}\eta)}{2\mathrm{i}}. \tag{3.56}$$

Together with $F_\ell$ and $G_\ell$ one may define also Coulomb distorted outgoing and ingoing waves

$$H_\ell^{\pm}(\eta,\rho) = G_\ell(\eta,\rho) \pm \mathrm{i}F_\ell(\eta,\rho), \tag{3.57}$$

where, in view of (3.2), the above functions must be proportional to $f_\ell^{\pm}$. By comparing the asymptotic forms of $H_\ell^{\pm}$ and $f_\ell^{\pm}$, we obtain the relation

$$H_\ell^{\pm}(\eta,\rho) = C_\ell(\epsilon\eta)k^{\ell}F_\ell^{\mp}(k)f_\ell^{\pm}(k,\,r). \tag{3.58}$$

Eliminating in (3.58) $H_\ell^{\pm}$ in favor of $(F_\ell, G_\ell)$ and, respectively, $f_\ell^{\pm}$ in favor of $(\phi_\ell, \theta_\ell)$, we arrive at two other important relations

$$F_\ell(\eta,\rho) - C_\ell(\epsilon\eta)\,k^{\ell+1}\,\phi_\ell(k,\,r); \tag{3.59a}$$

$$G_\ell(\eta,\rho) = \frac{\theta_\ell(k,\,r) + \frac{1}{2}[g^+(k)+g^-(k)]\phi_\ell(k,\,r)}{C_\ell(\epsilon\eta)\,k^{\ell}}, \tag{3.59b}$$

where $\frac{1}{2}(g_\ell^+ + g_\ell^-)$ occurring in (3.59b) is real for real $\eta$ and with the aid of (3.31) may be brought to the form

$$\tfrac{1}{2}[g_\ell^+(k) + g_\ell^-(k)] = \frac{(2\eta k)^{2\ell+1}}{[(2\ell+1)!]^2}\,A_\ell(\eta^2)\,h(\eta).$$

The function $F_\ell(\eta,\rho)$ may be then obtained by summation of a confluent hypergeometric series. The function $G_\ell(\eta,\rho)$ may be written as $G_\ell = \frac{1}{2}(H_\ell^+ + H_\ell^-)$ and in view of (3.57) will be a superposition of two Kummer functions whose series expansion is presented in (3.5).

What is often needed in many practical applications are the logarithmic derivatives of the Coulomb wave functions (i.e. $\phi_\ell'/\phi_\ell$ and $f_\ell^{\pm\prime}/f_\ell^{\pm}$) rather than the functions themselves and in general the computation of the former presents a much easier task. The most important advantage stems from the fact that for these ratios continued fraction expansions can be derived whose coefficients are given by simple algebraic forms. There are several efficient algorithms available for the progressive evaluation of the continued fractions to the required accuracies and this approach lies at the heart of all the modern numerical methods of computing the Coulomb wave functions. Indeed, given the three logarithmic derivatives, i.e. $\phi_\ell'/\phi_\ell$ and $f_\ell^{\pm\prime}/f_\ell^{\pm}$ supplemented by Wronskian condition in principle they provide together enough information and can be regarded as four equations which could be solved for a pair of two independent functions and their derivatives.

The regular solutions $\phi_\ell(k, r)$ for successive l-values and their derivatives are connected by recurrence relations which can be solved for the ratio $\phi'_\ell/\phi_\ell$ and the latter is given in the form of a continued fraction. The final formula (cf. [Barnett *et al.* (1977)]), is

$$r\frac{\phi'_\ell(k, r)}{\phi_\ell(k, r)} = \ell + 1 + \cfrac{\rho\eta}{\ell + 1 + \cfrac{a_{\ell+1}}{b_{\ell+1} + \cfrac{a_{\ell+2}}{b_{\ell+2} + \cfrac{a_{\ell+3}}{b_{\ell+3} + \cdots}}}} \tag{3.60}$$

where

$$\begin{aligned} a_{\ell+1} &= -\rho\left[(\ell+1)^2 + \eta^2\right](\ell+2)/(\ell+1), \\ a_j &= -\rho^2(j^2 - 1)(j^2 + \eta^2), \;\; j = \ell+2, \ell+3, ... \\ b_j &= (2j+1)\left[j(j+1) + \eta\rho\right], \;\; j = \ell+1, \ell+2, ... \end{aligned}$$

The continued fraction representation of the logarithmic derivative of the Jost solution (3.3) follows from the differentiation property of the Kummer function. The result (cf. [Barnett *et al.* (1977)]), is

$$\begin{aligned} &\mp \mathrm{i}\, r\,\frac{f_\ell^{\pm\prime}(k, r)}{f_\ell^{\pm}(k, r)} = \rho - \eta + \\ &+ \cfrac{(\pm\mathrm{i}\eta - \ell)(\pm\mathrm{i}\eta + \ell + 1)}{2(\rho - \eta \pm \mathrm{i}) + \cfrac{(\pm\mathrm{i}\eta - \ell + 1)(\pm\mathrm{i}\eta + \ell + 2)}{2(\rho - \eta \pm 2\mathrm{i}) + \cfrac{(\pm\mathrm{i}\eta - \ell + 2)(\pm\mathrm{i}\eta + \ell + 3)}{2(\rho - \eta \pm 3\mathrm{i}) + \cdots}}}, \end{aligned} \tag{3.61}$$

and this expansion is known to have strong convergence property throughout the complex k-plane exterior to a cut on the negative (positive) imaginary axis for the outgoing (ingoing) wave case. Although the right hand side of (3.61) converges in all four quadrants of the complex k-plane, as has been observed by [Tomphson *et al* (1986)], the expression (3.61) is a representation of $rf_\ell^{+\prime}/f_\ell^{+}$ ($rf_\ell^{-\prime}/f_\ell^{-}$) everywhere except when $k$ is in the III quadrant (II quadrant). The range of validity, however, is easily extended using time reversal invariance (3.7). Suppose that $k$ is in the III quadrant and we want to evaluate $rf_\ell^{+\prime}/f_\ell^{+}$. We first write

$$r\frac{f_\ell^{\pm\prime}(k,\,r)}{f_\ell^{\pm}(k,\,r)} = r\left[\frac{f_\ell^{\pm\prime}(-k^*,r)}{f_\ell^{\pm}(-k^*,r)}\right]^*, \tag{3.62}$$

and then use continued fraction representation for the expression in the square bracket in the right hand side of (3.62) which is legitimate as $-k^*$ will be in IV quadrant. Similarly we can get $rf_\ell^{-\prime}/f_\ell^{-}$ when $k$ is in the II quadrant. The continued fraction representation remains valid also in the limiting case of vanishing Coulomb field. However, in the latter case there is a simpler way of calculating logarithmic derivatives of the spherical Hankel functions. In the limiting case (3.9) , one has

$$\rho\frac{h_0^{\pm\prime}(\rho)}{h_0^{\pm}(\rho)} = \pm\mathrm{i}, \tag{3.63}$$

$$\rho\frac{h_\ell^{\pm\prime}(\rho)}{h_\ell^{\pm}(\rho)} = -\rho^2\left[\rho\frac{h_{\ell-1}^{\pm\prime}(\rho)}{h_{\ell-1}^{\pm}(\rho)} - \ell\right]^{-1} - \ell;\;\; \ell - 1, 2, \ldots \tag{3.64}$$

When $\phi_\ell(k,\,r)$ and $\phi_\ell'(k,\,r)$ are known, e.g. either by solving numerically the wave equation from $r = 0$ outwards, or by summing up the confluent hypergeometric function expansion (3.13), the remaining Coulomb functions are readily obtained by using the continued fraction representation. From the Wronskian (3.6), we have

$$\frac{f_\ell^{\pm}(k,\,r)}{F_\ell^{\pm}(k)} = \frac{r}{r\phi_\ell'(k,\,r) - \left[r\dfrac{f_\ell^{\pm\prime}(k,\,r)}{f_\ell^{\pm}(k,\,r)}\right]\phi_\ell(k,\,r)}; \tag{3.65}$$

where the logarithmic derivatives in the square bracket are calculated by continued fractions summations. This also gives $f_\ell^{\pm\prime}(k,\,r)$ once $f_\ell^{\pm}(k,\,r)$ is known. The irregular function $\theta_\ell(k,\,r)$ can be obtained by eliminating $f_\ell^{\pm}(k,\,r)$ between the equations (3.17) and (3.18). One has

$$\theta_\ell(k,\,r) = k^\ell C_\ell(\epsilon\eta)\,\frac{f_\ell^{\pm}(k,\,r)}{F_\ell^{\pm}(k)} - g_\ell^{\pm}(k)\phi_\ell(k,\,r), \tag{3.66}$$

where the first term is obtained from (3.65). Given $\theta_\ell(k,\,r)$ the derivative can be computed from the Wronskian (3.19). In order to avoid a possible loss of accuracy due to cancellations of the pole terms in (3.66) the plus or minus signs should be chosen in accordance with the sign of $Im(-k)$ so that the poles are absent.

The above review of Coulomb wave functions would not have been complete without mentioning the so called "physical" wave function $\psi_\ell(k,\, r)$. The momentum $k$ is taken to be "physical", (i.e. real, non-negative) and the physical wave function is a regular solution that asymptotically takes the familiar form

$$\lim_{r\to\infty} \psi_\ell(k,\, r) = \sin(kr - \eta \log 2kr - \pi\ell/2 + \sigma_\ell).$$

The function which satisfies these requirements, is

$$\psi_\ell(k,\, r) = k\left[F_\ell^+(k)\, F_\ell^-(k)\right]^{-1/2} \phi_\ell(k,\, r), \tag{3.67}$$

and comparison with (3.36) shows that the physical wave function is nothing else but $F_\ell(\eta, \rho)$. In the chargeless limit, we have

$$\lim_{\alpha Z\to 0} \psi_\ell(k,\, r) = kr\, j_\ell(kr)$$

and the functions $\psi_\ell(k,\, r)$ are orthogonal

$$\frac{2}{\pi}\int_0^\infty \psi_\ell(p, r)\, \psi_\ell(p', r)\, \mathrm{d}r = \delta(p - p'); \tag{3.68}$$

and form a complete set

$$\frac{2}{\pi}\int_0^\infty \psi_\ell(p, r)\psi_\ell(p, r')\, \mathrm{d}p + \sum_{N=0}^{\infty} u_{\ell N}(r)\, u_{\ell N}(r') = \delta(r - r'), \tag{3.69}$$

where

$$u_{\ell N}(r) = \frac{\phi_\ell(k_n, r)}{\left[\int_0^\infty \phi_\ell(k_n, r)^2\, \mathrm{d}r\right]^{1/2}}, \tag{3.70}$$

denotes the normalized bound state wave function. Of course, for repulsive Coulomb interaction the sum in (3.69) is absent.

The Coulomb scattering S-matrix $S_\ell^c(k)$ is obtained from (3.17) by using the asymptotic form for the Jost solutions and is given in terms of the Jost functions

$$S_\ell^c(k) = (-1)^\ell \frac{F_\ell^-(k)}{F_\ell^+(k)} = \mathrm{e}^{2\mathrm{i}\sigma_\ell(k)}. \tag{3.71}$$

Bound states are associated with poles of the S-matrix, i.e. with the zeros of the Jost function $F_\ell^+(k)$. The Jost function $F_\ell^+(k)$ for the attractive case has an infinite number of zeros located on the positive part of the imaginary axis and they correspond to the Coulomb bound states. As seen from (3.6), $F_\ell^+(k)$ vanishes at the poles of $\Gamma(\ell + 1 + \mathrm{i}\eta)$, i.e. when $\ell + 1 + \mathrm{i}\eta = -N$ with $N = 0, 1, 2, ...$, or for $\mathrm{i}\eta = -n$ where $n = \ell + 1 + N$ is the principal quantum number. This gives the locations of the Coulomb bound states $k_n$

$$k_n = \frac{\mathrm{i}\mu\alpha Z}{n}. \tag{3.72}$$

The bound states are the only discrete eigenvalues and for Coulomb potential the Gamow states are absent.

At the bound states $k = k_n$, the regular wave function $\phi_\ell(k_n, r)$ is proportional to the outgoing wave Jost solution and becomes normalizable. The hypergeometric function in (3.13) reduces to a Laguerre polynomial and the standard hydrogen like wave function is obtained.

# Chapter 4

# Coulomb Propagator and Scattering Operators

In relative coordinates the two non-interacting particle system is described by the free Hamiltonian $\widehat{H}_0 = \widehat{\mathbf{p}}^2/2\mu$. The propagator, or retarded Green's function for this system is

$$\widehat{G}_0^{(+)}(E) = \frac{1}{E - \widehat{H}_0 + \mathrm{i}\epsilon} \tag{4.1}$$

where $E = p^2/2\mu$ is the total energy. Both, $E$ and $p$ *throughout this chapter* are assumed to be real and non-negative quantities. When a complete set of plane wave eigenstates $|\,\mathbf{q}\rangle$ is inserted in the numerator, it becomes

$$\widehat{G}_0^{(+)}(E) = 2\mu \int \frac{\mathrm{d}^3 q}{(2\pi)^3} \frac{|\,\mathbf{q}\rangle\langle\mathbf{q}\,|}{\mathbf{p}^2 - \mathbf{q}^2 + \mathrm{i}\epsilon}. \tag{4.2}$$

The free propagation from an initial position $\mathbf{r}$ to a final position $\mathbf{r}'$ in coordinate space is therefore given by the function $\langle\mathbf{r}'\,|\widehat{G}_0^{(+)}(E)|\,\mathbf{r}\rangle$. Direct integration, gives

$$\langle\mathbf{r}'\,|\widehat{G}_0^{(+)}(E)|\,\mathbf{r}\rangle = 2\mu \int \frac{\mathrm{d}^3 q}{(2\pi)^3} \frac{\mathrm{e}^{-\mathrm{i}\mathbf{q}\cdot\mathbf{r}'}\mathrm{e}^{\mathrm{i}\mathbf{q}\cdot\mathbf{r}}}{\mathbf{p}^2 - \mathbf{q}^2 + \mathrm{i}\epsilon} = -\frac{2\mu}{4\pi} \frac{\mathrm{e}^{\mathrm{i}p|\mathbf{r}-\mathbf{r}'|}}{|\mathbf{r}-\mathbf{r}'|}. \tag{4.3}$$

Expanding the plane waves in (4.3) in Legendre polynomials, we obtain

$$\langle\mathbf{r}'\,|\widehat{G}_0^{(+)}(E)|\,\mathbf{r}\rangle = \frac{2\mu}{4\pi} \sum_{\ell=0}^{\infty} (2\ell+1) \frac{\langle r'\,|\mathrm{g}_\ell^{(+)}(p)|\,r\rangle}{rr'} P_\ell(\hat{\mathbf{r}}'\cdot\hat{\mathbf{r}}), \tag{4.4}$$

where

$$\frac{\langle r'\,|\mathrm{g}_\ell^{(+)}(p)|\,r\rangle}{rr'} = \frac{2}{\pi} \int_0^\infty \frac{j_\ell(qr')\, j_\ell(qr')\, q^2 \mathrm{d}q}{p^2 - q^2 + \mathrm{i}\epsilon} = -p\, h_\ell^{(+)}(pr_>)\, j_\ell(pr_<). \tag{4.5}$$

When the particles interact via the Coulomb potential $V_C$, the retarded and advanced Coulomb Green's operators are now

$$\widehat{G}_C^{(\pm)}(E) = \frac{1}{E - \widehat{H}_0 - \widehat{V}_C \pm i\epsilon}, \tag{4.6}$$

depending on the boundary conditions which are imposed at infinity. The above Coulomb resolvent is a solution of the integral equation $\widehat{G}_C^{(\pm)}(E) = \widehat{G}_0^{(\pm)}(E) + \widehat{G}_0^{(\pm)}(E)\widehat{V}_C\widehat{G}_C^{(\pm)}(E)$. After iteration, this equation gives the Coulomb propagator as an infinite sum of Feynman diagrams with zero, one, two and so on exchanged photons as shown in Fig. 4.1. The Schrödinger

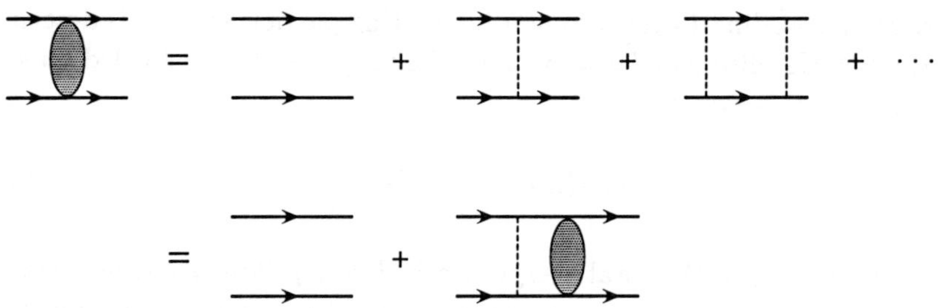

Fig. 4.1 Coulomb propagator as an infinite sum.

equation $(E - \widehat{H})|\chi\rangle = 0$ where $\widehat{H} = \widehat{H}_0 + \widehat{V}_C$ is the full Hamiltonian, has the corresponding incoming and outgoing solutions $|\chi_{\mathbf{p}}^{(\pm)}\rangle$. Formally, they can be expressed in terms of the free solutions as

$$|\chi_{\mathbf{p}}^{(\pm)}\rangle = [1 + \widehat{G}_C^{(\pm)}(E)\widehat{V}_C]|\mathbf{p}\rangle. \tag{4.7}$$

This means that $|\chi_{\mathbf{p}}^{(\pm)}\rangle = \widehat{G}_C^{(\pm)}(E)\widehat{G}_0^{(\pm)}(E)^{-1}|\mathbf{p}\rangle$, and these solutions must have the same normalization as the free solutions (plane waves), so that $\langle\chi_{\mathbf{q}}^{(\pm)}|\chi_{\mathbf{p}}^{(\pm)}\rangle = (2\pi)^3\delta(\mathbf{q} - \mathbf{p})$. The explicit solution of the Coulomb scattering problem in coordinate space has been obtained by solving the Schrödinger equation in parabolic coordinates and can be expressed in terms of confluent hypergeometric (Kummer) function $M(a, b, x)$ defined e.g. in [Abramowitz *et al.* (1965)]. The in-state solution with outgoing spherical waves in the future is

$$\chi_{\mathbf{p}}^{(+)}(\mathbf{r}) = \mathrm{e}^{-\frac{1}{2}\pi\eta}\,\Gamma(1 + \mathrm{i}\eta)\,\mathrm{e}^{\mathrm{i}\mathbf{p}\cdot\mathbf{r}}\,M(-\mathrm{i}\eta,\, 1,\, \mathrm{i}pr - \mathrm{i}\mathbf{p}\cdot\mathbf{r}), \tag{4.8}$$

whereas the corresponding out-state has incoming spherical waves in the distant past and is given by the wave function

$$\chi_{\mathbf{p}}^{(-)}(\mathbf{r}) = \mathrm{e}^{-\frac{1}{2}\pi\eta}\,\Gamma(1-\mathrm{i}\eta)\,\mathrm{e}^{\mathrm{i}\mathbf{p}\cdot\mathbf{r}}\,M(\mathrm{i}\eta,\,1,\,-\mathrm{i}pr-\mathrm{i}\mathbf{p}\cdot\mathbf{r}). \qquad (4.9)$$

These two solutions are related by time reversal which implies $\chi_{\mathbf{p}}^{(+)}(\mathbf{r}) = \chi_{-\mathbf{p}}^{(-)}(\mathbf{r})^{\dagger}$. The function $\chi_{\mathbf{p}}^{(+)}(\mathbf{r})$, similarly as the plane wave, can be expanded in partial waves

$$\chi_{\mathbf{p}}^{(+)}(\mathbf{r}) = \rho^{-1}\sum_{\ell=0}^{\infty}(2\ell+1)\,\mathrm{i}^{\ell}\,\mathrm{e}^{\mathrm{i}\sigma_\ell}\,F_\ell(\eta,\rho)\,P_\ell(\hat{\mathbf{p}}\cdot\hat{\mathbf{r}}), \qquad (4.10)$$

where $\sigma_\ell = \arg\Gamma(1+\ell+\mathrm{i}\eta)$ denotes the Coulomb phase shift, $F_\ell(\eta,\rho)$ is the regular Coulomb wave function defined in [Abramowitz *et al.* (1965)]

$$F_\ell(\eta,\rho) = C_\ell(\eta)\,\rho^{\ell+1}\mathrm{e}^{-\mathrm{i}\rho}\,M(\ell+1-\mathrm{i}\eta,\,2\ell+2,\,2\mathrm{i}\rho), \qquad (4.11)$$

and $C_\ell(\eta)$ is

$$C_\ell(\eta) = 2^{\ell}\,\mathrm{e}^{-\pi\eta/2}\frac{|\Gamma(\ell+1+\mathrm{i}\eta)|}{\Gamma(2\ell+2)}. \qquad (4.12)$$

The probability of finding two particles at zero separation can be immediately obtained as

$$|\chi_{\mathbf{p}}^{(\pm)}(0)|^2 = C_0(\eta)^2 = \mathrm{e}^{-\pi\eta}|\Gamma(1+\mathrm{i}\eta)|^2 = \frac{2\pi\eta}{\mathrm{e}^{2\pi\eta}-1} \qquad (4.13)$$

which is the well-known Sommerfeld factor (sometimes called Gamow factor). For repulsive Coulomb interaction, when the relative velocity between the particles goes to zero, this factor becomes exponentially small and the cross section is drastically reduced. By contrast, for attractive Coulomb interaction, when the parameter $\eta$ takes negative values, in the small momentum limit the Sommerfeld factor diverges increasing the cross section close to threshold.

With the Coulomb scattering wave function in hand, we are in the position to work out a more useful expression for the Green's functions (4.6). Since for Coulomb repulsion the scattering states form a complete set, we can use them similarly as in (4.2). Taking the matrix element in coordinate space, we are led to a spectral representation of the retarded Green's function

$$\langle\mathbf{r}'|\widehat{G}_C^{(+)}(E)|\mathbf{r}\rangle = 2\mu\int\frac{\mathrm{d}^3q}{(2\pi)^3}\frac{\chi_{\mathbf{q}}^{(+)}(\mathbf{r}')\chi_{\mathbf{q}}^{(+)*}(\mathbf{r})}{\mathbf{p}^2-\mathbf{q}^2+\mathrm{i}\epsilon}. \qquad (4.14)$$

Now consider oppositely charged particles (Coulomb attraction). In this case formula (4.14) must be supplemented by the contribution from the discrete spectrum (Coulomb bound states)

$$\langle \mathbf{r}' | \widehat{G}_C^{(+)}(E) | \mathbf{r} \rangle = 2\mu \int \frac{\mathrm{d}^3 q}{(2\pi)^3} \frac{\chi_{\mathbf{q}}^{(+)}(\mathbf{r}')\chi_{\mathbf{q}}^{(+)*}(\mathbf{r})}{\mathbf{p}^2 - \mathbf{q}^2 + \mathrm{i}\epsilon} + \\ + \sum_{n\ell m} \frac{\psi_{n\ell m}(\boldsymbol{r}')\psi^*_{n\ell m}(\boldsymbol{r})}{E - E_n}, \quad (4.15)$$

where $E_n = -\frac{1}{2}\mu\,(\alpha Z/n)^2$ are the Coulomb binding energies and

$$\psi_{n\ell m}(\boldsymbol{r}) = \left\{ \frac{(n-\ell-1)!}{2n\,(n+\ell)!} \left(\frac{2}{na_B}\right)^3 \right\}^{\frac{1}{2}} \mathrm{e}^{-\frac{1}{2}x}\, x^\ell\, L_{n-\ell-1}^{2\ell+1}(x)\, Y_{\ell m}(\theta,\phi), \quad (4.16)$$

with $x = (2r/na_B)$, is the bound state wave function corresponding to quantum numbers $n\ell m$ with $L^i_j(x)$ being the associated Laguerre polynomial.

In either case (Coulomb repulsion or attraction), it will be useful to expand (4.14) and (4.15) in spherical harmonics. Extending formula (4.5), the Coulomb Green's function can be then quite generally written as

$$\langle \mathbf{r}' | \widehat{G}_C^{(+)}(E) | \mathbf{r} \rangle = \frac{2\mu}{4\pi} \sum_{\ell=0}^{\infty} (2\ell+1) \frac{\langle r' | G_\ell^+(p) | r \rangle}{rr'} P_\ell(\hat{\mathbf{r}}' \cdot \hat{\mathbf{r}}), \quad (4.17)$$

where the explicit form of the angular projection $\langle r' | G_\ell^+(p) | r \rangle$ can be worked out by inserting the partial wave expansion in (4.17). For Coulomb repulsion, we obtain

$$\langle r' | G_\ell^{(+)}(p) | r \rangle = \frac{2}{\pi} \int_0^\infty \frac{F_\ell(\eta_q, qr')\, F_\ell(\eta_q, qr)\mathrm{d}q}{p^2 - q^2 + \mathrm{i}\epsilon} \quad (4.18)$$

where $\eta_q = -\mu\alpha Z/q$. Direct evaluation of the integral in (4.18) would be rather difficult but fortunately one can get around this problem. As a matter of fact, the Green's function $\langle r' | G_\ell^{(+)}(p) | r \rangle$ can be easily constructed by introducing the second, linearly independent radial Coulomb solution, i.e. the irregular Coulomb wave function $G_\ell(\eta, \rho)$ defined in [Abramowitz *et al.* (1965)]. Both functions $F_\ell(\eta, \rho)$ and $G_\ell(\eta, \rho)$ are solutions of the radial Schrödinger equation

$$\frac{\mathrm{d}^2 w(\rho)}{\mathrm{d}\rho^2} + \left[1 - \frac{2\eta}{\rho} - \frac{\ell(\ell+1)}{\rho^2}\right] w(\rho) = 0, \quad (4.19)$$

where $\rho > 0$, $-\infty < \eta < \infty$ and $w(\rho)$ can be either $F_\ell(\eta,\rho)$ or $G_\ell(\eta,\rho)$ but in the asymptotic region $\rho \to \infty$ these two functions assume different forms

$$F_\ell(\eta,\rho) = \sin(\rho - \eta \ln 2\rho - \tfrac{1}{2}\pi\ell + \sigma_\ell); \tag{4.20a}$$

$$G_\ell(\eta,\rho) = \cos(\rho - \eta \ln 2\rho - \tfrac{1}{2}\pi\ell + \sigma_\ell). \tag{4.20b}$$

It will prove useful in the following to introduce ingoing and outgoing waves by taking a linear combination of the functions (4.20)

$$H_\ell^{\pm}(\eta,\rho) = G_\ell(\eta,\rho) \pm \mathrm{i}\, F_\ell(\eta,\rho). \tag{4.21}$$

The partial Green's function $\langle r' \,| G_\ell^{(+)}(p) |\, r\rangle$ must be symmetric in $r$ and $r'$ and regular at the origin solution of the inhomogeneous wave equation

$$\left[\frac{\mathrm{d}^2}{\mathrm{d}r^2} + p^2 - \frac{2p\eta}{r} - \frac{\ell(\ell+1)}{r^2}\right] \langle r' \,| G_\ell^{(+)}(p) |\, r\rangle = \delta(r - r') \tag{4.22}$$

which in the asymptotic region has the form of an outgoing wave. The function which satisfies the above requirements will be sought in the form

$$\langle r' \,| G_\ell^{(+)}(p) |\, r\rangle = -c\, p^{-1}\, H_\ell^{+}(\eta, pr_>)\, F_\ell(\eta, pr_<), \tag{4.23}$$

where $c$ is a constant to be determined. For $r \neq r'$, by construction, the function (4.23) clearly satisfies the homogeneous wave equation (4.22). Evidently, $\langle r' \,| G_\ell^{(+)}(p) |\, r\rangle$ is continuous at $r = r'$ but its derivative is not. It is this discontinuity of the first derivative at $r = r'$ what makes the second derivative to become infinite at this point so that the inhomogeneous equation is satisfied for all $r$. In order to pin down the constant $c$ we insert (4.23) in (4.22) and subsequently we integrate this equation over $r$ in the interval from $r' - \epsilon$ to $r' + \epsilon$. In the limit $\epsilon \to 0$, only the second derivative term gives non-zero contribution on the left hand side and the result is

$$c\,\left[H_\ell^{+}(\eta,\rho)\, F_\ell'(\eta,\rho) - H_\ell^{+\prime}(\eta,\rho)\, F_\ell(\eta,\rho)\right] = 1, \tag{4.24}$$

where the prime denotes derivative with respect to the argument $\rho = pr'$. Using eq. (4.19), it can be easily checked that the derivative of the expression in the square bracket in (4.24) vanishes for all $\rho$ which means that in fact this expression is a constant independent of $\rho$. Consequently, the square bracket expression can be evaluated at any $\rho$, in particular for the asymptotic value of $\rho$ where formulae (4.20) hold and a simple calculation gives the ultimate answer

$$c = 1.$$

It should be stressed here that formula (4.23) is valid for both, repulsive and attractive Coulomb potentials. For vanishing Coulomb potential $\eta \to 0$, from (4.23) we recover the free propagation result (4.5).

When the strong interaction is added via the local potential operator $\widehat{V}_S$, we arrive at a two-potential problem and the corresponding Hamiltonian becomes $\widehat{H} = \widehat{H}_0 + \widehat{V}_C + \widehat{V}_S$. The full wave equation involving both, the Coulomb and the nuclear potential

$$(E - \widehat{H}_0 - \widehat{V}_C)|\psi_{\mathbf{p}}^{(\pm)}\rangle = \widehat{V}_S\,|\psi_{\mathbf{p}}^{(\pm)}\rangle$$

after multiplying by $(E - \widehat{H}_0 - \widehat{V}_C)^{-1}$ from the left, takes the form of an integral equation

$$|\psi_{\mathbf{p}}^{(\pm)}\rangle = |\chi_{\mathbf{p}}^{(\pm)}\rangle + \widehat{G}_C^{(\pm)}(E)\,\widehat{V}_S\,|\psi_{\mathbf{p}}^{(\pm)}\rangle, \tag{4.25}$$

in which the appropriate boundary condition has been imposed. The formal solution of this equation $|\psi_{\mathbf{p}}^{(\pm)}\rangle = [1-\widehat{G}_C^{(\pm)}(E)\,\widehat{V}_S]^{-1}\,|\chi_{\mathbf{p}}^{(\pm)}\rangle$, can be written as a perturbative power expansion with respect to $\widehat{V}_S$

$$|\psi_{\mathbf{p}}^{(\pm)}\rangle = \{1 + \sum_{n=1}^{\infty}[\widehat{G}_C^{(\pm)}(E)\widehat{V}_S]^n\}|\chi_{\mathbf{p}}^{(\pm)}\rangle. \tag{4.26}$$

It is worth noting that each term in this expansion contains Coulomb corrections to all orders. The scattering amplitude is given by the $S$-matrix element which is the overlap between an incoming state with momentum $\mathbf{p}$ and an outgoing state $\mathbf{p}'$. It takes the standard form

$$S(\mathbf{p}',\mathbf{p}) = \langle\psi_{\mathbf{p}'}^{(-)}\,|\psi_{\mathbf{p}}^{(+)}\rangle = (2\pi)^3\,\delta(\mathbf{p}'-\mathbf{p}) - 2\pi\mathrm{i}\,\delta(E'-E)\,T(\mathbf{p}',\mathbf{p}). \tag{4.27}$$

With the scattering states (4.26) the $T$-matrix element splits conveniently into two terms,

$$T(\mathbf{p}',\mathbf{p}) = T_C(\mathbf{p}',\mathbf{p}) + T_{SC}(\mathbf{p}',\mathbf{p}), \tag{4.28}$$

where

$$T_C(\mathbf{p}',\mathbf{p}) = \langle\mathbf{p}'\,|\widehat{V}_C|\,\psi_{\mathbf{p}}^{(+)}\rangle \tag{4.29}$$

denotes the purely Coulombic scattering amplitude, whereas

$$T_{SC}(\mathbf{p}',\mathbf{p}) = \langle\chi_{\mathbf{p}'}^{(-)}\,|\widehat{V}_S|\,\psi_{\mathbf{p}}^{(+)}\rangle \tag{4.30}$$

is the strong interaction scattering amplitude containing also the effect of the Coulomb distortion. The above result derives from the so called two-potential formula and can be seen as follows. The full T-matrix is defined as

$$T(\mathbf{p}',\mathbf{p}) = \langle \mathbf{p}'|\widehat{T}|\mathbf{p}\rangle = \langle \mathbf{p}'|\widehat{V}_S + \widehat{V}_C|\psi_{\mathbf{p}}^{(+)}\rangle,$$

but in view of (4.25), can be rewritten as a sum of four terms

$$T(\mathbf{p}',\mathbf{p}) = \langle \mathbf{p}'|\widehat{V}_C|\chi_{\mathbf{p}}^{(+)}\rangle + \langle \mathbf{p}'|\widehat{V}_S|\chi_{\mathbf{p}}^{(+)}\rangle + \\ + \langle \mathbf{p}'|\widehat{V}_C\widehat{G}_C^{+}(E)\widehat{V}_S|\psi_{\mathbf{p}}^{(+)}\rangle + \langle \mathbf{p}'|\widehat{V}_S\widehat{G}_C^{+}(E)\widehat{V}_S|\psi_{\mathbf{p}}^{(+)}\rangle. \tag{4.31}$$

Now, in the third term on the right hand side of (4.31) we use (4.7), replacing $\langle \mathbf{p}'|\widehat{V}_C\widehat{G}_C^{+}(E)$ by $(\langle\chi_{\mathbf{p}'}^{(-)}| - \langle \mathbf{p}'|)$ and similarly in the last term we invoke (4.25) to change $\widehat{G}_C^{+}(E)\widehat{V}_S|\psi_{\mathbf{p}}^{(+)}\rangle$ into $(|\psi_{\mathbf{p}}^{(+)}\rangle - |\chi_{\mathbf{p}}^{(+)}\rangle)$. As a result of cancellation, only two terms survive and formula (4.28) is thereby established.

Since the full Coulomb wave function $\chi_{\mathbf{p}}^{(+)}(\mathbf{r})$ is known one can calculate exactly the Coulomb scattering amplitude by taking the asymptotic limit $r \to \infty$ in (4.8). This yields the well known closed form expression

$$T_C(\mathbf{p}',\mathbf{p}) = \frac{4\pi}{2\mu}\frac{\eta}{2p\,\sin^2\frac{1}{2}\theta}\exp\left(2\mathrm{i}\sigma_0 - \mathrm{i}\eta\ln\sin^2\tfrac{1}{2}\theta\right) \tag{4.32}$$

where $\theta$ is the CM scattering angle. The T-matrix $T_C(\mathbf{p}',\mathbf{p})$ in which $|\mathbf{p}'|$ does not have to be equal to $|\mathbf{p}|$ has been known as the half-off-shell T-matrix and for the Coulomb case it can be obtained from (4.29). The integration is largely facilitated by introducing a convergence factor $\mathrm{e}^{-\lambda r}$ in the integrand. The integral can be then evaluated in analytic form, and the result is

$$T_C(\mathbf{p}',\mathbf{p}) = 4\pi\, p\eta\, C_0(\eta)\,\mathrm{e}^{\mathrm{i}\sigma_0} \lim_{\lambda\to 0}\frac{[p'^2 - (p+\mathrm{i}\lambda)^2]^{\mathrm{i}\eta}}{[(\mathbf{p}'-\mathbf{p})^2 + \lambda^2]^{1+\,\mathrm{i}\eta}}, \tag{4.33}$$

where the principal values of the powers are to be taken. If we now go on-shell, setting $p' = p$ and then take the limit $\lambda \to 0$, expression (4.33) gives (4.32) multiplied by an extra factor of $\Gamma(1-\mathrm{i}\eta)\exp\left[-\mathrm{i}\eta\ln(2p/\lambda)\right]$ indicating that the limiting procedures need to be performed in reversed order. Accordingly, taking the limit $\lambda \to 0$ *before* going on-shell, we get the on-shell amplitude (4.32) multiplied by an additional factor of $\mathrm{e}^{\pi\eta}$ if $p > p'$ and $\mathrm{e}^{-\pi\eta}$ if $p < p'$. The Coulomb scattering T-matrix has then a discontinuity on-shell and this anomalous behavior complicates seriously all two-potential calculations conducted in momentum space.

Formally, the on-shell Coulomb T-matrix (4.29) has the partial wave expansion

$$T_C(\mathbf{p}',\mathbf{p}) = \mathrm{i}\,\frac{\pi}{\mu p}\sum_{\ell=0}^{\infty}(2\ell+1)(\mathrm{e}^{2\mathrm{i}\sigma_\ell}-1)\,P_\ell(\cos\theta), \tag{4.34}$$

and when the full on-shell scattering amplitude $T(\mathbf{p}',\mathbf{p})$ is defined to have the phase shift $\sigma_\ell+\delta_\ell^c$, then the Coulomb distorted strong interaction scattering amplitude (4.30) has the partial wave expansion

$$T_{SC}(\mathbf{p}',\mathbf{p}) = \mathrm{i}\,\frac{\pi}{\mu p}\sum_{\ell=0}^{\infty}(2\ell+1)\,\mathrm{e}^{2\mathrm{i}\sigma_\ell}(\mathrm{e}^{2\mathrm{i}\delta_\ell^c}-1)\,P_\ell(\cos\theta). \tag{4.35}$$

The Coulomb interaction influences the strong interaction phase shift whose value differs from the Coulomb free phase shift $\delta_\ell$.

For spherically symmetric potentials $\widehat{V}_S$ the full wave function $\psi_{\mathbf{p}}^{(+)}(\mathbf{r})$ may be sought in the form of a partial wave expansion, similar to (4.10)

$$\psi_{\mathbf{p}}^{(+)}(\mathbf{r}) = (pr)^{-1}\sum_{\ell=0}^{\infty}(2\ell+1)\,\mathrm{i}^\ell\,\mathrm{e}^{\mathrm{i}\sigma_\ell}\,u_\ell(p,r)\,P_\ell(\hat{\mathbf{p}}\cdot\hat{\mathbf{r}}), \tag{4.36}$$

where $u_\ell(p,r)$ is the appropriate solution of the two-potential radial wave equation. When the latter function is known, the partial wave amplitudes $\mathrm{e}^{\mathrm{i}\delta_\ell^c}\sin\delta_\ell^c$ may be obtained by using (4.10) and (4.36) in (4.30). One has

$$f_\ell^c(p) \equiv \mathrm{e}^{\mathrm{i}\delta_\ell^c}\sin\delta_\ell^c = -\frac{2\mu}{p}\int_0^\infty F_\ell(\eta,pr)\,V_S(r)\,u_\ell(p,r)\,\mathrm{d}r \tag{4.37}$$

The radial wave function $u_\ell(p,r)$ satisfies also the integral equation

$$u_\ell(p,r) = F_\ell(\eta,pr) + 2\mu\int_0^\infty \langle r\,|G_\ell^{(+)}(p)|\,r'\rangle\,V_S(r')\,u_\ell(p,r')\,\mathrm{d}r', \tag{4.38}$$

which immediately follows when (4.36) and (4.17) are used in (4.25). For sufficiently large $r$ values, much bigger than the range of the strong interaction, from (4.38) we have

$$u_\ell(p,r) \sim F_\ell(\eta,pr) + f_\ell^c(p)\,H_\ell^+(\eta,pr) \tag{4.39}$$

with $f_\ell^c(p)$ given in (4.37). The analogy with the Coulomb free case is apparent. Outside the nuclear interaction region the wave function is given as a superposition of a plane wave plus an outgoing wave, both waves being now distorted by the Coulomb field.

In the analysis and interpretation of two-particle low-energy scattering data a central role plays the so called effective range function. It is well known from the simplest case of s-wave scattering by a short range potential that it is more advantageous to study the effective range function $p \cot \delta$ than the phase shift $\delta$. The former is an even function of the momentum $p$ and close to threshold can be expanded in a power series of $p^2$, known as the effective range expansion. The first two coefficients in this expansion: the scattering length and the effective range, respectively, carrying important information about the underlying interaction, are regarded as fundamental parameters in low-energy scattering phenomenology. The inclusion of the long range Coulomb interaction is a non-trivial complication because the latter changes dramatically the singularity structure of the scattering matrix. The Coulomb corrections to the effective range parameters are not only model dependent but can be also quite sizable. Especially sensitive is the scattering length where, strictly speaking, there is even no upper limit to the size of the Coulomb correction, i.e. the scattering length that is finite in absence of Coulomb interaction, could become infinite when Coulomb interaction is turned on. Perhaps the best illustration that in real life Coulomb correction can be quite large is provided by the nucleon-nucleon scattering. With exact isospin invariance the nn and pp $^1S_0$ scattering lengths in absence of Coulomb interaction should be equal but experimentally : $a_{nn} = -18.6 \pm 0.5$ fm and $a_{pp} = -7.828 \pm 0.008$ fm. It has been confirmed by model calculations that most of the difference in these two values may be attributed to the Coulomb interaction.

The effective range function, in contrast with the scattering amplitude which is a complex function, is real and is defined as the inverse of the K-matrix $\mathcal{K}_\ell(p)$. It is related to the scattering amplitude by means of a generalized Heitler formula

$$\frac{1}{\mathcal{K}_\ell(p)} = \frac{1}{\tau_\ell(p)} + g_\ell(p) \tag{4.40}$$

where the complex function $g_\ell(p)$ can be interpreted as a generalized barrier penetration factor. This function has been studied in several papers [Hull *et al.* (1959); Hamilton *et al.* (1973); Cornile *et al.* (1962); Humblet (1984); Lambert (1969); Kok *et al.* (1981)], and for the charge-less case, it is purely imaginary

$$g_\ell(p) = \mathrm{i}\, \frac{p^{2\ell+1}}{[(2\ell+1)!!]^2} \tag{4.41}$$

whereas for charged particles there must be also a real part attached and its form is

$$g_\ell(p) = \frac{C_\ell(\eta)^2}{C_0(\eta)^2}\, p^{2\ell+1}\, \{2\eta\, h(\eta) + \mathrm{i}\, C_0(\eta)^2\} \tag{4.42}$$

where

$$h(\eta) = \mathrm{Re}\,\psi(1+\mathrm{i}\eta) - \ln|\eta| = \sum_{j=1}^{\infty} \frac{\eta^2}{j(j^2+\eta^2)} - \gamma - \ln|\eta|,$$

$\psi$ denotes the logarithmic derivative of the gamma function and $\gamma$ is Euler constant. The function $h(\eta)$ is an even function of $\eta$ and for $p \to 0$, i.e. for $\eta \to \pm\infty$, is given by the asymptotic expansion

$$h(\eta) \sim \frac{1}{12\eta^2} + \frac{1}{120\eta^4} + \frac{1}{252\eta^6} + \cdots .$$

The function $\tau_\ell(p)$ for charge-less particles is equal to the scattering amplitude but for charged particles it is related to the Coulomb modified nuclear phase shift $\delta_\ell^c$ in the following way

$$\frac{1}{\tau_\ell(p)} = C_\ell(\eta)^2\, p^{2\ell}\, (p \cot\delta_\ell^c - \mathrm{i}\, p) \tag{4.43}$$

where

$$C_\ell(\eta)^2\, p^{2\ell} = C_0(\eta)^2\, \frac{(2\eta p)^{2\ell}}{[(2\ell+1)!]^2} \prod_{j=0}^{\ell} \left(1 + \frac{j^2}{\eta^2}\right). \tag{4.44}$$

Although for physical momenta $p$ the functions $1/\tau_\ell(p)$ and $g_\ell(p)$ are both complex, but their imaginary parts are equal in magnitude and have opposite signs, so that using (4.43) and (4.42) in (4.40), the imaginary parts are canceled and the resulting effective range function (4.40) is real, as it should be

$$\frac{1}{\mathcal{K}_\ell(p)} = \prod_{j=1}^{\ell} \left(1 + \frac{\eta^2}{j^2}\right) \frac{p^{2\ell+1}}{[(2\ell+1)!!]^2} \left[C_0(\eta)^2 \cot\delta_\ell^c + 2\eta h(\eta)\right]. \tag{4.45}$$

When the Coulomb potential is absent, the above formula reduces to the standard form

$$\frac{1}{\mathcal{K}_\ell(p)} \to \frac{1}{K_\ell(p)} = \frac{1}{[(2\ell+1)!!]^2}\, p^{2\ell+1} \cot\delta_\ell, \tag{4.46}$$

and in the particular case of s-wave scattering, from (4.45) we retrieve the well known formula

$$C_0(\eta)^2\, p\, \cot\delta_0^c + 2\eta p\, h(\eta) = \pm\frac{1}{a} + \tfrac{1}{2} r_0 p^2 + \cdots\,, \tag{4.47}$$

where $a$ and $r_0$ are, respectively, the scattering length and the effective range. The sign ambiguity in (4.47) reflecting the coexistence in the literature of two different definitions of the scattering length, has been resolved by adhering to the generally accepted convention that a $(-)$ sign applies for baryon-baryon scattering and a $(+)$ sign in all remaining cases.

# Chapter 5

# Two-Potential Scattering Formalism

When the strong interaction is present, the regular and Jost solutions and the Jost function can be defined similarly as for the pure Coulomb case. To distinguish the former from the pure Coulombic functions we introduce explicitly the parameter $s$ as the third argument while using the same symbols for the different wave functions. Hence, the Jost solutions $f_\ell^+(s,k,r)$ are specified by the boundary condition for $r \to \infty$ but since $v(r)$ is assumed to have a rapid fall-off for large $r$, they will go over into pure Coulomb functions

$$\lim_{r\to\infty} f_\ell^{\pm}(s,k,r) = f_\ell^{\pm}(k,r). \tag{5.1}$$

The regular solution $\phi_\ell(s,k,r)$ is specified by the boundary condition at $r = 0$, and we have

$$\lim_{r\to 0} \phi_\ell(s,k,r) = r^{\ell+1}. \tag{5.2}$$

The Wronskian of the Jost and the regular solution defines as before the Jost function $F_\ell^{\pm}(s,k)$, i.e.

$$F_\ell^{\pm}(s,k) = w\left[f_\ell^{\pm}(s,k,r), \phi_\ell(s,k,r)\right], \tag{5.3}$$

and the regular solution can be expressed as a linear combination of the Jost solutions :

$$\phi_\ell(s,k,r) = \frac{1}{2ik}\left[F_\ell^-(s,k)f_\ell^+(s,k,r) - F_\ell^+(s,k)f_\ell^-(s,k,r)\right]. \tag{5.4}$$

Since the boundary conditions (5.1) and (5.2) do not depend upon $s$ the regular solution, the Jost solutions and, consequently, also the Jost function (5.3) will be analytic functions of the parameter $s$. This follows from the Poincaré theorem which states that if a differential equation is analytic in some parameter $s$ then any solution thereof specified by boundary conditions independent of $s$ will be analytic functions of $s$.

Using the methods developed in [de Alfaro *et al.* (1965); Bertero *et al.* (1967)], it can be shown that when the strong interaction is present the regular and the Jost solutions satisfy the following Volterra equations

$$\phi_\ell(s,k,r) = \phi_\ell(k,r) - s\int_0^r \Gamma_\ell(k;r,r')v(r')\phi_\ell(s,k,r')\mathrm{d}r'; \tag{5.5}$$

$$f_\ell^\pm(s,k,r) = f_\ell^\pm(k,r) + s\int_r^\infty \Gamma_\ell(k;r,r')v(r')f_\ell^\pm(s,k,r')\mathrm{d}r'; \tag{5.6}$$

where the kernel $\Gamma_\ell(k;r,r')$ is an entire function of $k^2$

$$\Gamma_\ell(k;r,r') = \phi_\ell(k,r)\theta_\ell(k,r') - \theta_\ell(k,r)\phi_\ell(k,r'). \tag{5.7}$$

Multiplying (5.6) by $r^\ell$ and taking the $r \to 0$ limit, we obtain a formula for the Jost function

$$F_\ell^\pm(s,k) = F_\ell^\pm(k) - s\int_0^\infty \phi_\ell(k,r)v(r)f_\ell^\pm(s,k,r)dr, \tag{5.8}$$

and an alternative formula can be derived by taking the limit $r \to \infty$ in (5.3). Using (5.1) and (5.5), we obtain

$$F_\ell^\pm(s,k) = F_\ell^\pm(k) - s\int_0^\infty f_\ell^\pm(k,r)v(r)\phi_\ell(s,k,r)dr, \tag{5.9}$$

what implies equality of the integrals in (5.8) and (5.9). Iterating (5.6), it can be proved (cf. [de Alfaro *et al.* (1965)] ) that for potentials satisfying the condition

$$\int_0^\infty |V(r)|\mathrm{e}^{\lambda r} r\,\mathrm{d}r < \infty; \quad 0 \le \lambda < m$$

the corresponding Jost solution $f_\ell^+(s,k,r)$ will be analytic in the half plane $Im\ k > -\frac{1}{2}m$ with a cut from $k = -\frac{1}{2}im$ to $k = 0$ and $f_\ell^-(s,k,r)$, respectively, for $Im\ k < \frac{1}{2}m$ with a cut from $k = 0$ to $k = \frac{1}{2}im$. In both cases there is an essential singularity at $k = 0$. The branch point and the essential singularity at $k = 0$ are of Coulomb origin. In contrast with the purely Coulomb Jost function which is single valued, the full Jost function acquires a branch point at $k = 0$ on account of the integral term in (5.8). Consequently, the S-matrix will have a branch cut along the imaginary axis and essential singularity at $k = 0$.

Adding and subtracting to the right hand side of (5.5) the integral

$$\frac{s\phi_\ell(k,r)}{F_\ell^\pm(k)}\int_r^\infty f_\ell^\pm(k,r')v(r')\phi_\ell(s,k,r')\mathrm{d}r'$$

the Volterra integral equation can be converted into an integral equation of Fredholm type

$$\begin{aligned}\phi_\ell(s,k,r) = {} & \frac{F_\ell^\pm(s,k)}{F_\ell^\pm(k)}\phi_\ell(k,r) - \\ & - s\int_0^\infty \left\langle r|G_\ell^\pm(k)|r'\right\rangle v(r')\phi_\ell(s,k,r')\,\mathrm{d}r',\end{aligned} \tag{5.10}$$

where $\left\langle r|G_\ell^\pm(k)|r'\right\rangle$ is the Coulomb Green's function given in two equivalent forms

$$\left\langle r|G_\ell^\pm(k)|r'\right\rangle = -\frac{\phi_\ell(k,r_<)\,f_\ell^\pm(k,r_>)}{F_\ell^\pm(k)} = -\frac{F_\ell(\eta,\rho_<)\,H_\ell^\pm(\eta,\rho_>)}{k}, \tag{5.11}$$

where the symbols $r_<$, $r_>$ have the usual meaning, i.e. $r_< = \min(r,r')$, $r_> = \max(r,r')$. The Green's functions (5.11) are solutions of the inhomogeneous wave equation

$$\left[\frac{\mathrm{d}^2}{\mathrm{d}r^2} - \frac{\ell(\ell+1)}{r^2} + \frac{2\mu\alpha Z}{r} + k^2\right]\left\langle r|G_\ell^\pm(k)|r'\right\rangle = \delta(r-r'), \tag{5.12}$$

as can be directly checked by integrating equation (5.12) over $r$ in the interval $[r'-\epsilon, r'+\epsilon]$ and subsequently taking the limit $\epsilon \to 0$. Applying the operator in the square bracket on the left hand side of (5.12) to (5.10) we immediately recover the two-potential wave equation (2.3). The Green's function (5.11) has a well known spectral representation valid for $Im\, k > 0$ in which the symmetry in $r$ and $r'$ is manifest

$$\left\langle r|G_\ell^\pm(k)|r'\right\rangle = \frac{2}{\pi}\int_0^\infty \frac{\psi_\ell(p,r)\psi_\ell(p,r')}{k^2 - p^2 \pm \mathrm{i}\epsilon}\,\mathrm{d}p + \sum_{N=0}^\infty \frac{u_{\ell N}(r)\,u_{\ell N}(r')}{k^2 - k_n^2}, \tag{5.13}$$

where $n = \ell + N + 1$. Formula (5.13) could be proved by closing the contour of integration and using the residue theorem for calculating the integral. The contribution from the poles at $p = k_n$ will be cancelled with the discrete spectrum contribution and the pole at $p = k$ yields (5.11).

When the Coulomb potential vanishes we recover from (5.11) the usual expression for the free resolvent

$$\lim_{\alpha Z\to 0}\left\langle r|G_\ell^\pm(k)|r'\right\rangle = -krr'\,h_\ell^\pm(kr_>)\,j_\ell(kr_<).$$

The Fredholm determinant is equal to the ratio of the Jost functions $F_\ell^{\pm}(s,k)/F_\ell^{\pm}(k)$ so if this vanishes we are left with a homogeneous equation (5.10). However, if the Jost function $F_\ell^+(s,k)$ is equal to zero then in view of (5.4) there is no ingoing wave and the solution of the homogeneous equation satisfies Siegert boundary condition. We have to conclude that bound state wave function satisfy homogeneous Fredholm equation (5.10). When the Fredholm determinant is different from zero we have an inhomogeneous equation which is appropriate for the description of the scattering phenomena. This dichotomy is well known as the Fredholm alternative. By contrast, the Volterra equation (5.5) is always inhomogeneous and this form does not change comprising all situations which physically might be quite different.

In general the regular solution can be obtained only numerically, e.g by integrating the wave equation from $r = 0$ outwards using the boundary condition (5.2). Because $v(r)$ is assumed to have a short range when $r$ becomes sufficiently large the $sv(r)$ term in the wave equation is giving negligible contribution in comparison with the remaining terms. As a result in the asymptotic region the wave function could be approximated with arbitrary accuracy by a superposition of two independent Coulomb wave functions. This allows to calculate the S-matrix by matching at some large $r = R$ the regular solution to a superposition of ingoing and outgoing waves distorted by the Coulomb field. For $r \geq R$, we have

$$\phi_\ell(s,k,r) \sim f_\ell^-(k,r) - (-1)^\ell S_\ell(s,k) f_\ell^+(k,r), \tag{5.14}$$

where $S_\ell(s,k)$ denotes the S-matrix describing Coulomb and nuclear scattering. Using (5.4) together with (5.1), we obtain

$$S_\ell(s,k) = (-1)^\ell \frac{F_\ell^-(s,k)}{F_\ell^+(s,k)}, \tag{5.15}$$

where

$$F_\ell^{\pm}(s,k) = w\left[f_\ell^{\pm}(k,r), \phi_\ell(s,k,r)\right]. \tag{5.16}$$

It has become customary factorising the S-matrix $S_\ell(s,k)$ separating the pure Coulomb scattering part. We set

$$S_\ell(s,k) = S_\ell^c(k)\, S_\ell^{nuc}(s,k), \tag{5.17}$$

where $S_\ell^c(k)$ is the Coulomb S-matrix given by (3.71) and $S_\ell^{nuc}(s,k)$ denotes the so called nuclear S-matrix. We shall define the nuclear scattering phase shift $\delta_\ell^c(s,k)$, writing

$$S_\ell^{nuc}(s,k) = \mathrm{e}^{2\mathrm{i}\delta_\ell^c(s,k)}, \tag{5.18}$$

and for complex potentials the phase shift is also complex. Similarly as in the chargeless case, the nuclear T-matrix is related to the S-matrix by the familiar formula

$$kT_\ell^{nuc}(s,k) = \frac{1}{2\mathrm{i}}\left[S_\ell^{nuc}(s,k) - 1\right] = \mathrm{e}^{\mathrm{i}\delta_\ell^c(s,k)} \sin\delta_\ell^c(s,k).$$

The T-matrix can be obtained from the regular solution by imposing the following boundary condition at infinity

$$\phi_\ell(s,k,r) \sim \psi_\ell(k,r) + k\,T_\ell^{nuc}(s,k)\,H_\ell^+(\eta,\rho), \tag{5.19}$$

where $\psi_\ell(k,r)$ has been defined in (3.67). Matching at some large $r$ the logarithmic derivatives in the above equation, the T matrix is given as a ratio of two Wronskians

$$k\,T_\ell^{nuc}(s,k) = -\frac{w\left[\phi_\ell(s,k,r), \psi_\ell(k,r)\right]}{w\left[\phi_\ell(s,k,r), H_\ell^+(\eta,\rho)\right]}.$$

The T-matrix can be also obtained from Fredholm equation. To this end we introduce the wave function denoted as $\psi_\ell^T(s,k,r)$ regular at $r=0$ and for large $r$ satisfying the boundary condition (5.19). This function is a solution of the following integral equation

$$\psi_\ell^T(s,k,r) = \psi_\ell(k,r) - s\int_0^\infty \langle r|G_\ell^+(k)|r'\rangle\, v(r')\,\psi_\ell^T(s,k,r')\,\mathrm{d}r'. \tag{5.20}$$

Using the explicit form of the Green's function (5.11) and going subsequently to the asymptotic limit, we obtain

$$T_\ell^{nuc}(s,k) = \frac{s}{k^2}\int_0^\infty \psi_\ell(k,r)\,v(r)\,\psi_\ell^T(s,k,r)\,dr. \tag{5.21}$$

It is interesting to note that for *repulsive* Coulomb interaction formula (5.20) could be cast into a form resembling chargeless Lippmann–Schwinger equation. To demonstrate that we insert the spectral representation for the Green's function (5.13) in (5.21) and define the half-off-shell T-matrix

$$\langle p|T_\ell(k)|k\rangle = \frac{s}{pk}\int_0^\infty \psi_\ell(p,r)\,v(r)\,\psi_\ell^T(s,k,r)\,dr \tag{5.22}$$

together with the appropriate potential matrix

$$\langle p|sv|k\rangle = \frac{s}{pk}\int_0^\infty \psi_\ell(p,r)\,v(r)\,\psi_\ell(k,r)\,dr. \tag{5.23}$$

Multiplying (5.20) by $(s/pk)\psi_\ell(p,r)v(r)$ and integrating over $r$, we end up with the equation

$$\langle p|T_\ell(k)|k\rangle = \langle p|sv|k\rangle - \frac{2}{\pi}\int_0^\infty \langle p|sv|p'\rangle\,\frac{p'^2\,\mathrm{d}p'}{k^2-p'^2+\mathrm{i}\epsilon}\,\langle p'|T_\ell(k)|k\rangle\,, \tag{5.24}$$

whose form is identical with the familiar Lippmann–Schwinger equation derived for the scattering of chargeless particles. When this equation has been solved, the scattering amplitude is obtained in the usual way, by taking the on-shell value of the T-matrix

$$T_\ell^{nuc}(s,k) = \lim_{p\to k}\langle p|T_\ell(k)|k\rangle\,.$$

The above Coulomb representation of the T-matrix which yields (5.24) was introduced many years ago by [Harrington (1965)] and rediscovered recently by [Elster *et al.* (1993)]. For attractive Coulomb interaction the spectral representation of the Green's function (5.13) contains the discrete spectrum contribution the presence of which precludes the possibility of writing down a simple equation for the T-matrix. The importance of the Fredholm equation approach derives from the fact that (5.20) could be immediately extended to arbitrary non-local interactions and in such a case this equation remains the only sensible alternative for solving the scattering problem.

In many problems instead of the S-matrix it is more convenient to work with the K-matrix. To obtain the latter we use as our two independent Coulomb functions the regular and the irregular functions and for $r \geq R$ formula (5.14) is replaced by the expression

$$\phi_\ell(s,k,r) \sim \phi_\ell(k,r) + \mathcal{K}_\ell(s,k)\theta_\ell(k,r), \tag{5.25}$$

where $\mathcal{K}_\ell(s,k)$ denotes the K-matrix which can be obtained by matching the logarithmic derivatives of (5.25). We have

$$\mathcal{K}_\ell(s,k) = -\frac{w\left[\phi_\ell(k,r),\phi_\ell(s,k,r)\right]}{w\left[\theta_\ell(k,r),\phi_\ell(s,k,r)\right]}. \tag{5.26}$$

The K-matrix can be also obtained from Fredholm equation after converting

(5.5) to the form

$$\phi_\ell(s,k,r) = \left[1 - s\int_0^\infty \theta_\ell(k,r'')\,v(r'')\,\phi_\ell(s,k,r'')\,\mathrm{d}r''\right]\phi_\ell(k,r) - \\ -s\int_0^\infty \langle r|G_\ell(k)|r'\rangle\, v(r')\,\phi_\ell(s,k,r')\,\mathrm{d}r', \tag{5.27}$$

where the Green's function without a superfix is free from singularities in any finite part of the complex k-plane and can be written as

$$\langle r|G_\ell(k)|r'\rangle = -\phi_\ell(k,r_<)\,\theta_\ell(k,r_>) \tag{5.28}$$

The above Green's function can be expressed in terms of the $G_\ell^\pm(k)$ Green's functions

$$\langle r|G_\ell(k)|r'\rangle = \frac{g_\ell^+(k)\,\langle r|G_\ell^-(k)|r'\rangle - g_\ell^-(k)\,\langle r|G_\ell^+(k)|r'\rangle}{g_\ell^+(k) - g_\ell^-(k)}. \tag{5.29}$$

In the limit when the Coulomb interaction goes to zero the above formula reduces to the well known result that the principal value Green's function equals to the arithmetic mean of the ingoing and outgoing Green's functions. From (5.27) we obtain the following expression for the K-matrix

$$\mathcal{K}_\ell(s,k) = \frac{s\displaystyle\int_0^\infty \phi_\ell(k,r)v(r)\phi_\ell(s,k,r)\,\mathrm{d}r}{1 - s\displaystyle\int_0^\infty \theta_\ell(k,r)v(r)\phi_\ell(s,k,r)\,\mathrm{d}r}. \tag{5.30}$$

Similarly as before Fredholm alternative applies to the integral equation (5.27),i.e. at certain specific values of $k$ the Fredholm determinant vanishes and the equation becomes homogeneous. As seen from (5.30) at these values of $k$ the K-matrix has poles.

Fredholm equation approach is greatly simplified when the boundary condition (5.25) is incorporated into the integral equation. We define for this purpose a new wave function regular at $r = 0$ which will be denoted as $\psi_\ell^K(s,k,r)$ and which satisfies the following integral equation

$$\psi_\ell^K(s,k,r) = \phi_\ell(k,r) - s\int_0^\infty \langle r|G_\ell(k)|r'\rangle\; v(r')\,\psi_\ell^K(s,k,r')\,\mathrm{d}r'. \tag{5.31}$$

Going to the asymptotic limit, we obtain a simple formula

$$\mathcal{K}_\ell(s,k) = s\int_0^\infty \phi_\ell(k,r)\,v(r)\,\psi_\ell^K(s,k,r)\,\mathrm{d}r. \tag{5.32}$$

Similarly as before in the case of the T-matrix, the potential $v(r)$ can be replaced by a general non-local interaction in (5.31)–(5.32).

The K-matrix can be also obtained from a first order differential equation. In order to demonstrate that, let us introduce two auxiliary functions

$$\mathcal{S}_\ell(s,k,r) = s \int_0^r \phi_\ell(k,r')v(r')\phi_\ell(s,k,r')\,\mathrm{d}r'; \tag{5.33}$$

$$\mathcal{C}_\ell(s,k,r) = 1 - s \int_0^r \theta_\ell(k,r')v(r')\phi_\ell(s,k,r')\,\mathrm{d}r'; \tag{5.34}$$

and also their ratio

$$\mathcal{R}_\ell(s,k,r) = \frac{\mathcal{S}_\ell(s,k,r)}{\mathcal{C}_\ell(s,k,r)}. \tag{5.35}$$

The latter function is directly connected to the K-matrix and by comparing (5.30) and (5.35), we have

$$\mathcal{K}_\ell(s,k) = \lim_{r\to\infty} \mathcal{R}_\ell(s,k,r). \tag{5.36}$$

Using the Volterra equation (5.5) we can express the wave function $\phi_\ell(s,k,r)$ in terms of the functions defined in (5.33) and (5.34)

$$\phi_\ell(s,k,r) = \phi_\ell(k,r)\,\mathcal{C}_\ell(s,k,r) + \theta_\ell(k,r)\,\mathcal{S}_\ell(s,k,r). \tag{5.37}$$

Inserting the expression (5.37) into (5.33) and (5.34), we end up with a set of two integral equations which can be then converted into differential equations by differentiation with respect to the independent variable $r$. As the result, we obtain two first order differential equations

$$\begin{aligned} \mathcal{S}'_\ell(s,k,r) &= sv(r)\,\phi_\ell(k,r)\,\times \\ &\times\,\Big[\phi_\ell(k,r)\,\mathcal{C}_\ell(s,k,r) + \theta_\ell(k,r)\,\mathcal{S}_\ell(s,k,r)\Big]; \end{aligned} \tag{5.38}$$

$$\begin{aligned} \mathcal{C}'_\ell(s,k,r) &= -sv(r)\,\theta_\ell(k,r)\,\times \\ &\times\,\Big[\phi_\ell(k,r)\,\mathcal{C}_\ell(s,k,r) + \theta_\ell(k,r)\,\mathcal{S}_\ell(s,k,r)\Big]; \end{aligned} \tag{5.39}$$

where the prime in these two equations denotes derivative with respect to $r$ and the boundary conditions at $r = 0$ are

$$\mathcal{S}_\ell(s,k,0) = 0; \qquad\qquad \mathcal{C}_\ell(s,k,0) = 1;$$

The two equations (5.38) and (5.39) can be combined into a single differential equation for the function $\mathcal{R}_\ell(s,k,r)$ defined in (5.35). Indeed, multiplying (5.38) by $C_\ell(s,k,r)$ and (5.39) by $S_\ell(s,k,r)$ and subtracting (5.39) from (5.38), we obtain the following non-linear equation

$$\mathcal{R}'_\ell(s,k,r) = sv(r)[\phi_\ell(k,r) + \theta_\ell(k,r)\mathcal{R}_\ell(s,k,r)]^2, \tag{5.40}$$

and the appropriate boundary condition is $\mathcal{R}_\ell(s,k,0) = 0$. This is a well known Ricatti differential equation and similar equations have been derived in the literature in the variable phase approach to scattering theory. This subject has been thoroughly discussed in the books by [Calogero (1967)] and by [Babikov (1976)]. The necessity to evaluate the Coulomb wave functions at each integration step is a complication and therefore equations (5.38)–(5.39) are less attractive for numerical step-wise integration than the Schrödinger equation. In the latter case the Coulomb wave functions are needed only once in the matching condition (5.26). It is interesting, however, that the equations (5.38)–(5.39) can be readily solved using perturbative approach. Representing $S_\ell(s,k,r)$ and $C_\ell(s,k,r)$ as power series in $s$, we can write

$$S_\ell(s,k,r) = \sum_{n=0}^{\infty} s^n a_n(\ell,k,r); \tag{5.41}$$

$$C_\ell(s,k,r) = \sum_{n=0}^{\infty} s^n b_n(\ell,k,r); \tag{5.42}$$

where the expansion coefficients $a_n(\ell,k,r)$ and $b_n(\ell,k,r)$ are obtained from the recurrences

$$a_n(\ell,k,r) = \int_0^r \mathrm{d}r' v(r')\phi_\ell(k,r') \times \times [\phi_\ell(k,r')b_{n-1}(\ell,k,r') + \theta_\ell(k,r')a_{n-1}(\ell,k,r')]; \tag{5.43}$$

$$b_n(\ell,k,r) = -\int_0^r \mathrm{d}r' v(r')\theta_\ell(k,r') \times \times [\phi_\ell(k,r')b_{n-1}(\ell,k,r') + \theta_\ell(k,r')a_{n-1}(\ell,k,r')]; \tag{5.44}$$

with $a_0(\ell,k,r) = 0$ and $b_0(\ell,k,r) = 1$. The convergence of the series in (5.41) and (5.42) could be accelerated by introducing Padé approximants . It should be noted that the perturbative scheme presented above results in a rational approximation for the K-matrix as function of $s$.

The K-matrix has the following rather advantageous features:

- *(i)* for real potentials and physical momenta K-matrix is real;
- *(ii)* K-matrix is a single valued function of $k^2$;

- *(iii)* except poles K-matrix has no other singularities in any finite part of the k-plane.

To relate the K-matrix with the S-matrix it is sufficient to eliminate in (5.25) the functions $\phi_\ell(k,r)$ and $\theta_\ell(k,r)$ in favor of the Jost solutions $f_\ell^\pm(k,r)$ (cf.(3.17) and (3.18)) and writing the resulting expression in the form (5.14). Subsequently this can be solved for $S_\ell$, and, we obtain

$$S_\ell(s,k) = (-1)^\ell \frac{F_\ell^-(k)}{F_\ell^+(k)} \frac{1 - g_\ell^-(k)\,\mathcal{K}_\ell(s,k)}{1 - g_\ell^+(k)\,\mathcal{K}_\ell(s,k)} \tag{5.45}$$

and comparing this formula with (5.17) we can identify the last factor on the right hand side of (5.45) with the nuclear S-matrix $S_\ell^{nuc}(s,k)$. Using (5.45) we can find the poles of the S-matrix and it is important to note that the nuclear interaction shifts the positions of the Coulombic bound states. Although the Jost function $F_\ell^+(k)$ vanishes at $k = k_n$ where $k_n$ correspond to Coulomb bound states but the function $g_\ell^+(k)$ present in the denominator of (5.45) has poles at $k_n$ so that the zeros are exactly cancelled and the S-matrix goes to a finite limit when $k \to k_n$. On the other hand, the S-matrix does have poles whenever the second denominator in (5.45) vanishes, i.e at the roots of the equation

$$g_\ell^+(k) = \frac{1}{\mathcal{K}_\ell(s,k)}. \tag{5.46}$$

This equation plays a central role in this book and will be extensively used to locate the bound states in the two-potential problem under consideration. Solving equation (5.45) for $\mathcal{K}_\ell(s,k)$ we can express the K-matrix in terms of the nuclear phase shift. We can also calculate the effective range function, and, the result is

$$\frac{1}{\mathcal{K}_\ell(s,k)} = -\mathrm{i}\tfrac{1}{2}\left[g_\ell^+(k) - g_\ell^-(k)\right] \cot\delta_\ell^c + \tfrac{1}{2}\left[g_\ell^+(k) + g_\ell^-(k)\right], \tag{5.47}$$

and in the limit of vanishing Coulomb field, we have

$$\lim_{\alpha Z\to 0} \frac{1}{\mathcal{K}_\ell(s,k)} = \frac{1}{K_\ell(s,k)} = \frac{k^{2\ell+1}}{[(2\ell+1)!!]^2}\cot\delta_\ell, \tag{5.48}$$

where $K_\ell(s,k)$ and $\delta_\ell$ are, respectively the K-matrix and the Coulomb-free nuclear phase shift.

In the following we shall need a formula relating the T-matrix with the K-matrix. From (5.45), we have

$$\frac{1}{\mathcal{K}_\ell(s,k)} = \frac{1}{F_\ell^+(k)F_\ell^-(k)} \frac{1}{T_\ell^{nuc}(s,k)} + g_\ell^+(k),$$

and this equation can be viewed as a generalization of Heitler equation accounting for the presence of the Coulomb field.

# Chapter 6

# Bound States and Low-Energy Scattering

Far and away the best tool for studying the low-energy scattering phenomena is the effective range approximation. On a phenomenological level it constitutes an economic parametrisation of the S-matrix ensuring that the latter will be always unitary. Theoretically, this approximation is also very important and in particular it provides simple means to continue analytically the S-matrix below threshold allowing thereby to pin down the near by poles corresponding to shallow bound states. In the following we shall make extensive use of this possibility and therefore it is essential to develop a reliable scheme for calculating the coefficients of the various low-energy expansions. Actually, in a two-potential scattering problem when both the nuclear and the Coulomb potential are present the derivation of the low-energy scattering parameters presented in the literature has been unnecessarily complicated often involving unjustified approximations. The difficulties stem from the fact that the standard Coulomb wave functions $F_\ell(\eta, \rho)$, $G_\ell(\eta, \rho)$ used in these calculations are not analytic functions of $k^2$ what makes them ill-chosen for studying expansions in powers of $k^2$. We are going to show that by using Coulomb wave functions which are analytic in $k^2$ the low-energy scattering parameters can be immediately obtained in a straightforward way without approximations.

## 6.1 Effective Range Approximation

Since $1/\mathcal{K}_\ell(s,k)$ and $1/K_\ell(s,k)$ are even functions of $k$, for small $|k|$ it should be possible to expand these functions in powers of $k^2$ in a Taylor series around zero

$$\frac{1}{\mathcal{K}_\ell(s,k)} = \frac{1}{A_\ell(s)} + \tfrac{1}{2}R_\ell(s)k^2 + O(k^4), \tag{6.1}$$

$$\frac{1}{K_\ell(s,k)} = \frac{1}{a_\ell(s)} + \tfrac{1}{2}r_\ell(s)k^2 + O(k^4), \tag{6.2}$$

where the effective range approximation, based on the supposition of a fast convergence, consists in keeping only the first two terms in the series. The leading expansion coefficients $A_\ell(s)$ and $R_\ell(s)$ are known as Coulomb corrected scattering length and effective range, respectively while $a_\ell(s)$ and $r_\ell(s)$ denote their Coulomb-free counterparts. For complex potentials these parameters also become complex. The radius of convergence of the expansions in powers of $k^2$ (6.1) or (6.2) is determined by the singularity located nearest to the origin. For a given strength $s$ this singularity can be easily determined by finding zeros of the K-matrix which, as seen from (5.26), will be located at the roots of the equation

$$w\left[\phi_\ell(k,R), \phi_\ell(s,k,R)\right] = 0.$$

The effective range approximation was first introduced for s-waves (cf. ref. [Landau *et al.* (1944a)] ) in which case $A_\ell(s)$ and $R_\ell(s)$ had indeed the dimension of length, however, for arbitrary $\ell$ (cf. ref. [Landau *et al.* (1944b)]) their dimensions are $(length)^{2\ell+1}$ for the scattering length and $(length)^{1-2\ell}$ for the effective range.

In practise, the scattering length and the effective range can be obtained by setting $k = 0$ and integrating the wave equation (2.3) from $r = 0$ outwards to $r = R$ with $R$ chosen to be big enough so that the strong interaction term in (2.3) gives negligible contribution as compared with the remaining terms. Denoting the regular solution of (2.3) as $\phi_\ell(s,k,r)$, for $r = R$ we have

$$\phi_\ell(s,0,R) \sim \phi_\ell(0,R) + A_\ell(s)\theta_\ell(0,R), \tag{6.3}$$

and writing similar formula for the derivative, we obtain

$$A_\ell(s) = -\frac{w\left[\phi_\ell(s,0,R), \phi_\ell(0,R)\right]}{w\left[\phi_\ell(s,0,R), \theta_\ell(0,R)\right]}. \tag{6.4}$$

Using (3.43)–(3.45), the scattering length can be expressed in terms of Bessel functions

$$\frac{A_\ell(s)}{R^{2\ell+1}} = \frac{1}{\pi}\left[\frac{(2\ell+1)!}{(x/2)^{2\ell+1}}\right]^2 \frac{J_{2\ell+1}(x)\left[\ell+1-\mathcal{L}_\ell(s)\right] - (x/2)J_{2\ell+2}(x)}{Y_{2\ell+1}(x)\left[\ell+1-\mathcal{L}_\ell(s)\right] - (x/2)Y_{2\ell+2}(x)}; \tag{6.5}$$

for Coulomb attraction, and

$$\frac{A_\ell(s)}{R^{2\ell+1}} = -\frac{1}{2}\left[\frac{(2\ell+1)!}{(x/2)^{2\ell+1}}\right]^2 \frac{I_{2\ell+1}(x)\left[\ell+1-\mathcal{L}_\ell(s)\right]+(x/2)I_{2\ell+2}(x)}{K_{2\ell+1}(x)\left[\ell+1-\mathcal{L}_\ell(s)\right]-(x/2)K_{2\ell+2}(x)}; \tag{6.6}$$

for Coulomb repulsion. In the above equations (6.5) - (6.6) $x = 2|2\mu\alpha ZR|^{1/2}$ and $\mathcal{L}_\ell(s)$ denotes the logarithmic derivative of the wave function

$$\mathcal{L}_\ell(s) = R\frac{\phi'_\ell(s,0,R)}{\phi_\ell(s,0,R)}. \tag{6.7}$$

For vanishing Coulomb field, i.e. for $x \to 0$ both (6.5) and (6.6) are reduced to the following Coulomb-free expression

$$\lim_{\alpha Z\to 0}\frac{A_\ell(s)}{R^{2\ell+1}} = \frac{a_\ell(s)}{R^{2\ell+1}} = (2\ell+1)\frac{\ell+1-\mathcal{L}_\ell(s)}{\ell+\mathcal{L}_\ell(s)}. \tag{6.8}$$

To obtain the formula for the effective range $R_\ell(s)$, we subtract the equation for $\phi_\ell(s,0,r)$ from the equation for $\phi_\ell(s,k,r)$ and after integrating from 0 to $R$, we have

$$w\left[\phi_\ell(s,k,R),\phi_\ell(s,0,R)\right] = k^2\int_0^R \phi_\ell(s,0,r)\phi_\ell(s,k,r)\mathrm{d}r. \tag{6.9}$$

For large $R$, the left hand side can be evaluated using the asymptotic expressions (5.25) and (6.3). Expanding (6.9) in powers of $k^2$ and neglecting terms $O(k^4)$, we end up with the ultimate formula for the effective range

$$\begin{aligned}\tfrac{1}{2}R_\ell(s) = {} & \frac{w\left[\dot\phi_\ell(0,R),\phi_\ell(0,R)\right]}{A_\ell(s)^2} + 2\frac{w\left[\dot\theta_\ell(0,R),\phi_\ell(0,R)\right]}{A_\ell(s)} \\ & + w\left[\dot\theta_\ell(0,R),\theta_\ell(0,R)\right] - I(s)\left\{\frac{\phi_\ell(0,R)}{A_\ell(s)} + \theta_\ell(0,R)\right\}^2,\end{aligned} \tag{6.10}$$

where

$$I(s) = \frac{w\left[\dot\phi_\ell(s,0,R),\,\phi_\ell(s,0,R)\right]}{\phi_\ell(s,0,R)^2} = \int_0^R \frac{\phi_\ell(s,0,r)^2}{\phi_\ell(s,0,R)^2}\,\mathrm{d}r \tag{6.11}$$

and dot means derivative with respect to $k^2$. The three Wronskians in (6.10) can be easily calculated with the aid of (3.43)–(3.45). In the limit of vanishing Coulomb field, formula (6.10) takes the Coulomb-free form

$$\tfrac{1}{2}r_\ell(s) = \frac{R^{1-2\ell}}{(2\ell+1)^2}\left[\frac{\xi^2}{2\ell+3} + \xi - \frac{1}{2\ell-1} - (\xi+1)^2\frac{I(s)}{R}\right], \tag{6.12}$$

where $\xi = (2\ell+1)R^{2\ell+1}/a_\ell(s)$. It should be noted that the above expressions for the effective range are exact. Although the matching radius $R$ is arbitrary but, in fact, the scattering length and the effective range do not depend on $R$, and, as could be directly checked, the derivatives of (6.8) and (6.12) with respect to $R$ vanish.

For real potentials in the Coulomb-free case the scattering length $a_\ell(s)$ (6.8) is known to be a rapidly varying function of the potential strength and has interlacing zeros and poles when considered to be a function of the argument $s$. It has been shown quite rigorously that the scattering length $a_\ell(s)$ can be represented conveniently as an infinite product (cf. ref. [Deloff (1978)])

$$a_\ell(s) = \frac{\int_0^\infty v(r)\, r^{2\ell+2}\, \mathrm{d}r}{\int_0^\infty v(r)\, r\, \mathrm{d}r} \frac{s}{1 - s/s_{\ell 0}} \prod_{N=1}^{\infty} \frac{1 - s/z_{\ell N}}{1 - s/s_{\ell N}}, \tag{6.13}$$

where the sequences of zeros $z_{\ell N}, N = 1, 2, 3, ...$ and poles $s_{\ell N}, N = 0, 1, 2...$ are just numbers depending upon the shape of the potential $v(r)$ (note that there is always a zero at $s = 0$). For fixed $\ell$ both sequences increase monotonically quite rapidly so that for $s$ not exceedingly large the product in (6.13) can be truncated providing rather accurate rational approximation. When the Coulomb potential is present, we obtain a similar expression for the scattering length $A_\ell(s)$

$$A_\ell(s) = \frac{\int_0^\infty v(r)\, \phi_\ell(0,r)^2\, \mathrm{d}r}{\int_0^\infty v(r)\, r\, \mathrm{d}r} \frac{s}{1 - s/\tilde{s}_{\ell 0}} \prod_{N=1}^{\infty} \frac{1 - s/\tilde{z}_{\ell N}}{1 - s/\tilde{s}_{\ell N}}, \tag{6.14}$$

in which Coulomb shifted zeros $\tilde{z}_{\ell N}$ and poles $\tilde{s}_{\ell N}$ replace, respectively, $z_{\ell N}$ and $s_{\ell N}$. Actually, formula (6.14) is a direct consequence of (6.4). By Poincaré theorem the regular wave function $\phi_\ell(s, 0, r)$ entering (6.4) is necessarily an analytic function of $s$ what leads to the conclusion that $A_\ell(s)$ must be a meromorphic function of $s$. The coefficient in front of (6.14) is fixed and can be unambiguously adjusted by observing that in the limit $s \to 0$ we should recover the first order Born approximation. The zeros and poles in (6.14) can be quite easily located by solving the following two transcendental equations with respect to the variable $s$ :

$$w\left[\phi_\ell(s,0,R), \phi_\ell(0,R)\right] = 0, \tag{6.15}$$

$$w\left[\phi_\ell(s,0,R), \theta_\ell(0,R)\right] = 0, \tag{6.16}$$

so that the roots of the first equation (6.15) yield $\tilde{z}_{\ell N}$ whereas the roots of the second equation (6.16) yield $\tilde{s}_{\ell N}$.

In the past many attempts had been made to express $A_\ell(s)$ in terms of $a_\ell(s)$ because people had hoped that the Coulomb corrections would be independent upon the underlying potential. If this had been the case one could have used the experimentally measured scattering lengths for two charged members of an isospin multiplet (e.g. proton-proton) to infer the scattering length for the two neutral members of the same multiplet (e.g. neutron-neutron). Although by eliminating $s$ between (6.13) and (6.14) it would be possible to express $A_\ell(s)$ in terms of $a_\ell(s)$ but the resulting expression would still depend upon the shape of the potential $v(r)$, primarily via the integrals occurring in (6.13) and (6.14).

As seen from (6.12) and (6.10) the effective range takes the form of a second order polynomial in the inverse of the scattering length and therefore in general exhibits a second order pole at the zero of the scattering length. If the coefficient of the quadratic term vanishes the polynomial is of the first order and the effective range has only a simple pole. This accidental degeneracy occurs if $I(s) = a\phi_\ell(0, R)^2$ for the charged case and, respectively, if $I(s) = R/(2\ell + 3)$ for the neutral case. The sign of the effective range depends upon i the value of $\ell$ and the shape of the underlying potential which is here represented by the integral $I(s)$. The behavior of the effective range for $\ell > 0$ is different from that for $\ell = 0$, as can be easily established by examining the roots of the above mentioned polynomial: for $\ell = 0$ the polynomial in (6.12) has real roots if $I(s) \geq \frac{1}{4}R$ whereas for $\ell > 0$ this condition becomes $I(s) \leq \frac{1}{4}R$. We can summarize all possibilities:

$\ell = 0$:

if $I(s) \leq \frac{1}{4}R$ then $r_\ell(s) \geq 0$;

if $\frac{1}{4}R \leq I(s) < \frac{1}{3}R$ then $r_\ell(s) > 0$ becoming negative between the roots;

if $I(s) = \frac{1}{3}R$ degenerate case, interlacing single poles and zeros;

if $I(s) > \frac{1}{3}R$ then $r_\ell(s) < 0$ becoming positive between the roots;

$\ell > 0$:

if $I(s) \geq \frac{1}{4}R$ then $r_\ell(s) \leq 0$;

if $\frac{1}{2\ell+3}R < I(s) \leq \frac{1}{4}R$ then $r_\ell(s) \leq 0$ becoming positive between the roots;

if $I(s) = \frac{1}{2\ell+3}R$ degenerate case, interlacing single poles and

zeros;
if $I(s) < \dfrac{1}{2\ell+3}R$ then $r_\ell(s) \geq 0$ becoming negative between the roots;

The charged case can be analyzed in the same way and all the qualitative features of the effective range are found to be similar to the neutral case. A typical behavior of the scattering length and the effective range is displayed in Fig. 6.1 for $\ell = 2$ and attractive Coulomb potential.

For illustrative purposes in this work we have been using as the nuclear interaction the potential representing the phenomenological strong interaction of negatively charged K-meson with the $^{32}$S nucleus which is assumed to be of the same shape as the nuclear charge density $\rho(r)$. For traditional reasons it has been customary expressing the nuclear potential in the form

$$2\mu V(r) = -\bar{a}A\left(1 + \frac{m_K}{m}\right)\rho(r) \tag{6.17}$$

where $\bar{a}$ is a complex parameter of the dimension of length representing the strength of the interaction, $A$ is mass number, $m_K$ and $m$ denote, respectively, the masses of the kaon and the nucleon. The function $\rho(r)$ has been taken in the form of a two parameter Fermi distribution normalized to one

$$\rho(r) = \frac{3/c^3}{\left[1 + (\frac{\pi a}{c})^2\right]\left\{\exp[(r-c)/a] + 1\right\}}, \tag{6.18}$$

where $c = 3.2\,fm$ and $a = 0.59\,fm$. Using our normalisation, the nuclear potential is written as $s\,v(r)$ with

$$s = \bar{a}\,A\,\frac{3}{c^3}\,\frac{1 + (m_K/m)}{1 + (\pi a/c)^2}\,\frac{1}{v_0}; \tag{6.19}$$

and

$$v(r) = \frac{v_0}{\exp[(r-c)/a] + 1}, \tag{6.20}$$

where $v_0$ is given as

$$\frac{1}{v_0} = \int_0^\infty \frac{r\,\mathrm{d}r}{\exp[(r-c)/a] + 1} = \frac{1}{2}c^2 + a^2\left[\frac{\pi^2}{6} + \sum_{n=1}^{\infty}\frac{(-1)^n}{n^2}\mathrm{e}^{-nc/a}\right],$$

and the sum rapidly converges. With the adopted values of the parameters we obtain $v_0^{-1} = 5.691067885\,fm^2$ and $s = 19.04822663\,fm^{-1}\,\bar{a}$ .

In Table 6.1 we present the values of the strength s at which the scattering length has a pole computed for the potential (6.17) for the three lowest partial waves. When Coulomb attraction is present less attraction is

Table 6.1 Values of the strength $s$ of the potential (6.17) for which the scattering length has a pole for $\ell = 0, 1, 2$. The upper part of the table contains the values of $s_{\ell N}$ from formula (6.13), whereas the lower part contains $\tilde{s}_{\ell N}$ from formula (6.14).

Without Coulomb

| N | $\ell = 0$ | $\ell = 1$ | $\ell = 2$ |
|---|---|---|---|
| 0 | 1.29426 | 5.57216 | 12.0430 |
| 1 | 9.98779 | 18.6802 | 29.5111 |
| 2 | 25.6735 | 38.6413 | 53.7663 |
| 3 | 48.2873 | 65.4863 | 84.8689 |

With Coulomb

| N | $\ell = 0$ | $\ell = 1$ | $\ell = 2$ |
|---|---|---|---|
| 0 | -2.87457 | 3.21110 | 9.26730 |
| 1 | 4.77458 | 15.7286 | 25.6160 |
| 2 | 19.5569 | 35.4307 | 48.8005 |
| 3 | 41.5437 | 62.2426 | 78.8781 |

needed to make $1/A_\ell(s)$ vanish and the poles of $A_\ell(s)$ are shifted towards smaller values of s comparing with the poles of $a_\ell(s)$. We note that for $\ell = 0$ the first pole of $A_0(s)$ has been pushed towards negative s. At the values of s at which the scattering length $a_\ell(s)$ has a pole we have a zero-energy bound states. Since the smallest value of s at which the Coulomb corrected scattering length $A_0(s)$ has a pole corresponds to a *repulsive* potential, the relation between the poles of $A_\ell(s)$ and zero-energy bound states in the two-potential problem cannot be sustained any more. We shall resume this subject in the next Section.

In Fig. 6.1 we present the inverse of the scattering length and the effective range as functions of s calculated for the potential (6.20). As seen from Fig. 6.1 the effective range takes only negative values and a similar behavior is observed for all partial waves with $\ell > 0$. The case $\ell = 0$, however, is

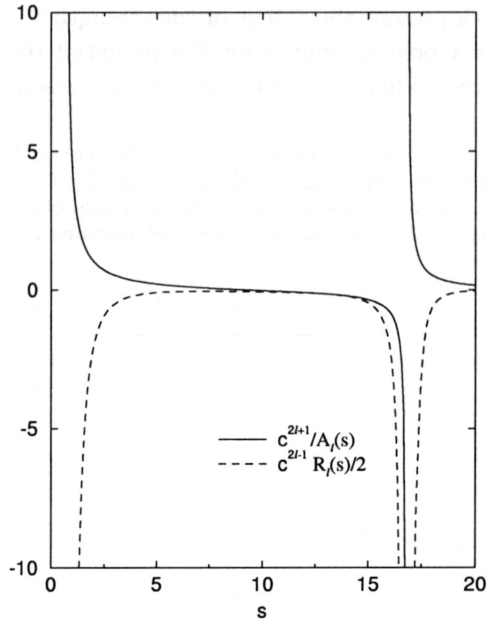


Fig. 6.1 Low-energy scattering parameters versus $s$ for $\ell = 2$. The inverse of the scattering length $1/A_\ell(s)$ and the effective range $R_\ell(s)$ have been made dimensionless by scaling them with the appropriate power of the range $c$ of the nuclear potential, i.e. with $c^{-(2\ell+1)}$ and $\frac{1}{2}c^{1-2\ell}$, respectively. We used the potential (6.20) together with a point-like Coulomb potential.

different and the effective range is always positive taking infinite values at zeros of the scattering length.

The validity of the product formulae (6.13) and (6.14) can be immediately extended to arbitrary complex $s$ values by analytic continuation. Since the poles $s_{\ell N}$ and $\tilde{s}_{\ell N}$ reside on the real axis, for absorptive potentials close to the pole we observe a rapid resonant like behavior but the scattering length always remains finite in contrast with the real $s$ case. More insight into the qualitative behavior of $a_\ell(s)$ and $A_\ell(s)$ for complex $s$ can be gained by rewriting (6.13) as a sum of pole terms

$$a_\ell(s) = s \sum_N \frac{C_{\ell N}}{s_{\ell N} - s}, \tag{6.21}$$

where the residues $C_{\ell N}$ are real and positive numbers (cf. ref. [Deloff (1978)]) and similar expression follows also for the charged case, just by replacing in (6.21) $C_{\ell N}, s_{\ell N}$ , respectively, by $\tilde{C}_{\ell N}, \tilde{s}_{\ell N}$. The presence of the Coulomb field brings no additional complications so that the two cases can be jointly discussed. When $s$ is complex $s = x + iy$ with real $x, y$ such that $x$ is close to $s_{\ell N}$ and $y \ll s_{\ell N}$, we can keep only the contribution from the dominant pole in (6.21), writing the real and imaginary part of the scattering length in the form

$$Re\, a_\ell(s) \approx C_{\ell N} \frac{x(s_{\ell N} - x) - y^2}{(s_{\ell N} - x)^2 + y^2}, \tag{6.22}$$

$$Im\, a_\ell(s) \approx C_{\ell N} \frac{s_{\ell N} y}{(s_{\ell N} - x)^2 + y^2}. \tag{6.23}$$

For fixed $y$, $Re\, a_\ell$ and $Im\, a_\ell$ show very simple behavior as functions of $x$: $Re\, a_\ell$ has a maximum and a minimum passing through zero in between, while $Im\, a_\ell$ exhibits a Lorentzian peak at $x = s_{\ell n}$. The zero of $Re\, a_\ell$ falls very close to $s_{\ell N}$ and from (6.22) we immediately infer that the zero is located at $s_{\ell N}(1 - y^2/s_{\ell N}^2)$. Taking into account the fact that in general the scattering length is given as a sum of pole terms, we conclude that for fixed absorption both $Re\, a_\ell$ and $Im\, a_\ell$ will be oscillating functions of $Re\, s$. The parameter $Im\, s$ controls the height of the amplitude of these oscillations and the oscillations are damped when $Im\, s$ increases.

A typical example is presented in Fig. 6.2 where we consider the kaonic Sulphur case with the nuclear potential given in (6.20) for $\ell = 2$. We regard the imaginary part of $s$ as a fixed parameter while $Re\, s$ is our independent variable and the real and imaginary part of the scattering length $A_\ell(s)$ are displayed in Fig. 6.2 as functions of $Re\, s$. We can see that for $Re\, s$ in the vicinity of the pole $\tilde{s}_{\ell N}$ the rapid variation in both $Re\, A_\ell$ and $Im\, A_\ell$ is in good accord with our approximate formulae (6.22) and (6.23). The "resonant" behavior of the scattering length can be seen also from the Argand diagram presented in Fig. 6.3 where $Im\, A_\ell$ is plotted against $Re\, A_\ell$ for the same physical situation as before. Similarly as in Fig. 6.2 $Im\, s$ is regarded as a fixed parameter and varying $Re\, s$ from 8.0 to 10.0 causes $A_\ell$ to move in the complex plane along certain trajectory. The presence of poles in the scattering length induces characteristic anti-clockwise looping of the trajectory resembling the familiar behavior of the scattering amplitude close to a resonance. With increasing absorption the loops get tighter and tighter, as illustrated in Fig. 6.3, and in consequence the trajectories become confined to a smaller and smaller region of the complex plane.

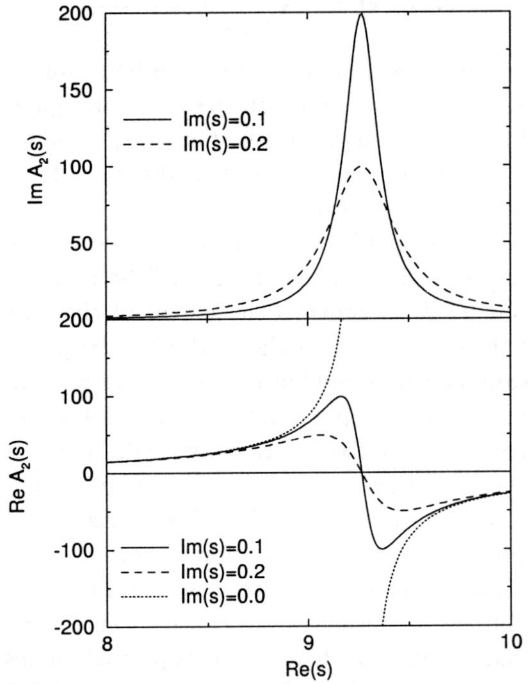


Fig. 6.2 Real and imaginary part of the complex scattering length $A_\ell(s)$ in $a_B^{2\ell+1}$ units versus $Re\, s$ when to the nuclear potential (6.20) is supplemented by the Coulomb interaction for $\ell = 2$.

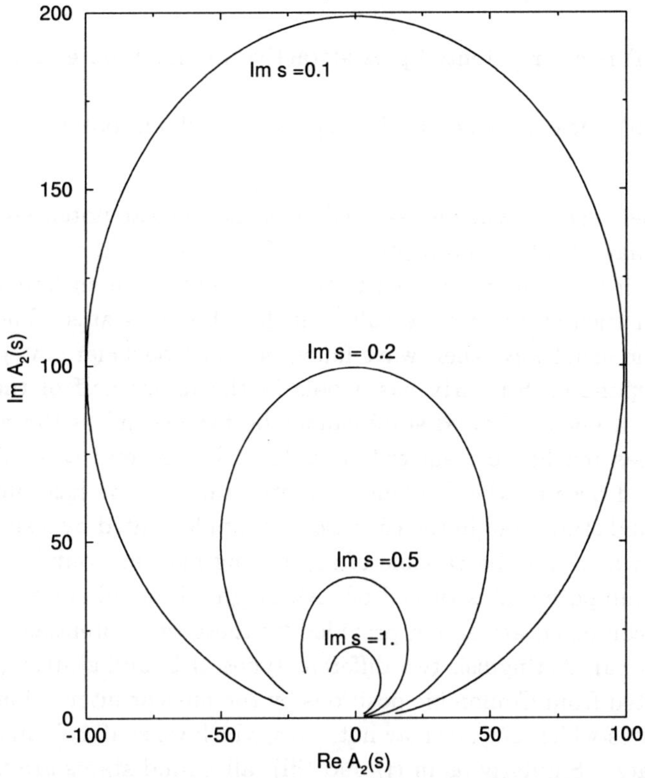


Fig. 6.3 Contour plots for the complex scattering length $A_\ell(s)$ in $a_B^{2\ell+1}$ units when the nuclear potential (6.20) is supplemented by the Coulomb interaction for $\ell = 2$. The values of $Re\,s$ have been varied in the range [8.0,10.0] and the fixed values of $Im\,s$ have been given at the top of each curve.

## 6.2 Nuclear and Quasi-Nuclear Bound States

We are going now to consider in some detail the bound state problem and it will be appropriate to examine separately the following three choices of the interaction in the two-potential Hamiltonian:

- (i) nuclear potential without Coulomb interaction;
- (ii) nuclear potential plus attractive Coulomb potential,
- (iii) nuclear potential plus repulsive Coulomb potential.

In all three cases it will be assumed that the nuclear potential is attractive and may also have absorptive part. Obviously, in the first and in the third case if the bound state exists then it ought to be entirely of nuclear origin and such states will be called nuclear bound states. They are defined in the usual way: their wave functions must be square integrable and the corresponding S-matrix has a pole in the upper half of the k-plane. Of course, a nuclear bound state cannot be formed unless the nuclear attraction has reached certain critical value. The second case (ii) is more complicated because the Coulomb potential taken alone has infinite number of bound states and in the (ii) case it is supplemented by a short ranged potential which may be capable of supporting nuclear bound states when the Coulomb potential is turned off. Although all bound states result then from the combined attraction provided by these two potentials but nevertheless we can distinguish two different types of bound states: (a) atomic levels shifted from Coulombic positions by the nuclear interaction, and, (b) bound states which may, or may not, exist which we shall call quasi-nuclear bound states. Similarly as in (i) and (iii), all bound states are necessarily associated with poles of the S-matrix that are located in the upper half of the k-plane. In the case (ii), however, when the strength of the nuclear potential is reduced to zero ($s \to 0$) the bound state poles either tend to occupy positions on the imaginary axis as appropriate for the unperturbed Coulomb spectrum, or move into the lower half of the k-plane. If in the limit $s \to 0$ the pole resumes a position on the imaginary axis we have to do with the case (a) and if it moves into the lower half of the k-plane we have a quasi-nuclear bound state (b). It should be noted that knowing just the position of the pole is not enough to tell to which category it belongs either (a), or (b). To answer this question the limiting procedure $s \to 0$ has to be effected in order to examine the motion of the pole in the k-plane.

In the last section we derived the bound state condition (5.46) connecting the location of the bound state pole in the k-plane with the corresponding strength $s$ of the effective potential. For fixed $s$ formula (5.46) can be regarded as an equation with respect to $k$ whose solutions yield all the bound states of our two-potential problem. In the following we shall be faced also with the inverse problem when the position of the pole is known and we want to find all $s$ which generate such a bound state. Then, it is the variable $k$ in (5.46) that is fixed and we obtain an equation with respect to $s$. In order to resolve the problem of quasi-nuclear bound states we shall make extensive use of formula (5.46). To begin with, let us consider first the case when absorption is absent assuming for the time being that the strength $s$ is real. For the chargeless case (i) the potential is capable of supporting a nuclear bound state only if $s$ exceeds certain critical value which could be determined from (5.46). We know from (2.7) that for real $s$ the nuclear bound state poles must be necessarily located on the positive imaginary axis of the k-plane and the critical $s$ corresponds to a bound state pole situated at the origin of the k-plane. To obtain the strength $s$ necessary to produce a zero-energy nuclear bound state in the chargeless case (i), we set $k = 0$ and $\alpha Z \to 0$ in (5.46), which immediately yields the well known condition

$$\frac{1}{a_\ell(s)} = 0. \tag{6.24}$$

As seen from (6.24), the poles of the scattering length provide the sought for values of the strength necessary for supporting zero-energy nuclear bound states. Indeed, at these values of $s$ the pole of the S-matrix resumes the position at the origin $k = 0$ and we have a zero-energy nuclear bound state. With further increase of the potential strength the S-matrix pole enters the the positive imaginary axis in the complex k-plane and the force is capable of supporting a bound state. With still bigger attraction we can produce second, third, and more bound states in a similar way.

When the Coulomb potential is present the analytic structure of the S-matrix changes dramatically and $k = 0$ becomes a highly singular point because a branch point singularity together with an essential singularity happen to reside right at the origin. Nevertheless, we can examine the possibility of zero-energy bound states by taking real values of $k$ setting $k \to 0$ in (5.46). For repulsive Coulomb potential, i.e. for the case (iii), from (3.37) and (5.46), we obtain

$$\frac{1}{A_\ell(s)} = 0, \tag{6.25}$$

which is the same result as for the Coulomb-free case (6.24). Reiterating, the poles of the Coulomb corrected scattering length $A_\ell(s)$ provide the values of the strength necessary for supporting zero-energy nuclear bound states. However, for the case (ii), i.e. for an attractive Coulomb potential, using (3.39), we obtain completely different answer, arriving at the condition

$$\frac{1}{A_\ell(s)} = -\mathrm{i}\pi \frac{(2\mu\alpha Z)^{2\ell+1}}{[(2\ell+1)!]^2}, \tag{6.26}$$

in which on the right hand side zero has been replaced by an imaginary constant. Since for real $s$ the scattering length $A_\ell(s)$ must necessarily be real it is quite clear that this condition can never be satisfied. As a result the quasi-nuclear bound state pole cannot enter the positive part of the imaginary axis because passing through the origin is prohibited according to (6.26). We conclude that if we have an attractive Coulomb potential and a real short ranged nuclear potential then this system always has only Coulomb bound states whose positions become shifted owing to the presence of the nuclear potential. It is not possible to produce a quasi-nuclear bound state, no matter how strongly attractive we make the nuclear potential. We have arrived at the remarkable saturation theorem.

**Theorem 1** *Coulomb spectrum is saturated: by supplementing attractive Coulomb potential with a real short range potential a distorted Coulomb spectrum is obtained but the resulting infinite denumerable set of bound states will remain in a one-to-one correspondence with the unperturbed Coulomb spectrum.*

The essence of the saturation theorem is that for real potentials the quasi-nuclear bound states do not exist. Since the additional attraction provided by a real nuclear potential can not increase the number of levels, the only effect is a restructuring of the Coulomb spectrum so that each Coulombic level becomes shifted, in general by a different amount. The proof of the saturation theorem is a corollary of the following no-crossing theorem, valid for real potentials.

**Theorem 2** *The energy levels in the two-potential problem where the Coulomb potential is combined with a real short range potential are not* *allowed to cross.*

The eigenmomenta $k_n$ can be regarded as functions of the real parameter $s$ representing the strength of the nuclear potential. When $s$ is varied the momenta $k_n(s)$ will be moving in the k-plane along certain trajectories and the different trajectories cannot cross each other in the upper half of the k-plane. To prove this theorem let us suppose the opposite, i.e. that at some $s = s_0$ two trajectories did cross. This means that as $s$ is varied in the vicinity of $s_0$ two bound state poles of the S-matrix come close to each other. At $s = s_0$ they would coincide and as a result we would have a second order pole. This, however, is not possible because, as well known [Newton (1982); de Alfaro *et al.* (1965)], for real potentials all bound state poles are necessary simple. Indeed, at a double pole we would need

$$F_\ell^+(s,k) = \dot{F}_\ell^+(s,k) = 0 \tag{6.27}$$

where the dot denotes derivative with respect to $k$. But the derivative of the Jost function can be also written as

$$\dot{F}_\ell^+(s_0,k_0) = -\frac{4ik_0^2}{F_\ell^-(s_0,k_0)}\int_0^\infty \phi_\ell(s_0,k_0,r)^2\,dr, \tag{6.28}$$

where $k_0$ is the position of the pole. Since for real potentials the wave function is real, the normalisation integral in (6.28) can not vanish and it is evident that $\dot{F}_\ell^+(s_0,k_0)$ must be different from zero. Thus, the supposition that levels could cross leads to a contradiction proving thereby the no-crossing theorem.

We are now in the position to prove the saturation theorem by *reductio ad absurdum.* Let us assume then that the nuclear potential is strong enough to support a bound state on its own. For the sake of the argument we make a supposition that after adding the Coulomb potential we obtain an infinite number of levels but one of them would be quasi-nuclear while the remaining levels are atomic. In order to find out whether this is possible, we let the strength of the nuclear potential go to zero. The poles associated with atomic levels all tend then to a well prescribed limit, namely to the unperturbed Coulombic positions on the positive part of the imaginary axis in the k plane. By definition, as a result of the above limiting procedure, the pole associated with the quasi-nuclear level would be forced to travel along the imaginary axis towards zero, reaching eventually the origin at certain value of the strength when the nuclear potential becomes too weak to sustain binding. Since the atomic spectrum is infinite, wherever we choose to put the quasi-nuclear pole, there will be still infinite number of atomic poles between zero and the chosen location of the quasi-nuclear pole. Therefore, the quasi-nuclear pole before reaching the origin would have to collide with at least one of the atomic poles. This, however, is not possible without violating the no-crossing theorem. The assumed existence of a quasi-nuclear level leads to a contradiction and the proof of the saturation theorem is complete.

The conclusion that in a two-potential problem – where the Coulomb potential is combined with a *real* nuclear potential – no quasi-nuclear bound states can be produced, irrespectively of how strongly attractive is the nuclear potential, goes perhaps against common prejudices and at first sight might even appear to be wrong. Therefore, in order to shed some more light on this unexpected phenomenon, we are going to provide an illustrative example in which the calculations can be done analytically. Let us consider then a toy model of a hadronic hydrogen (Z=1), where an unspecified hadron and a proton are bound thanks to a point-like Coulomb potential. For simplicity, we choose to simulate the strong hadron-proton interaction by a delta-shell potential, putting $2\mu V(r) = -(s/R)\,\delta(r-R)$, where $\mu$ is the reduced mass of the system and the two real parameters s and R represent, respectively, the strength and the range of our nuclear potential. The exact bound state condition for this two-potential model, takes the form ( cf. (11.15) for particulars)

$$s = \rho\frac{F'_\ell(\eta,\rho)}{F_\ell(\eta,\rho)} - \rho\frac{H_\ell^{+\prime}(\eta,\rho)}{H_\ell^{+}(\eta,\rho)} \tag{6.29}$$

where $\rho = kR$ and the logarithmic derivatives occurring in this formula may be readily computed by using the appropriate continued fraction representations (cf. (3.60)–(3.61)). In a particular case when $s = 0$, (6.29) reduces to the familiar condition for Coulomb bound states. The bound state condition (6.29) remains also valid when there is no Coulomb interaction ($\eta$=0) and in such a case the solution of (6.29) with respect to k, for an assigned and large enough value of s, gives the nuclear bound state energy. It is a specific feature of the delta-shell potential that in each partial wave this potential can support at most a single nuclear bound state. In the following we take $\ell$=1, and adopt the Coulomb energy of the 2p level, $E_{2p}$, for setting the "atomic" energy scale. We have now three dimensionless parameters in the game, namely $y = E/E_{2p}$, s and $\xi = R/a_B$. Since a typical nuclear potential is of a short range, $\xi$ must be a small number, i.e $\xi \ll 1$ and hereafter we set rather arbitrarily $\xi = 0.1$. As a matter of fact, to obtain

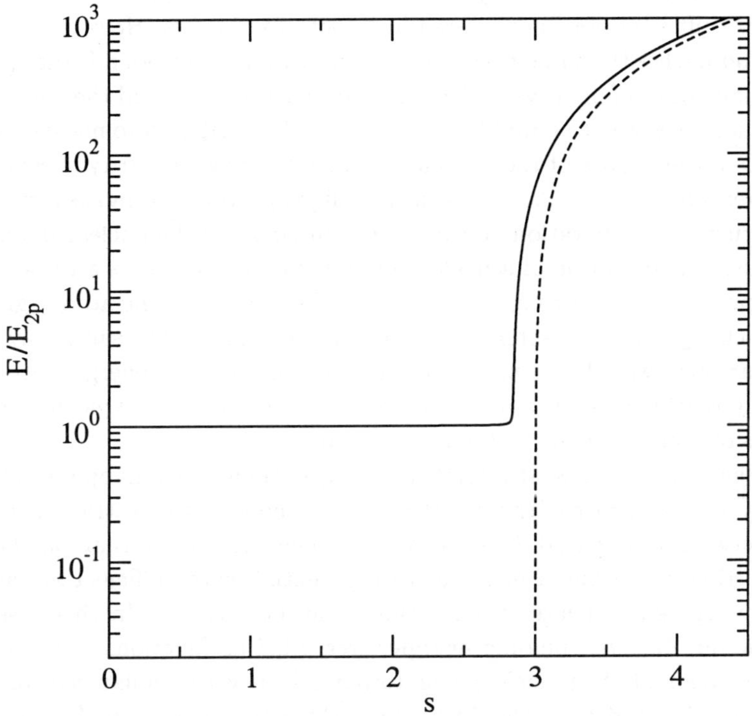


Fig. 6.4 Binding energy of the atomic 2p level in a hadronic hydrogen as a function of strength of the nuclear potential (continuous curve). The dashed curve represents the energy of the nuclear 2p bound state formed in the same hadron-proton system in absence of Coulomb interaction.

the binding energy y for a given value of s, there is no need to solve (6.29) for y. With the aid of (6.29) it is much easier to calculate s as a function of y and then plot y versus s. The results of our computation are presented in Fig. 6.4. We begin with the nuclear bound state which is obtained by switching off the Coulomb interaction and then using (6.29). The nuclear binding energy, depicted by the dashed curve in Fig. 6.4, exhibits a typical behavior, i.e. binding taking place only above certain minimal value of the strength (here, s=3) and for increasing s (s>3) we note a rapid increase of the binding energy changing by several orders of magnitude. Next, we turn to the hadronic atom problem where both potentials are present. In contrast with the nuclear case, the atomic bound state, depicted by the full curve in Fig. 6.4, exists for all values of s. For s>0 but smaller than 3 the increase of the binding energy is very slow which is on a par with the small energy shift situation encountered in X-ray experiments. However, as soon as s approaches the value necessary to support a nuclear bound state (s=3), the shape of the full curve undergoes a dramatic change and the small shift behavior gives way to a rapid increase normally attributed to nuclear levels. For s> 3 the shapes of the two curves show striking similarity except that for fixed s it is the atomic state that has slightly more binding, as a result of constructive collaboration between the Coulomb and the nuclear potential. Because the formation of a nuclear bound state is barred, as a response to the application of a strongly attractive nuclear potential the atomic ground state changes its characteristics behaving very much the same way as a nuclear level would. In particular, the atomic binding energies are very rapidly reaching values appropriate for the nuclear scale, typically several orders of magnitude above the atomic scale.

When s is increased still further (s≫ 3) nothing much happens but this is only a consequence of the fact that our toy model is too restrictive in that the delta-shell potential does not permit to generate more than one bound state. However, using a more realistic potential, with sufficient attraction the 3p, 4p, *etc.* nuclear bound states may be created. In this case the atomic 3p, 4p, *etc.* binding energies, regarded as functions of s, imitate the behavior of their nuclear counterparts in a very similar way to that presented if fig. 6.4. It should be also added that the above features are by no means particular to the p-wave and in qualitative terms a similar behavior is observed in all partial waves. We shall return to that in the next chapter.

We wish to reiterate that for the above theorems to be true it was absolutely crucial that the potentials were taken to be real. Now, we wish to relax this limitation admitting complex nuclear potentials. As we shall see, this extension has profound consequences. In particular, for complex nuclear potentials the above theorems are no longer valid. As regards the saturation theorem, the condition for a zero-energy quasi-nuclear bound state (6.26) can be viewed as an equation in which $s$ is the unknown quantity. We know that for real s this equation has no solution but when $s$ is extended into the complex domain a solution always exists. Besides that, with complex $s$ the bound state poles are no longer confined to the imaginary axis. Therefore, for complex potentials, it becomes perfectly possible to generate quasi-nuclear bound state poles. We note that when $s$ is complex, then the appropriate regular solution of the wave equation will also be complex and vanishing of the normalisation integral in (6.28) is no longer impossible. Therefore, the bound state poles of the S-matrix are not necessarily simple and the corresponding energy levels are allowed to cross.

Summarizing, when the potential is real the hadronic atom spectrum is purely atomic and only when the nuclear potential is complex we may have additional quasi-nuclear levels. Varying $s$ causes all poles to move in the k-plane along certain trajectories and the limiting procedure $s \to 0$ reveals the origin of the bound state. The behavior of pole trajectories has been studied extensively by [Nussenzveig (1972)] for the square well potential, by [Ferreira *et al.* (1966)] for a cut-off Coulomb potential, and, by [Bang *et al.* (1980)] for a Saxon–Woods potential but all the above authors assumed real potentials and did not include Coulomb potential. On the other hand, pole trajectories for Coulomb interaction combined with a complex, separable potential of rank one have been extensively investigated in a number of papers (cf. [Kok *et al.* (1981); van Haeringen (1977); Kok *et al.* (1982)]). The general case of a Coulomb plus local complex potential in its full complexity presents a non-trivial numerical problem when it comes to Gamow states primarily because the corresponding wave functions blow up at large separations. We are not aware of any paper which would examine pole trajectories for that case and therefore we are going to fill this gap. However, before we do that some comments on pole collision phenomenon are in order.

It has to be emphasized that the no-crossing theorem given above is valid only for bound states. In the lower half of the k-plane this theorem is not applicable any more and even for real potentials the pole trajectories are allowed to cross there, or using equivalent parlance, the poles may collide. As a matter of fact, as follows from numerical studies, this is a not an uncommon occurrence in the motion of the poles.

For complex potentials the poles can collide anywhere in the k-plane. We recall that the poles of interest to us arise from zeros of the Jost function $F_\ell^+(s,k)$ and although the latter by Poincaré theorem is analytic function of $s$ but the solution of the equation $F_\ell^+(s,k) = 0$, i.e. the zero of the Jost function $k_n(s)$ is not analytic function of $s$. This is immediately seen by invoking the Feynman–Hellmann theorem for calculating the derivative of $k^2$ with respect to $s$. We obtain

$$\frac{\mathrm{d}k^2}{\mathrm{d}s} = \frac{\int_0^\infty \phi_\ell(s,k,r)^2 v(r)\mathrm{d}r}{\int_0^\infty \phi_\ell(s,k,r)^2 \mathrm{d}r}; \tag{6.30}$$

but at the crossing point of two trajectories in the upper half of the k-plane the norm integral vanishes causing the derivative (6.30) to become infinite. This implies that the function $k_n(s)$ has a branch point singularity there. Actually, this is true also for Gamow states i.e. when crossing takes place in the lower half of the k-plane in which case it does not matter whether $s$ real or complex. This general property is a consequence of the following theorem (cf. [McVoy (1967)]).

**Theorem 3** *If by varying $Re\, s$ for fixed $Im\, s$, two poles collide in the k-plane, the line of escape is at right angle to the line of approach.*

To demonstrate that suppose that for $s = s_0$ two poles collide at $k = k_0$. The Jost function can be expanded about these values in a double Taylor series. We write

$$F_\ell^+(s,k) = F_\ell^+(s_0,k_0) + \dot{F}_\ell^+(s_0,k_0)(k-k_0) + F_\ell^{+\prime}(s_0,k_0)(s-s_0) + \\ + \ddot{F}_\ell^+(s_0,k_0)(k-k_0)^2 + \cdots, \tag{6.31}$$

where prime denotes derivative with respect to $s$. Since by assumption at $(s_0, k_0)$ there is a crossing point, the first two terms in (6.31) are bound to vanish, and to the lowest order the motion of the zero in the k-plane is given as

$$k(s) \approx k_0 \pm \sqrt{\frac{2(s-s_0)F_\ell^{+\prime}(s_0,k_0)}{\ddot{F}_\ell^+(s_0,k_0)}}, \tag{6.32}$$

and this formula exhibits the branch point singularity just mentioned. The signs $\pm$ correspond to the two trajectories crossing at $k_0$. When $s$ is varied, the two $k$ values approach towards each other along a line passing through $k_0$. When the collision occurs $s - s_0$ passes from positive to negative values and this produces the factor of i before the square root in (6.32) so that the line of recession acquires a phase factor $e^{i\pi/2}$. Consequently, the line of escape is rotated by right angle with respect to the line of approach.

It is interesting to note that the no-crossing theorem is a direct consequence of the above just proven theorem [McVoy (1967)]. Indeed, for real potentials bound state poles are necessarily confined to the positive imaginary axis. However, in this case a collision of two poles must be prohibited because otherwise after collision both poles would be bound to leave the imaginary axis.

Resuming our discussion of the pole trajectories in the k-plane, the motion of the S-matrix poles very close to threshold could be studied by making analytic continuation of the effective range expansion. However, each of the three cases listed above (i.e, (i) zero-Coulomb, (ii) attractive Coulomb and (iii) repulsive Coulomb potential) is different and requires a separate treatment. Therefore, we shall deal with them in sequence beginning with the well known Coulomb-free case.

• *(i) Coulomb-free case*

The S-matrix has poles when $\cot\delta_\ell = \mathrm{i}$ and replacing $\cot\delta_\ell$ by the effective range formula (6.2), the poles that are located close to the origin, are obtained from the equation

$$\frac{1}{a_\ell(s)} + \tfrac{1}{2} r_\ell(s)\, k^2 - \mathrm{i}\, \frac{k^{2\ell+1}}{[(2\ell+1)!!]^2} = 0, \tag{6.33}$$

in which the last term represents the centrifugal barrier penetration factor. The centrifugal barrier plays a very important role introducing a marked difference in the behavior of the $\ell = 0$ trajectories as compared with $\ell > 0$ case. Since for real potentials T-invariance holds the poles in the lower half of the k-plane come in pairs for which the imaginary axis is the axis of symmetry. Each pair can be labeled by the radial quantum number $N$ and in the limit $s \to 0$ they are found to be located at equidistant points $k_N = (\pm N\pi/2 - \mathrm{i}\infty)/R$ where $R$ is some scaling factor roughly proportional to the range of the strong interaction and $N$ is even when $\ell$ is even and is odd for odd $\ell$. As has been described in many text-books [Newton (1982); Nussenzveig (1972)], for $\ell = 0$ when the attraction is gradually increased the ground state pole moves from $-\mathrm{i}\infty$ along the negative imaginary axis towards the origin and when the attraction is increased beyond certain critical value, the pole eventually passes through the origin moving into the positive part of the imaginary axis so that it becomes a bound pole. Meanwhile, the other poles move also closer to the physical region. As mentioned above they come in pairs and with sufficient amount of attraction for $\ell$=0 they approach vertically the negative imaginary axis from both sides colliding at some point $-\mathrm{i}/R$ where $R$ is proportional to the range of the potential. In result of collision one of them moves up the imaginary axis to become a bound state after passing through the origin while the other moves down in the opposite direction. Further strengthening of the potential brings new pairs of poles close to the physical region to be promoted to become bound states in a very similar way. The collision point for $\ell = 0$ is always on the negative imaginary axis but not necessarily at the same place

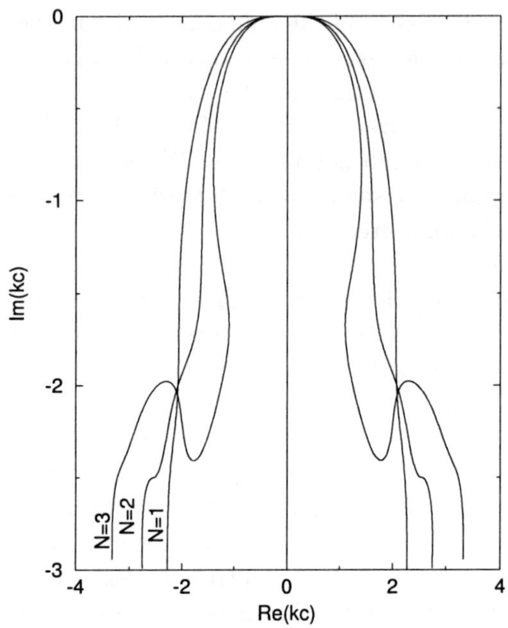


Fig. 6.5 S-matrix pole trajectories in the complex k-plane for $\ell = 2$ when only nuclear potential is present. The shape of nuclear potential is specified in (6.20) and the momentum $k$ on the graph is given in the units of the inverse range $c$. The real strength $s$ of the nuclear potential was used as a parameter whose variation caused the motion of the poles. The poles collide at $k = 0$ and after collision one of them enters the positive imaginary axis while the other moves along the negative imaginary axis in the opposite direction.

(e.g. for square well potential the crossing point happens to be always the same). The $\ell > 0$ case is different as the centrifugal barrier causes two poles to approach the origin tangent to the real axis so that the collision point is fixed at the origin for all $\ell$. After collision one of the poles immediately becomes a bound state entering the positive half of the imaginary axis while the other escapes in the opposite direction. This scenario is essentially universal for all bound states with $\ell > 0$. In Fig. 6.5 we have shown some typical trajectories for the potential given in (6.20).

It is of interest to provide further insight into these phenomena by an analytic approach based on formula (6.33). For small $|k|$, the centrifugal term in (6.33) is important only for $\ell = 0$ whereas for $\ell > 0$ it is of higher order in $k$ than the effective range term and could be dropped. For $\ell = 0$, neglecting the effective range term, we obtain the solution of (6.33)

$$k \approx -\frac{\mathrm{i}}{a_0(s)}; \tag{6.34}$$

and this solution illustrates the motion of the s-wave pole. When the strength is close to binding $1/a_0(s)$ changes sign passing through zero from positive to negative values. This behavior causes the pole to move on the imaginary axis from negative to positive values. For $\ell > 0$, neglecting centrifugal barrier term, the solution of (6.33) is

$$k \approx \pm\sqrt{-\frac{2}{r_\ell(s)}\, a_\ell(s)}. \tag{6.35}$$

As seen from (6.12), the effective range $r_\ell(s)$ at zero of $1/a_\ell(s)$ is necessarily negative and is slowly varying function of $s$ so that $k$ in (6.35) will be determined by the square root of the scattering length. The $\pm$ signs correspond to two different solutions moving towards each other along the real axis when $1/a_\ell(s)$ is positive and decreasing. When $1/a_\ell(s)$ passes through zero the solutions (6.35) collide at the origin and finally when $1/a_\ell(s)$ changes sign the solutions become imaginary and we have a bound state.

For a complex potential the s-wave pole is not confined any more to the imaginary axis and with enough attraction enters the upper half of the k-plane crossing the real axis at some distance from the origin. This behavior again could be traced down from the approximate formula (6.34). When the scattering length is complex this formula can be written as

$$k \approx -\frac{Im\, a_0(s) + \mathrm{i}Re\, a_0(s)}{|a_0(s)|^2}$$

and for absorptive potentials flux conservation implies that $Im\, a_0(s) > 0$ so that in this case we have $Re\, k < 0$. The $\ell > 0$ case is slightly more complicated because for complex $s$ the imaginary axis is no longer the axis of symmetry and for each pair of poles coming from $-\mathrm{i}\infty$ the fate of the two members will be different. For absorptive potentials with sufficient attraction the pole with $Re\, k < 0$ crosses the real axis at some negative value of $Re\, k$ and enters the upper half of the k-plane. At the same time the second pole with $Re\, k > 0$ at first approaches towards the real axis imitating the behavior of the first pole but then makes a sharp U-turn moving next in the opposite direction. Unlike the real potential case the two poles never collide. When the potential is emissive the two poles interchange their roles and it is the pole with $Re\, k > 0$ that becomes a bound state and the other makes U-turn. Since it is more to it than just the overall sign change it is clear that the approximate formula (6.35) is inadequate to describe the motion of the two poles. It turns out that of the two solutions in (6.35) only one choice of sign is physically correct and only the pole making U-turn can be accounted for in this simple approximation.

• *(ii) attractive Coulomb potential*

We turn now to the case when an attractive Coulomb potential has been added to the nuclear potential. In this case the analytic approach based on effective range expansion still can be used but our starting point is formula (5.46) in which we approximate $1/\mathcal{K}_\ell(s,k)$ by the effective range expansion (6.1). The barrier penetration function $g_\ell^+(k)$ is approximated by using the small $k$ expansion (cf. (3.39)). In the small $|k|$ limit, neglecting terms $O(k^4)$, the poles are obtained from the equation

$$\frac{1}{A_\ell(s)} + \tfrac{1}{2} R_\ell(s)\, k^2 \approx -\frac{(2\mu\alpha Z)^{2\ell+1}}{[(2\ell+1)!]^2} \left[ \mathrm{i}\pi + \frac{k^2}{(\mu\alpha Z)^2} (\frac{1}{12} + \mathrm{i}\pi\alpha_\ell) \right] . \quad (6.36)$$

This equation may be regarded as a generalization of (6.33) accounting for Coulomb effects and as far as the dependence upon $k$ is concerned it appears to be even simpler than (6.33). We wish to recall, however, that the expansion (3.39) is only asymptotic which means that there is no guarantee that the lowest two-term approximation will always be sufficiently accurate, even for the smallest $|k|$ well within the radius of convergence of the effective range expansion.

Let us assume first that $s$ is real. In accordance with the saturation theorem the poles from the lower half of the k-plane are not allowed to cross into the positive imaginary semi axis. Therefore, it is interesting to know what happens to them and where would they go when the nuclear attraction was strengthened arbitrarily. Although we still have reflection symmetry with respect to the imaginary axis, so that the poles come in pairs, but by contrast with the Coulomb-free case the poles from the III quadrant cannot collide with the poles from the IV quadrant owing to the presence of the logarithmic branch cut of Coulomb origin running along the negative imaginary axis.

In order to examine the influence of the Coulomb field on the behavior of pole trajectories we consider once more the previous example of the kaonic Sulphur potential (6.20). The pole trajectories for $\ell = 0, 1, 2, 3$ are displayed in Figs. 6.6-6.9 and all poles even for $\ell = 0$ come now in pairs and far from the physical region the trajectories are almost the same as those for the Coulomb-free case. With increasing nuclear attraction, each pole

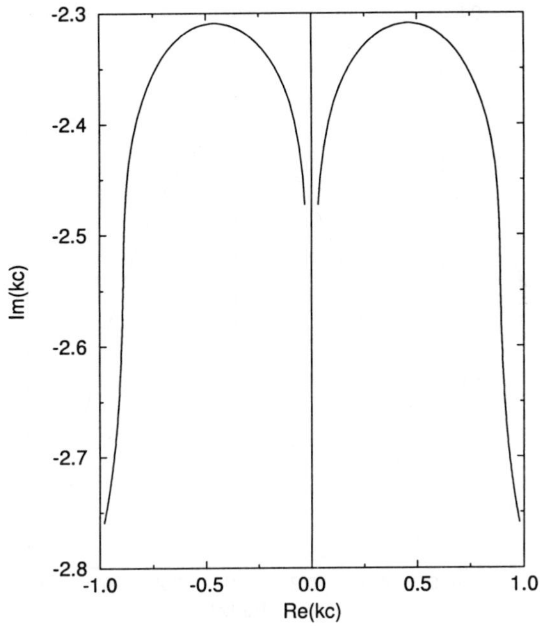


Fig. 6.6 S-matrix pole trajectories corresponding to nuclear plus Coulomb interaction in the complex k-plane for $\ell = 0$. The nuclear potential is specified in (6.20) and the momentum $k$ on the graph is given in the units of the inverse range $c$. The real strength $s$ of the nuclear potential was used as a parameter whose variation caused the motion of the poles. Owing to the presence of Coulomb potential the collision of the poles at $k = 0$ is prohibited. Instead, with increasing $s$ the poles come closer and closer to the Coulomb branch cut running along the negative imaginary axis. After crossing through the cut the poles eventually enter another Riemann sheets.

and its mirror partner approach from opposite sides a fixed point on the imaginary k-axis not far from the origin. However, the collision never takes place, since after eventually hitting the imaginary axis both poles leave the principal sheet and each of them enters through the cut a different Riemann sheet. The point of escape depends upon the orbital momentum: the bigger is the value of $\ell$ the closer this point is located to the origin (cf. Fig. 6.10). It is quite clear that the Coulomb potential strongly affects pole trajectories, with the biggest distortion to be expected for $\ell = 0$. For large $\ell$ the Coulomb potential looses its importance so the pole trajectories resemble very much those known for the Coulomb-free case. Only very close to the origin instead of colliding at $k = 0$, the poles miss the origin crossing through the cut into different sheets.

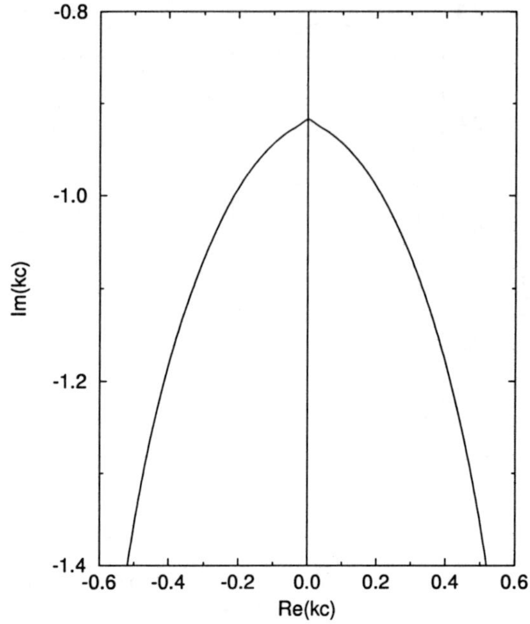


Fig. 6.7 S-matrix pole trajectories corresponding to nuclear plus Coulomb interaction in the complex k-plane for $\ell = 1$. The nuclear potential is specified in (6.20) and the momentum $k$ on the graph is given in the units of the inverse range $c$. With increasing strength $s$ two poles come from opposite sides towards a point on the negative imaginary axis. Since there is a Coulomb branch cut running along negative imaginary axis the poles do not collide but enter different Riemann sheets adjacent to the principal sheet shown on the graph.

The individual trajectories may be labeled by the excitation (or, nodal) quantum number (N=0,1,2,...), which in the Coulomb-free case numbers the zero-energy bound states in order of appearance when the nuclear attraction is increased arbitrarily. When $\ell$ is fixed and $N$ is varied instead all the different trajectories seem to be passing through the same fixed point on the cut before leaving the principal sheet. This is illustrated in Fig. 6.8 for $\ell = 2$ and should be compared with Fig. 6.5 where the Coulomb potential was absent. It is not clear, however, whether this is a general property.

The solution of equation (6.36) has the same form for all $\ell$

$$ka_B \approx \pm Z \frac{\left\{-1 + \dfrac{\mathrm{i}}{\pi A_\ell(s)} \dfrac{[(2\ell+1)!!]^2}{(2\mu\alpha Z)^{2\ell+1}}\right\}^{1/2}}{\left\{\alpha_\ell + \dfrac{1}{12\pi\mathrm{i}} + \dfrac{1}{8} R_\ell(s) \dfrac{[(2\ell+1)!!]^2}{(2\mu\alpha Z)^{2\ell-1}}\right\}^{1/2}}, \tag{6.37}$$

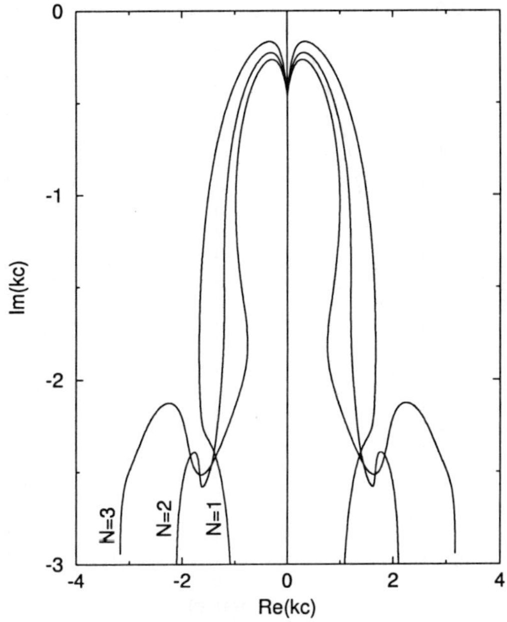


Fig. 6.8 S-matrix pole trajectories corresponding to nuclear plus Coulomb interaction in the complex k-plane for $\ell = 2$. The nuclear potential is specified in (6.20) and the momentum $k$ on the graph is given in the units of the inverse range $c$. The present case should be compared with that displayed in Fig. 6.2 where we left out Coulomb interaction. In contrast with Fig. 6.2 where the poles collided at $k = 0$, here with increasing strength $s$ two poles approach from opposite sides a common point on the negative imaginary axis. Since there is a Coulomb branch cut running along negative imaginary axis the poles do not collide but enter different Riemann sheets adjacent to the principal sheet shown on the graph.

where $a_B = 1/\mu\alpha$ denotes the Bohr radius. For real $s$ the right hand side of the above expression never vanishes and the poles given by (6.37) must necessarily be complex. By invoking the saturation theorem we can resolve the sign ambiguity in (6.37) because we know that the physical solution has $Im\, k < 0$ while the other solution must be rejected. The poles are generated in pairs and the second pole trajectory is obtained by reflecting the physical solution about the imaginary axis. Of course, these two trajectories never cross because of the branch cut along the negative part of the imaginary axis. In the approximate solution (6.37) we have a square root branch point instead of a logarithmic singularity and the solution (6.37) may be used only off the imaginary axis.

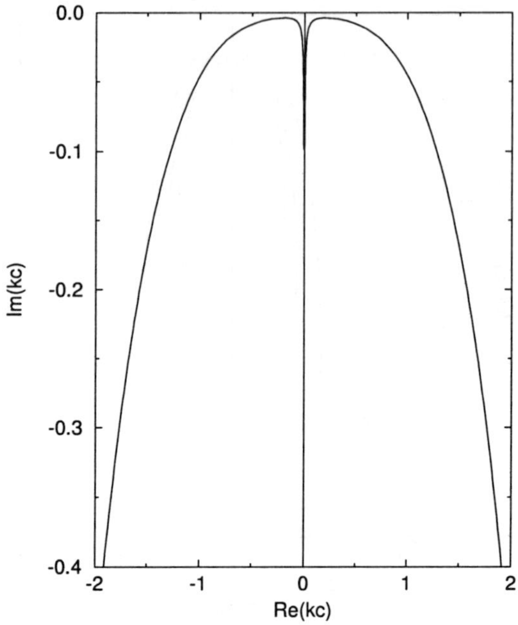


Fig. 6.9 S-matrix pole trajectories corresponding to nuclear plus Coulomb interaction in the complex k-plane for $\ell = 3$. The nuclear potential has the shape specified by (6.20) and the momentum $k$ is given on the graph in units of the inverse range $c$. With increasing strength $s$, two poles approach from opposite sides a point on the negative imaginary axis. Since there is a Coulomb branch cut running along negative imaginary axis the poles do not collide but enter different Riemann sheets.

We have examined the accuracy of the analytic formula (6.37) by comparing the approximate solution with the complete solution of the eigenvalue problem obtained numerically which is here referred to as "exact". As seen from Fig. 6.10 for s-waves all the $|k|$ values along the trajectory are quite large and the conditions under which (6.37) has been derived are certainly not likely to be satisfied so that this formula is not applicable for s-waves and gives nonsensical results. On the other hand, for higher waves the performance of (6.37) is rapidly improving with increasing $\ell$ as can be seen in Figs. 6.11-6.13. When the orbital momentum increases the trajectories are shifted closer to the origin. As a result, the convergence rate of the expansions used in the derivation of (6.37) improves and this formula becomes then remarkably accurate.

Finally, we are going to discuss the most interesting complex $s$ case. Then, the imaginary axis is no longer the axis of symmetry, and, as can

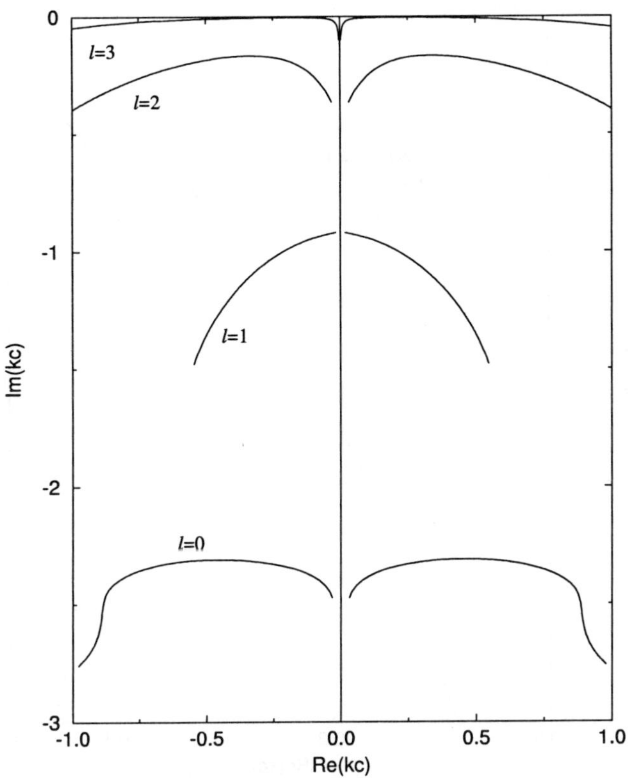


Fig. 6.10 Relative positions of the different pole trajectories in the complex k-plane for $\ell = 0, 1, 2, 3$ We consider here a two-potential problem in which attractive Coulomb interaction is combined with the nuclear potential (6.20). On the graph the momentum $k$ is given in the units of the inverse range $c$ of the nuclear potential. For large values of $\ell$ the trajectories are seen to approach very close to the origin.

be seen from Fig. 6.14, the trajectories in the III and IV quadrant are different: when the absorption increases the trajectories in the III quadrant move upwards while those from the IV quadrant are shifted in the opposite direction. By increasing *Re s* all the curves presented in Fig. 6.14 are forced to make a U-turn and eventually disappear through the Coulomb branch cut from the principal Riemann sheet. This picture, however, does not exhaust all the possibilities since, as we already know, for complex $s$ it is possible to generate the quasi-nuclear bound states and when this happens the corresponding pole trajectories will show a different behavior from that apparent in Fig. 6.14. We are now going to examine closer this most interesting case. It is rather obvious that for generating a quasi-nuclear bound

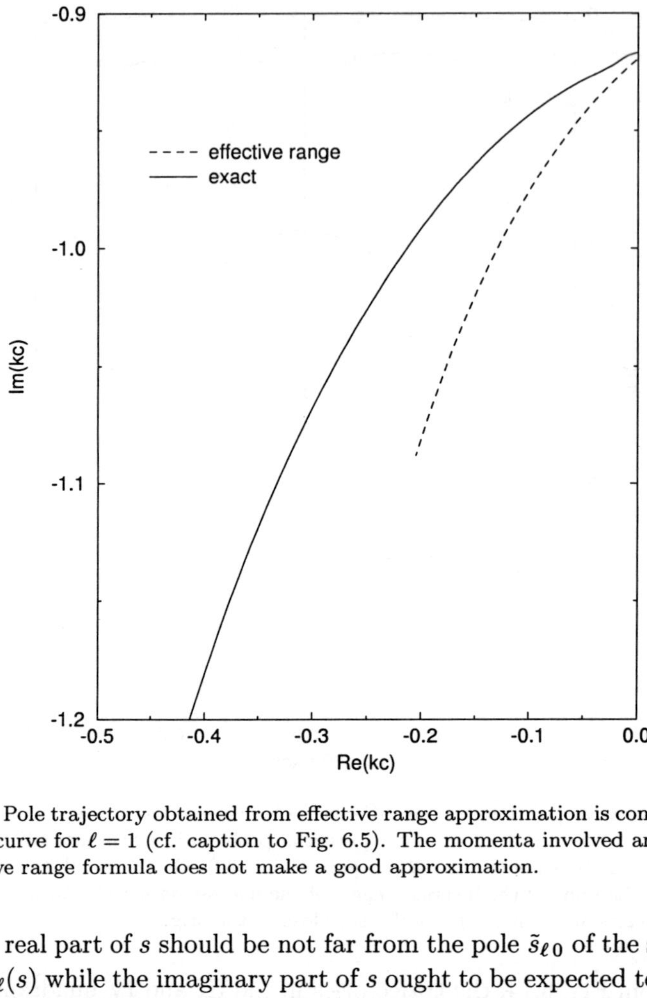


Fig. 6.11 Pole trajectory obtained from effective range approximation is compared with the exact curve for $\ell = 1$ (cf. caption to Fig. 6.5). The momenta involved are large and the effective range formula does not make a good approximation.

state the real part of $s$ should be not far from the pole $\tilde{s}_{\ell 0}$ of the scattering length $A_\ell(s)$ while the imaginary part of $s$ ought to be expected to be small in comparison with the real part. For such values of $s$ the scattering length can be approximated by a single pole term and using the general product formula (6.14), we have

$$\frac{1}{A_\ell(s)} \frac{[(2\ell+1)!]^2}{(2\mu\alpha Z)^{2\ell+1}} \approx \frac{1}{C_\ell}\left(\frac{1}{s} - \frac{1}{\tilde{s}_{\ell 0}}\right), \tag{6.38}$$

where $C_\ell$ is a constant given by the formula

$$C_\ell = \frac{\int_0^\infty J_{2\ell+1}(2\sqrt{2\mu\alpha Z r})^2\, v(r) r\, \mathrm{d}r}{\int_0^\infty v(r) r\, \mathrm{d}r}.$$

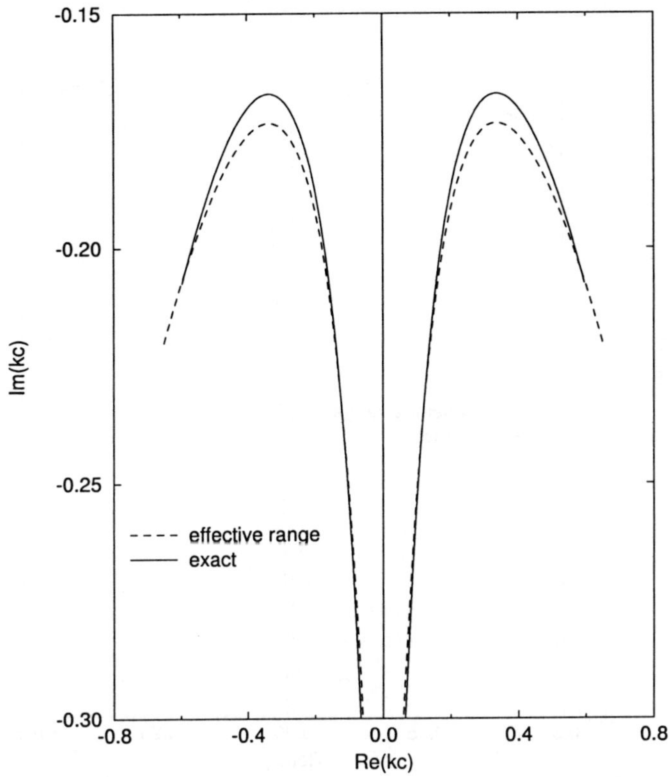


Fig. 6.12 Pole trajectory obtained from effective range approximation is compared with the exact curve for $\ell = 2$ (cf. caption to Fig. 6.6). This trajectory involves small momenta and the effective range approximation is more accurate than for $\ell = 1$ case shown in Fig. 6.9.

Using the single pole approximation (6.38) in (6.37), we can explicitly calculate the critical $s$ value $s = s_c$ for which the numerator of (6.37) vanishes so that the corresponding trajectory is bound to go through the origin. We obtain

$$s_c = \frac{\tilde{s}_{\ell 0}}{1 - \mathrm{i}\pi\, C_\ell \tilde{s}_{\ell 0}}, \tag{6.39}$$

and formula (6.37) takes the form

$$ka_B \approx \pm Z \left\{ \frac{\mathrm{i}}{\pi \alpha_\ell C_\ell} \left( \frac{1}{s} - \frac{1}{s_c} \right) \right\}^{1/2}, \tag{6.40}$$

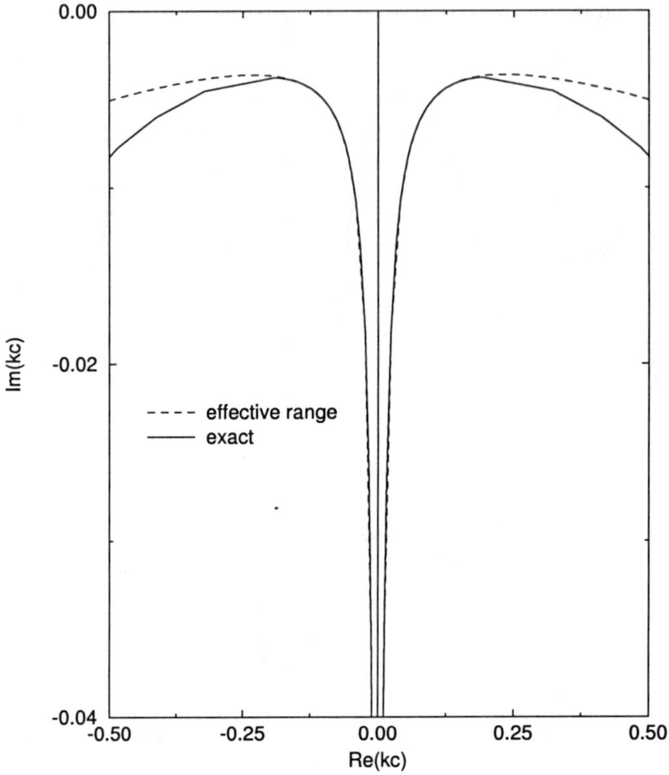


Fig. 6.13 Pole trajectory obtained from effective range approximation is compared with the exact curve for $\ell = 3$ (cf. caption to Fig. 6.7).

where to simplify matters as much as possible we have retained only the $\alpha_\ell$ term in the denominator of formula (6.37) (the effective range is small for $s$ close to $\tilde{s}_{\ell 0}$). It could be established by trial and error that the sign in front of (6.40) is determined by the sign of $Im\, s$ in such a way that minus (plus) sign should be chosen for absorption (emission). Formula (6.40) explains when and how the quasi-nuclear bound states can be formed. To make it even more transparent, we expand the expression under the square root in (6.40) around $s_c$ retaining only terms linear in $s - s_c$. Leaving out all uninteresting constants, we have

$$k\, a_B \sim -\sqrt{Im(s - s_c) - \mathrm{i}\, Re(s - s_c)},$$

and it is now evident that for generating a quasi-nuclear bound state ($Im\, k > 0$) it is not enough to supply sufficiently strong attraction, i.e.

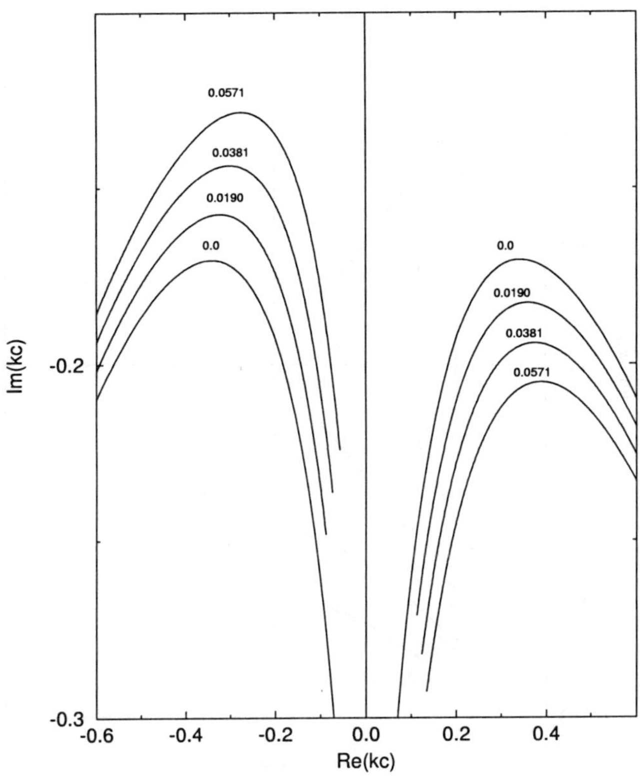


Fig. 6.14 Pole trajectories for a complex Saxon–Woods potential (6.20) in the presence of the Coulomb potential for $\ell = 2$. The values of $k$ on the axes are given in the units of the inverse range $c$ of the nuclear potential. To obtain these curves the absorption has been held fixed and $Re\, s$ has been varied to obtain a trajectory. The fixed value of $Im\, s$ is specified at the top of each curve. When absorption is present the imaginary axis is no longer the axis of symmetry.

such that $Re\, s > Re\, s_c$, as it was the case when the Coulomb potential was absent. Now, two conditions have to be fulfilled because additionally we have to ensure that also $Im\, s > Im\, s_c$. If the latter condition is not satisfied but $Re\, s > Re\, s_c$, the pole is pushed through the branch cut into the second Riemann sheet and even with the largest attraction a quasi-nuclear bound state will never be formed. The pole trajectories obtained from formula (6.40) are plotted in Fig. 6.15. We have used here again the potential (6.20), for $\ell = 2$. For each of the presented curves the value of $Im\, s$ has been kept fixed and equal to $\lambda\, Im\, s_c$ where the parameter $\lambda$ assumes values around unity and our choice was $\lambda = 0.8, 0.9, 1.0, 1.1, 1.2$. Keeping $Im\, s$

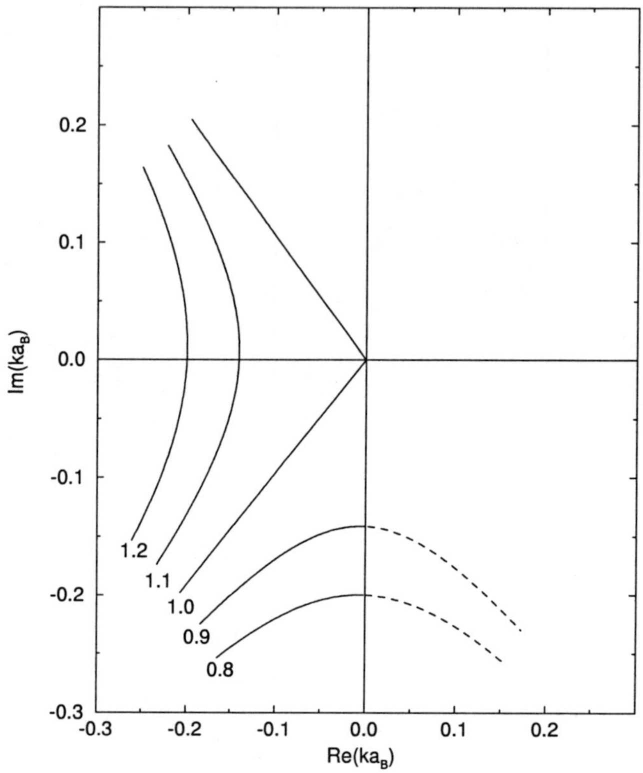


Fig. 6.15 Formation of a quasi-nuclear bound state. Pole trajectories which are presented here have been obtained from formula (6.40). The imaginary part of $s$ has been set equal to $\lambda\, Im\, s_c$ and the values of $\lambda$ are given at each curve. For $\lambda \geq 1$ a quasi-nuclear bound state is formed and in this case the trajectories cross into upper half of the k-plane. The Coulomb cut runs along the negative part of the imaginary axis. Trajectories depicted by dashed lines are in the second Riemann sheet.

fixed, by varying $Re\, s$ from $0.95\, Re\, s_c$ to $1.05\, Re\, s_c$, we obtained the five curves displayed in Fig. 6.15. It is apparent from this figure that when the absorption is too small, i.e. $Im\, s < Im\, s_c$, the corresponding trajectories never enter the upper half of the k-plane. Consequently, the quasi-nuclear bound state cannot be formed and by increasing the attraction the trajectory eventually disappears from the principal Riemann sheet through the branch cut which in Fig. 6.15 has been depicted by the dashed line. It is only when the absorption reaches certain critical value $Im\, s_c$, the trajectory will be pushed through the origin into the upper half of the k-plane. In this case we have a quasi-nuclear bound state. Increasing the absorption above the critical value has the effect that the crossing point is not at zero but becomes shifted away from the origin similarly as in the Coulomb-free case. For comparison, the exact (numerically) solution of the very same problem is presented in Fig. 6.16 and the curves show very similar pattern to those resulting from our simplified treatment based on formula (6.40). The numbers above the curves in Fig. 6.16 indicate the values of $Im\, s$ which have been kept fixed for each of the trajectories.

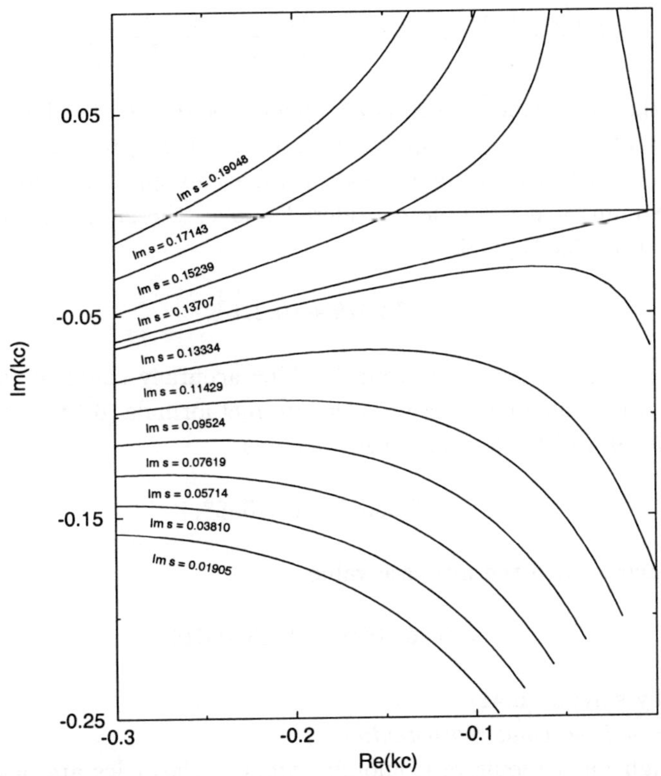


Fig. 6.16 Pole trajectories for a complex Saxon–Woods potential (6.20) in the presence of the Coulomb potential for $\ell = 2$. The values of $k$ on the axes are given in the units of the inverse range $c$ of the nuclear potential. To obtain these curves, the absorption has been held fixed, the value of $Im\, s$ being specified at the top of each curve, while $Re\, s$ has been varied to obtain a trajectory. When $Im\, s$ exceeds certain critical value, the trajectory enters II quadrant and this corresponds to a formation of a quasi-bound state. The curves displayed here have partners in the IV quadrant (not shown) which sink down with increasing absorption as can be seen in Fig. 6.12.

In a general case, the critical value of the strength parameter necessary for supporting a zero-energy quasi-nuclear bound state can be very precisely computed by solving equation (6.26) with respect to s. This task may be greatly facilitated by using for $A_\ell(s)$ the product representation (6.14). Keeping only a finite number of factors, we are left with an algebraic equation which is far easier to solve. In this method all that is needed are the locations of the zeros and the poles in the product formula. If they are not available, one can always use (6.5) in (6.26), arriving at the following equation with respect to $s$

$$R\frac{\phi_\ell'(s,0,R)}{\phi_\ell(s,0,R} = \ell + 1 - \frac{x}{2}\,\frac{Y_{2\ell+2}(x) + \mathrm{i}\,J_{2\ell+2}(x)}{Y_{2\ell+1}(x) + \mathrm{i}\,J_{2\ell+1}(x)}, \tag{6.41}$$

in which the right hand side is independent of $s$ and the left hand side for each value of $s$ must be computed by numerical integration of the wave equation. We note that formula (6.41) involves no approximation.

In the numerical example considered above, the simplest single pole approximation (6.39) yields

$$s_c = 9.2616 + \mathrm{i}\,0.2281,$$

but this estimate is not very accurate. The accuracy could be improved, however, by including more terms in the product formula (6.14). Truncating the product in (6.14) after six terms, one has

$$s_c = 9.26695 + \mathrm{i}\,0.13888,$$

to be compared with the ultimate value

$$s_c = 9.2669838 + \mathrm{i}\,0.13706703$$

obtained by solving (6.41).

• *(iii) repulsive Coulomb potential*

Although for a repulsive Coulomb potential the poles are determined from (5.46), similarly as for Coulomb attraction, but the barrier penetration functions $g_\ell^+(k)$ are markedly different in these two cases. Using the effective range expansion and the asymptotic form for $g_\ell^+(k)$ (cf. (3.37)), the poles are obtained from the equation

$$\frac{1}{A_\ell(s)} + \tfrac{1}{2}R_\ell(s)\,k^2 \approx \frac{|2\mu\alpha Z|^{2\ell-1}}{[(2\ell+1)!]^2}\,\frac{k^2}{3}, \tag{6.42}$$

in which terms of the order $O(k^4)$ have been neglected. The solution of (6.42) is

$$ka_B \approx \pm \left\{ \frac{12}{A_\ell(s)} \frac{[(2\ell+1)!]^2}{|2\mu\alpha Z|^{2\ell+1}} \right\}^{1/2} \left\{ 1 - \frac{3}{2} R_\ell(s) \frac{[(2\ell+1)!]^2}{|2\mu\alpha Z|^{2\ell-1}} \right\}^{-1/2}, \quad (6.43)$$

where in the first approximation the effective range term can be dropped. The expression (6.43) becomes then qualitatively very similar to the Coulomb free case with $\ell > 0$ and the quasi-nuclear bound states can be formed for real potentials. Indeed, for real $s < \tilde{s}_{\ell 0}$ the scattering length is real and positive and the poles (6.43) are located on the real axis. The bound state is formed when $1/A_\ell(s)$ passes through zero and when this happens the two poles coming from opposite directions along the real axis collide and after collision one of them enters the positive imaginary axis. The second pole leaves the principal Riemann sheet through the cut along the negative imaginary axis. For absorptive potentials, i.e. for complex $s$ the pole crosses the real axis entering the upper half of the k-plane at some negative value. Qualitatively, this is a very similar scenario as in the Coulomb-free case and now even the behavior of the s-wave pole trajectory is no different from the $\ell > 0$ case.

# Chapter 7

# Atomic Spectrum

Recapitulating the results of the preceding chapter, in a two-potential problem when we have an attractive Coulomb potential plus a short ranged complex nuclear potential we can distinguish between the atomic and the quasi-nuclear bound states. To decide whether a given bound state is atomic or quasi-nuclear is quite complicated as it requires tracing down the provenance of the associated pole in the complex momentum plane. In practise this may be effected by reducing gradually to zero the strength $s$ of the nuclear potential investigating the trajectory associated with the bound state pole in the complex k-plane. If, as a result of this limiting procedure for $s = 0$ the pole resumes the Coulombic location on the positive imaginary axis, then the bound state is atomic. By contrast, if the limiting procedure forces the pole to cross from II quadrant into III quadrant (absorptive potentials) then we have to do with a quasi-nuclear bound state. We have examined in some detail the motion of the quasi-nuclear poles and in particular we have shown that for real nuclear potentials no quasi-nuclear bound state can be formed. In this chapter we are going to investigate the atomic part of the spectrum.

## 7.1 Real Nuclear Potentials

We begin our discussion by assuming that the nuclear potential is real. In accordance with the saturation theorem the whole spectrum is in that case purely atomic. In other words, the nuclear potential can only deform the Coulombic spectrum and it is quite obvious that the strongest distortion will show up at the lowest orbital momenta. The centrifugal potential plays a crucial role here for owing to its presence the total interaction potential takes a conspicuous two-well shape: the inner well, deep and narrow, may be associated with the nuclear interaction while the outer well, broad and shallow, is typical for atomic systems. The height of the barrier separating the two wells is proportional to the angular momentum $\ell$. For small $\ell$ there is a substantial overlap between the nuclear and the atomic wave functions and with increasing $s$ Coulombic levels will rapidly acquire large shifts typical for the nuclear scale of the binding energies, often several orders of magnitude bigger than the corresponding Coulomb energies. With increasing $\ell$ the barrier separating the two wells eventually gets so high that the mixing between nuclear and atomic wave functions becomes strongly suppressed and the level shifts will be dramatically reduced. The absolute value of the energy shift due to the nuclear forces is then typically several orders of magnitude smaller than the Coulomb binding energy. An increase of $\ell$ by one could well cause a reduction of the energy shift by several orders of magnitude. The above qualitative considerations have been confirmed by explicit computations carried out for the Saxon–Woods potential (6.20). In Figs. 7.1-7.3 we present the dependence of the binding energy for our two-potential problem upon the strength of the nuclear potential for three lowest angular momenta $\ell = 0, 1, 2$. Higher values of $\ell$ are less interesting because the emerging curves are qualitatively very similar to those obtained for $\ell = 2$ and we felt that nothing new could be learnt from them. Each plot shows six curves corresponding to the six lowest eigenenergies labeled by the excitation quantum number $N$. As seen from Fig. 7.1, for $\ell = 0$ by increasing s the nuclear potential almost immediately wins the competition with the Coulomb potential and the energy curves show parabolic behavior typical for a shallow nuclear bound state what can be easily demonstrated from the scattering length approximation and the product representation. It should be noted that all the quasi-nuclear poles are prohibited from entering the upper half of the k-plane and yet it is quite remarkable that the purely atomic states behave in the same way as the nuclear states would. We observed this phenomenon before for a toy model presented in Fig. 6.29. The $\ell = 1$ case illustrated in Fig. 7.2 is slightly more complicated since the levels result from a collaboration between Coulomb and nuclear potential with a large degree of mixing between the two wells. Nevertheless for sufficiently large $s$ the nuclear potential eventually takes over and the energy shows roughly a linear dependence upon $s$ in accordance with the scattering length approximation. For $\ell = 2$ the energies shown in Fig. 7.3 exhibit a different pattern. The energies either stay close to the purely Coulomb positions or make jointly a rapid, almost collective transition to the next deeper Coulomb level. This remarkable restructuring

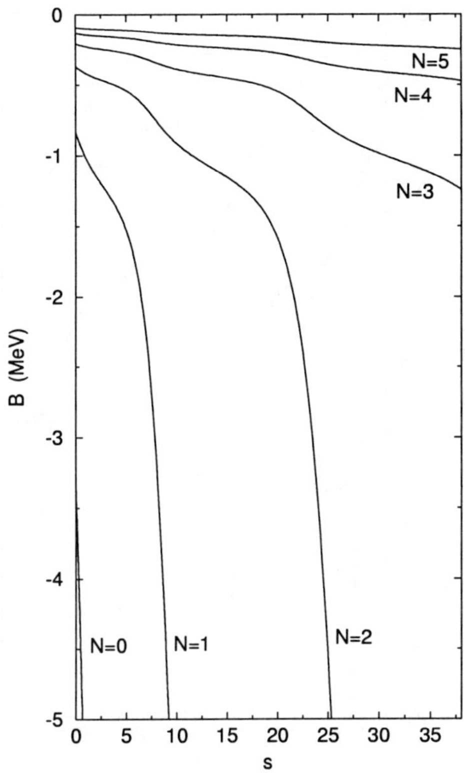


Fig. 7.1 Binding energy for a real Saxon–Woods potential (6.20) and Coulomb interaction present versus the dimensionless strength $s$ for $\ell = 0$. The different curves correspond to the six lowest levels labeled by the nodal quantum number $N = 0, 1, 2, 3, 4, 5$.

of the atomic spectrum has been known as Zeldovich effect as it was apparently first observed by [Zel'dovich (1959)] in a different context many years ago, and subsequently rediscovered by others (cf. [Karnakov *et al.* (1987); Karnakov *et al* (1985); Badalyan *et al.* (1982)]). It is of interest to provide a general quantum mechanical explanation of this effect and we are going to take on this issue. When the barrier separating the nuclear well from the Coulomb well becomes sufficiently high, there is very little mixing between the levels corresponding to each well taken separately and the solution of the two-potential problem can be sought as a superposition of two single well problems. The energy of the system might be determined variationally using a trial function of the form

$$\Psi = C_1 \psi_c + C_2 \psi_s, \tag{7.1}$$

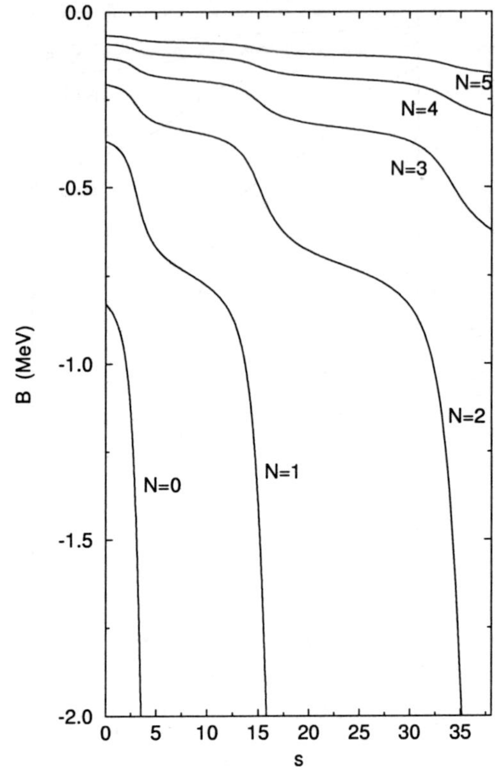


Fig. 7.2 Binding energy for a real Saxon–Woods potential (6.20) and Coulomb interaction present versus the dimensionless strength $s$ for $\ell = 1$. The different curves correspond to the six lowest levels labeled by the nodal quantum number $N = 0, 1, 2, 3, 4, 5$.

where $C_1, C_2$ are two adjustable parameters, $\psi_c$ is a purely Coulombic wave function while $\psi_s$ is a purely nuclear wave function. These wave functions are obtained by solving the corresponding single potential wave equations, *viz.*

$$(H_0 + V_c - E_c)\psi_c = 0, \tag{7.2}$$

$$(H_0 + V_s - E_s)\psi_s = 0, \tag{7.3}$$

where $H_0$ denotes kinetic energy, the indices $c$ and $s$ distinguish, respectively, Coulomb and nuclear components in the potentials and binding energies. The index $s$ not only stands for the strong interaction but is also meant to indicate the dependence upon the strength parameter $s$. It is understood that $s$ is large enough so that $V_s$ is capable of supporting purely

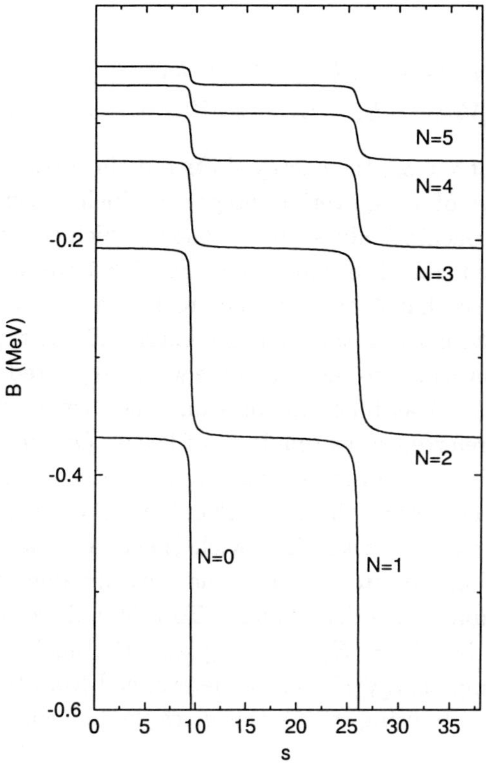


Fig. 7.3 Binding energy for a real Saxon–Woods potential (6.20) and Coulomb interaction present versus the dimensionless strength $s$ for $\ell = 2$. The different curves correspond to the six lowest levels labeled by the nodal quantum number $N = 0, 1, 2, 3, 4, 5$. This plot illustrates the so called Zeldovich effect: for $s \approx 10$ all levels with $N > 0$ make a collective transition in result of which a level with the nodal quantum number $N$ resumes position appropriate for the undisturbed Coulomb level with nodal quantum number equal $N - 1$. Similar transition occurs also for $s \approx 26$.

nuclear bound states. We assume that the wave functions $\psi_c$ and $\psi_s$ are normalized but they do not need to be mutually orthogonal and their overlap $\langle \psi_c | \psi_s \rangle$ will be denoted as $\xi$. Standard diagonalization procedure of the two-potential Hamiltonian yields a quadratic equation for the binding energy and we obtain two solutions for the eigenenergy

$$E_{\pm}(s) = E_c + \frac{U \pm \sqrt{U^2 - (1 - \xi^2)W}}{1 - \xi^2}, \tag{7.4}$$

where

$$U = \tfrac{1}{2}\left(\langle\psi_c|V_s|\psi_c\rangle + \langle\psi_s|V_c|\psi_s\rangle + E_s - E_c\right) - \xi\,\langle\psi_s|V_s|\psi_c\rangle\,; \tag{7.5}$$
$$W = \langle\psi_c|V_s|\psi_c\rangle\,\left(E_s - E_c + \langle\psi_s|V_c|\psi_s\rangle\right) - \langle\psi_s|V_s|\psi_c\rangle^2\,. \tag{7.6}$$

For most values of $s > s_{\ell 1}$ the energy shifts remain small. However, for certain critical values of $s$ the overlap integral $\xi$ which is usually rather small can increase approaching values close to one in which case the eigenenergies will be strongly enhanced in magnitude. For $\ell \geq 1$ the wave functions are pushed outward so that $\xi$ has a chance to became large when the variation of $s$ brings $E_s$ close to the Coulomb energy $E_c$. A typical situation is illustrated schematically in Fig. 7.4 where we have plotted the two eigenenergies energies $E_{\pm}(s)$ as functions of $s$ and the displayed phenomenon is well known and referred to as repulsion of levels. One may argue that the repulsion of levels is the only alternative to level crossing since the latter is forbidden by no-crossing theorem. Therefore, the repulsion of levels appears to be the essence of the Zeldovich effect, as is also apparent from Fig. 7.4. It is of interest to calculate the critical value of $s$ at which the energy shift exhibits a sudden change. To this end we use the condition for maximal overlap $E_c = E_s$ assuming that the nuclear level is shallow so that the binding energy $E_s$ can be determined from the effective range expansion. For $\ell > 0$, using (6.35), this gives the following equation for the critical strength

$$r_\ell(s)\,a_\ell(s) \approx 2a_B^2(N+\ell+1)^2, \tag{7.7}$$

where $a_B$ as before denotes the classical Bohr's radius. Since the critical $s$ is quite close to $s_{\ell 1}$, we can use the product representation for $a_\ell(s)$, retaining in (6.13) only the single pole term. The effective range $r_\ell(s)$ varies slowly so that it can be regarded as constant which will be evaluated at $s = s_{\ell 1}$. With these approximations equation (7.7) can be explicitly solved for the critical $s$, and, the result is

$$s_{crit} \approx s_{\ell 1}\left[1 - \frac{\int_0^\infty v(r) r^{2\ell+2} \mathrm{d}r}{\int_0^\infty v(r) r \mathrm{d}r}\,\frac{r_\ell(s_{\ell 1})}{2a_B^2(N+\ell+1)^2}\right], \tag{7.8}$$

where $r_\ell(s_{\ell 1})$ must necessarily be a negative quantity as may be inspected from (6.12). As a matter of fact, formula (7.7) gives for fixed $\ell$ the critical strength for all Coulomb levels $N = 0, 1, 2, \ldots$ explaining the slow decrease of $s_{crit}$ with increasing $N$. Ultimately, for large $\ell$ the critical strength becomes practically equal to $s_{\ell 1}$ and does not depend on $N$ any more. In

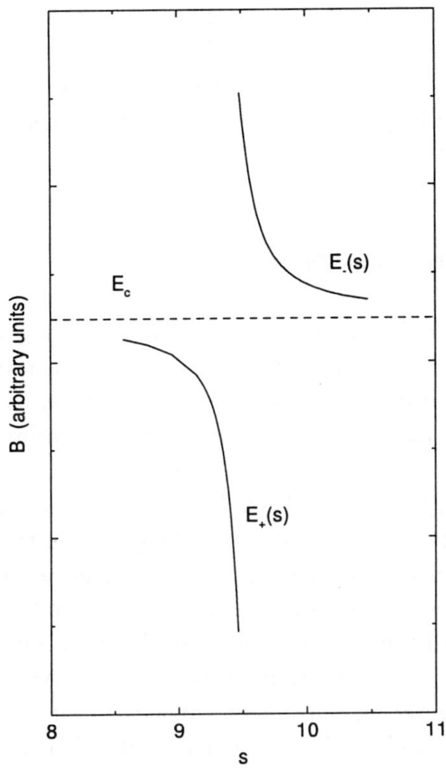


Fig. 7.4 Repulsion of levels phenomenon is displayed for the Saxon–Woods potential (6.20) for the $\ell = 2$ case. The undisturbed Coulomb level is depicted by the dashed line. When the strength $s$ is close to the value required for supporting a nuclear bound state there is a strong mixing between the atomic and the nuclear component of the total wave function resulting in a strong enhancement of the eigenenergies $E_{\pm}(s)$.

this case we have to do with a collective transition of all levels at the same value of $s$, that has been observed in numerical studies.

Summarizing, Zeldovich effect results as a consequence of the no-crossing and saturation theorems. Since the quasi-nuclear level is not allowed to exist the atomic levels form a complicated pattern in which they simulate its presence, and in particular the rapid $s$ dependence represented by the vertical segments of the curves in Fig. 7.3. The simulation, however, is not complete as these segments cannot be joined into a single curve without level crossing. Since level crossing is forbidden, the repulsion of levels mechanism sets in.

Actually, Zeldovich effect is not restricted to Coulomb interactions and the latter may be replaced by any shallow potential whose range is bigger than that of the nuclear potential. The only difference it would make is the possibility of increasing the total number of bound states. Additional bound states could be produced as the saturation theorem is valid only for Coulomb potential.

## 7.2 Complex Nuclear Potential

We are going now to lift the restrictions introduced in the preceding section and consider the general case when the strength of the nuclear potential is complex. For real potentials the bound state poles were confined to the positive imaginary axis in the k-plane and their motion caused by varying $s$ was possible only in one dimension. With complex $s$ ($Im\, s > 0$) the poles go off the imaginary axis into II quadrant and the motion of the poles will be two dimensional. The poles moving along certain trajectories may now collide anywhere in the k-plane. Furthermore, in addition to the atomic spectrum, quasi-nuclear levels may exist. The theoretical background necessary to explain all these phenomena has already been presented and now we shall concentrate our attention on some illustrative examples. Similarly as before our computations have been carried out for the Saxon–Woods potential (6.20). Although we have now two real independent variables ($Re\, s$ and $Im\, s$) which ought to be treated on equal footing but for practical reasons we will continue using $Re\, s$ as our sole variable, freezing the value of $Im\, s$ and regarding it as if it was a fixed parameter. To find out how things are changing with increasing absorption all computations have been done repeatedly increasing successively the value of $Im\, s$. For every fixed value of $Im\, s$ we are going to present two plots: in the first, similarly as before, we show the binding energies for the lowest states versus $Re\, s$, whereas in the second plot we display the trajectories of the corresponding bound state poles whose motion results form varying $Re\, s$.

When $Im\, s$ is very small the energy plots are almost the same as those for real potentials. Zeldovich effect is still clearly visible and the only difference is seen in the pole trajectories whose linear shapes for $Im\, s = 0$ have been deformed into little arches starting and ending very close to purely Coulombic positions. When the absorption increases the trajectories are further deformed in such a way that the convex parts of one trajectory and the concave part of another trajectory come very close together. At the energy plot this behavior is reflected through the formation of little spikes pointing upward on one curve and downward on another curve so that it looks as if their ends barely escaped from meeting (Fig. 7.5 and Fig. 7.6). For some critical value of $Im\, s$ the two trajectories associated with the two

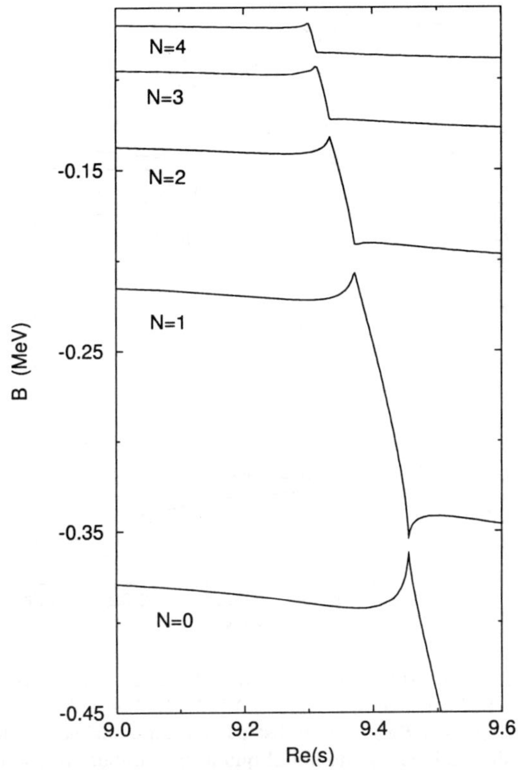


Fig. 7.5 Binding energy versus the real part of the dimensionless potential strength $s$ for the potential (6.20) and for $\ell = 2$. The imaginary part of $s$ has been held fixed $Im\, s = 0.132195$. The plot shows five curves corresponding to the lowest lying levels labeled by the nodal quantum number $N$. The Zeldovich effect is still visible but on the $N = 0$ and $N = 1$ curves conspicuous spikes are observed.

lowest levels cross each other so that we observe a collision of poles. On the energy plot the two spikes have been welded into a single curve and we observe a level crossing phenomenon (Fig. 7.7 and Fig. 7.8). It should be noted that although the remaining levels do not cross yet but it is quite clear that also for them crossing may be imminent. The absorption which has been thus far considered is still rather small and insufficient yet for a formation of a quasi-nuclear bound state. Therefore, the quasi-nuclear pole

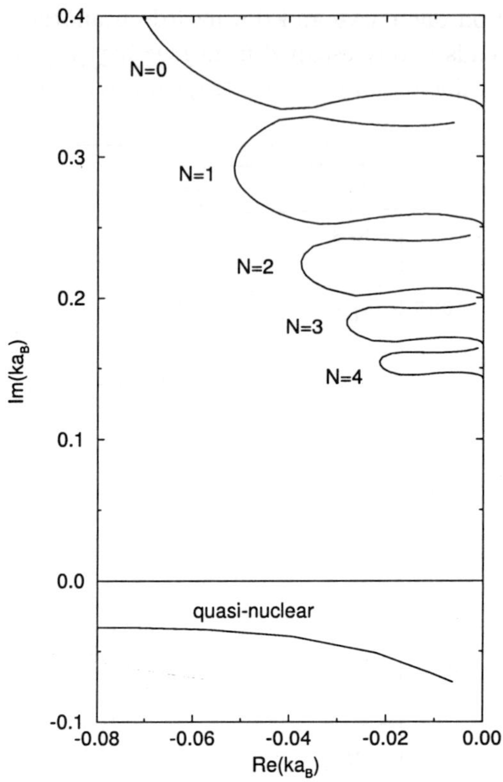


Fig. 7.6 The pole trajectories for the potential (6.20) and $\ell = 2$ calculated for fixed value of the imaginary part of the strength $Im\, s = 0.132195$. The different curves correspond to trajectories starting with $Re\, s = 0$ almost at the unperturbed Coulomb positions on the imaginary axis determined by the nodal quantum number. In the lower half we also show the trajectory connected with a quasi-nuclear level which comes quite close to the origin but owing to insufficient absorption never enters the upper half of the k-plane. The $N = 0$ and $N = 1$ trajectories form a bottle neck which suggests the possibility of crossing.

trajectory still remains in III quadrant. By strengthening the attraction a quasi-nuclear bound state is not formed but the pole is pushed away from the physical region and eventually crosses through the cut on the negative imaginary axis into a different Riemann sheet, as shown before in Fig. 6.16.

In the next phase achieved by a tiny increase of $Im\, s$ we can see what happens after collision (cf. Fig. 7.9 and Fig. 7.10). The two trajectories associated with the two lowest states become disconnected again but the ground state trajectory makes a loop whereas the $N = 1$ trajectory takes

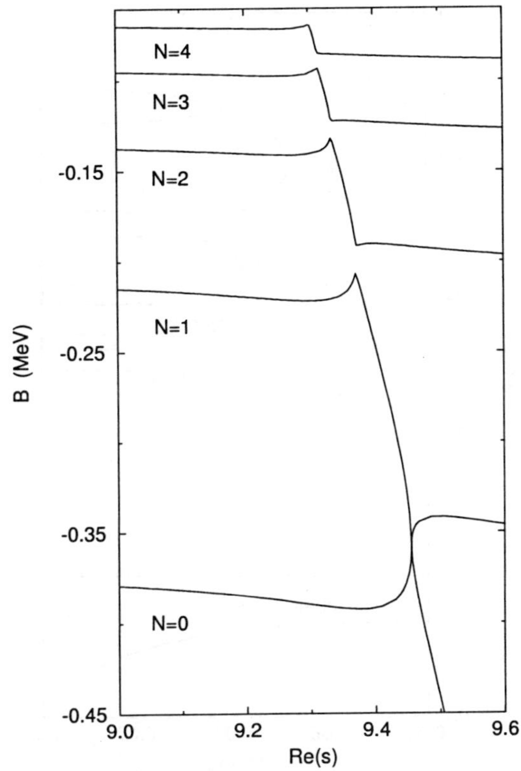


Fig. 7.7 Binding energy versus the real part of the dimensionless potential strength $s$ for the potential (6.20) and for $\ell = 2$. The imaginary part of $s$ has been held fixed $Im\, s = 0.132204$. The plot shows five curves corresponding to the lowest lying levels labeled by the nodal quantum number $N$. The spikes on the $N = 0$ and $N = 1$ curves seen if Fig. 7.3 have now been welded and we observe a level crossing phenomenon. The other curves do not cross yet but have prominent spikes

over its role moving fast toward large $Im\, k$ values. Looping of a pole trajectory gives rise to oscillatory behavior of both the binding energy and the level width, a phenomenon referred to as [Krell (1971)] oscillations. Despite the fact that on the energy plot the two curves still cross, there is in fact no level crossing any more since this would require crossing of the binding energies and the widths at the same value of $Re\, s$. This would imply crossing of pole trajectories but as our graph shows it does not take place here.

By going to even larger $Im\, s$ values, a collision between poles with higher N takes place and subsequently one by one these poles become successively

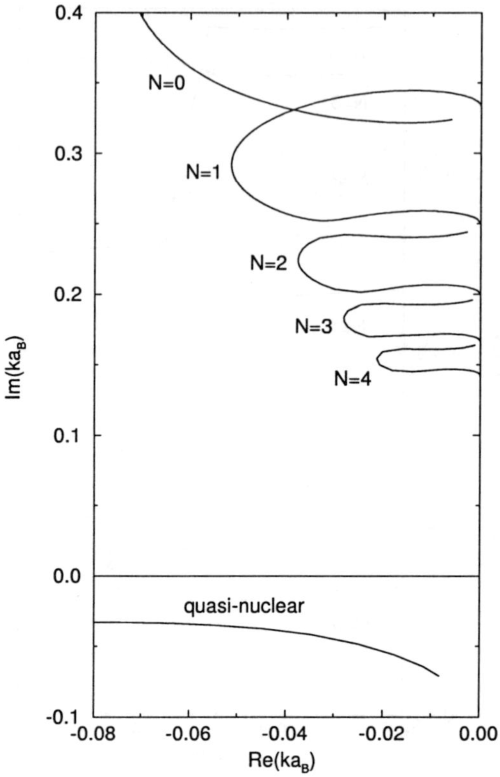


Fig. 7.8 The pole trajectories for the potential (6.20) and $\ell = 2$ calculated for fixed value of the imaginary part of the strength $Im\, s = 0.132204$. The different curves correspond to trajectories starting with $Re\, s = 0$ almost at the unperturbed Coulomb positions on the imaginary axis determined by the nodal quantum number. In the lower half we also show the trajectory connected with a quasi-nuclear level which comes quite close to the origin but owing to still insufficient absorption never enters the upper half of the k-plane. The $N = 0$ and $N = 1$ trajectories are seen to cross each other

engaged in a looping motion. At the same time one trajectory always stays apart and is almost linear. As $Im\, s$ increases this trajectory will be fed by states with higher and higher N and the trajectory comes closer and closer to the origin $k = 0$. Finally, for certain critical value of $Im\, s$ the trajectory reaches this limit and then for the same value of absorption, the quasi-nuclear pole trajectory in III quadrant for certain value of $Re\, s$ passes through the origin so that these two trajectories can be joined together to form a single curve. We have then a quasi-nuclear bound state. As has been shown in the preceding section, the critical value of $s$ at which zero-energy

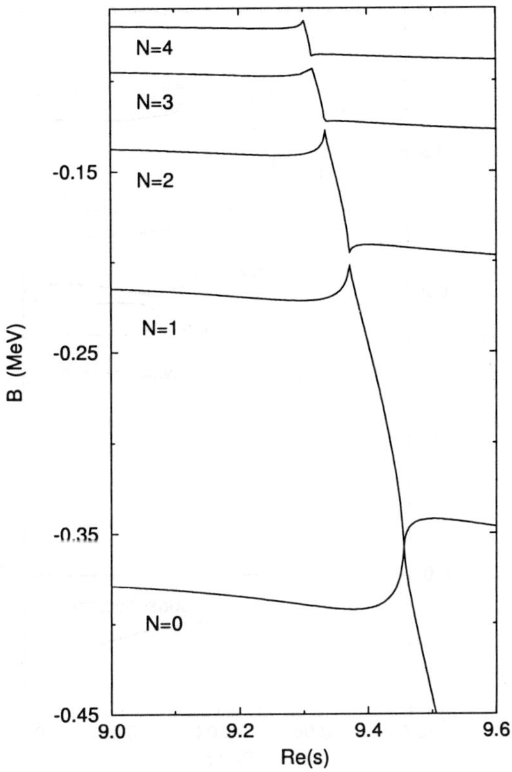


Fig. 7.9 Binding energy versus the real part of the dimensionless potential strength $s$ for the potential (6.20) and for $\ell = 2$. The imaginary part of $s$ has been held fixed $Im\, s = 0.13333759$. The plot shows five curves corresponding to the lowest lying levels labeled by the nodal quantum number $N$. Although the $N = 0$ and $N = 1$ curves do cross there is no level crossing because the widths of these levels are different. This will become apparent from Fig. 7.8

bound state is formed can be determined as a solution of the equation (6.26) and this is precisely how we calculated $Im\, s$ necessary to obtain the curves presented in Figs. 7.11-7.12. It should be noted that at this specific value of $Im\, s$ all atomic trajectories form loops whereas the quasi-nuclear trajectory has been completely disconnected from them and has a markedly different shape. With still further increasing of absorption the loops representing the atomic trajectories become tighter while the quasi-nuclear trajectory is pushed away from the imaginary axis so that it crosses the real axis at some negative value of $Re\, k$ as shown in Fig. 6.16. Also more attraction is needed to create a quasi-nuclear bound state in this case.

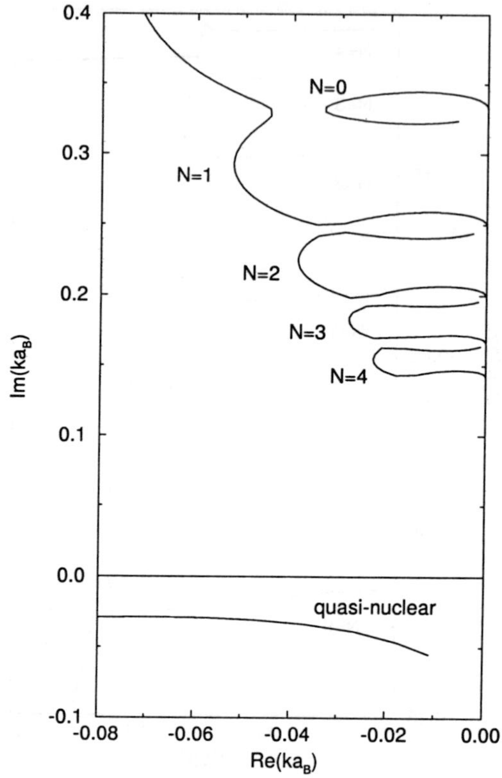


Fig. 7.10 The pole trajectories for the potential (6.20) and $\ell = 2$ calculated for fixed value of the imaginary part of the strength $Im\, s = 0.13333759$. The different curves correspond to trajectories starting with $Re\, s = 0$ almost at the unperturbed Coulomb positions on the imaginary axis determined by the nodal quantum number. In the lower half we also show the trajectory connected with a quasi-nuclear level which comes quite close to the origin but owing to still insufficient absorption never enters the upper half of the k-plane. The $N = 0$ and $N = 1$ trajectories are separated and the ground state trajectory makes a loop while the $N = 1$ trajectory takes over from the former going toward large binding energies

In Figs. 7.5-7.12 for the sake of clarity of the presentation we deliberately did not use $Re\, s$ values bigger than 10. It is nevertheless interesting to know what happens if the attraction is pushed beyond this limit. It turns out that essentially history will repeat itself in an almost cyclic fashion. When absorption is small as in Fig. 7.12, after completing a single loop each trajectory takes the form of an arc joining two adjacent Coulombic positions what corresponds to Zeldovich effect connected with the second quasi-nuclear level. Only the ground state trajectory is different as the

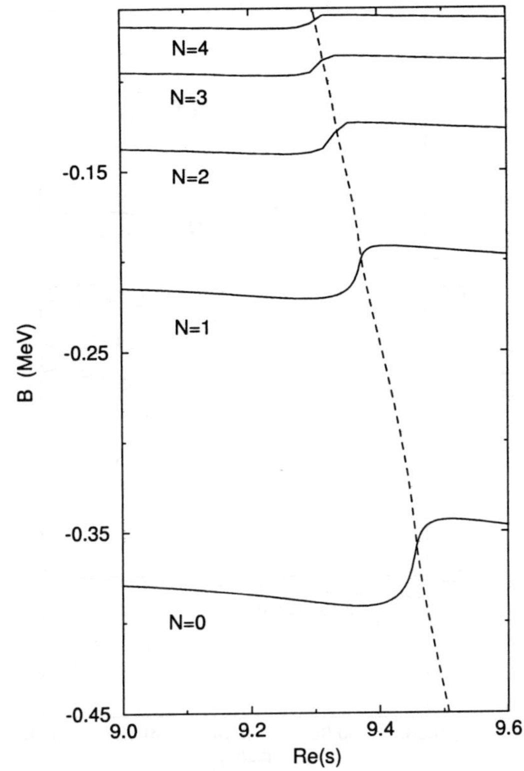


Fig. 7.11 Binding energy versus the real part of the dimensionless potential strength $s$ for the potential (6.20) and for $\ell = 2$. The imaginary part of $s$ has been held fixed $Im\, s = 0.137067$. The plot shows five curves corresponding to the lowest lying levels labeled by the nodal quantum number $N$. The quasi-nuclear level has been depicted by a dashed line. There is now no trace of Zeldovich like behavior, instead the atomic curves display slow oscillations

arc is quickly deformed into a straight line. If the loops were not present we would have a repetition of the situation presented in Fig. 7.6. We can reach the next phase by increasing absorption so that different pairs of poles collide and eventually a second quasi-nuclear bound state will be created. By the time it happens the first quasi-nuclear trajectory has been shifted to the left and the atomic trajectories take the form of double loops. The level shifts and widths both display ondulatory behavior. Thus, the second cycle has been completed and the essence of each cycle is the formation of a new quasi-nuclear bound state.

Comparing the different energy plots in Figs. 7.5 and 7.11 we can see

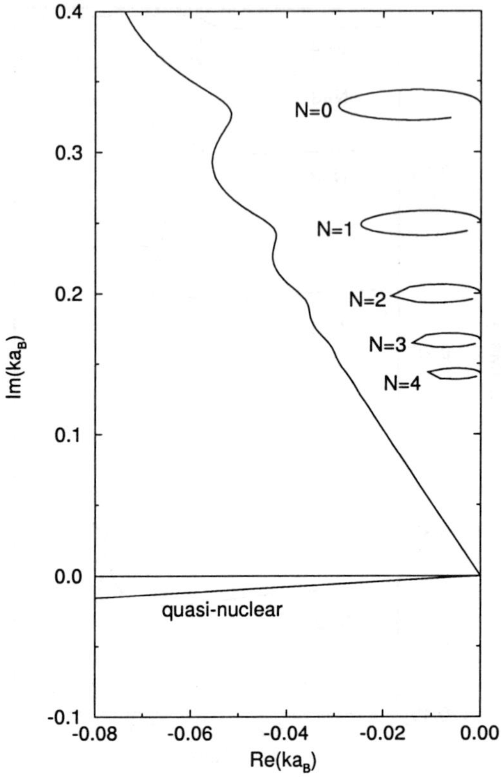


Fig. 7.12 The pole trajectories for the potential (6.20) and $\ell = 2$ calculated for fixed value of the imaginary part of the strength $Im\, s = 0.137067$. The different curves correspond to trajectories starting with $Re\, s = 0$ almost at the unperturbed Coulomb positions on the imaginary axis determined by the nodal quantum number. We also show the trajectory connected with a quasi-nuclear level which comes from the lower half of the k-plane passes through the origin and finally enters the upper half. The trajectories connected with atomic levels show all a very pronounced looping behavior

that Zeldovich effect at certain value of the strength gives way to the oscillatory behavior. The transition takes place when two poles collide and as a result we have a second order pole, i.e. the Jost function has a double zero. The condition for a double zero has been already given (cf. equations (6.27)) but a simultaneous solution of these two equations using a direct approach presents in general a difficult task. It seems therefore instructive and practically useful to develop an interpolative approximation scheme. As has already been noted the Jost function is analytic in both variables so that it could be locally approximated by polynomials. If we have at our disposal a calculational scheme which allows to compute the value of the Jost function for any values of the independent variables $s$ and $k$ then it is a simple matter to construct for $F_\ell(s,k)$ an interpolating polynomial. When such a polynomial has been determined a double zero of the Jost function can be evaluated directly using elementary algebraic methods and if need arises this procedure could be plugged into an iterative scheme. It is of interest to pursue this approach in some more detail so let us assume that we have located two single zeros of the Jost function at $(s_1,k_1)$ and $(s_2,k_2)$, respectively, positioned not very far apart and we want to find a double zero. The most general polynomial of the second order in $s$ and $k$ vanishing at $(s_1,k_1)$ and $(s_2,k_2)$, takes the form

$$\begin{aligned} F_\ell(s,k) \approx (s-s_1)\left[A(s-s_2)+B(k-k_2)\right] + \\ + (k-k_1)\left[C(s-s_2)+D(k-k_2)\right], \end{aligned} \tag{7.9}$$

where the four constants $A, B, C, D$ can be adjusted by providing the values of the Jost function at four points preferably in the neighborhood of the two zeros. After simple algebra, we obtain

$$A = -4\left[F_\ell(\bar{s},k_1) - \tfrac{1}{2}F_\ell(s_2,k_1)\right]/(s_1-s_2)^2; \tag{7.10}$$

$$B = -F_\ell(s_2,k_1)/\left[(s_1-s_2)(k_1-k_2)\right]; \tag{7.11}$$

$$C = -F_\ell(s_1,k_2)/\left[(s_1-s_2)(k_1-k_2)\right]; \tag{7.12}$$

$$D = -4\left[F_\ell(s_1,\bar{k}) - \tfrac{1}{2}F_\ell(s_1,\bar{k})\right]/(k_1-k_2)^2, \tag{7.13}$$

where $\bar{k} = \frac{1}{2}(k_1+k_2)$ and $\bar{s} = \frac{1}{2}(s_1+s_2)$. Since the interpolating polynomial has been completely determined we can express analytically $k$ in terms of $s$ and in particular calculate the critical value of $s$ at which a double zero occurs. Denoting this strength as $s_c$, we find two solutions

$$s_c = \left[-b \pm (b^2 - ac)^{1/2}\right]/a, \tag{7.14}$$

where

$$a = (B+C)^2 - 4AD; \tag{7.15}$$

$$b = -(B+C)(2\bar{k}D + Cs_2 + Bs_1) + 2D(Ck_1 + Bk_2 + 2A\bar{s}); \tag{7.16}$$

$$c = (2D\bar{k} + Cs_2 + Bs_1)^2 - 4D\left[k_1(Dk_2 + Cs_2) + s_1(Bk_2 + As_2)\right]. \tag{7.17}$$

The solution which is close to the two zeros $s_1$ and $s_2$ is the physical $s_c$ while the other one is usually far away from the region where our interpolation could be valid and will be denoted as $\tilde{s}_c$. The explicit expression for the two trajectories reads

$$k_{\pm}(s) = \bar{k} + \frac{B(s_1 - s) + C(s_2 - s) \pm \sqrt{a(s - s_c)(s - \tilde{s}_c)}}{2D}. \tag{7.18}$$

So far in the presented plots the values of $Im\, s$ have been quite small and it is interesting to know what happens when the absorption is increased arbitrarily. It turns out that the loops displayed in Fig. 7.12 become then tighter and tighter. Further increase of absorption deforms them into a twist and finally into a wiggle. In consequence, all the effects described above in this section are washed out. In principle it is still possible to generate a quasi-nuclear bound state but this might require very strong attraction. The quasi-nuclear pole trajectory shown in Fig. 7.11 is forced to cross the negative real k-axis further and further away from the origin which means that these bound states will be rather pathological with large width and positive $Re\, E$. Although it is true that by increasing the attraction the binding energy could be made negative but this would also push the width to still bigger values diminishing the chances that the quasi-nuclear bound states could be observable.

## 7.3 Small Shift Approximation (SSA)

When the orbital momentum is large enough the level shift in a hadronic atom is usually quite small in comparison with the Coulomb energy. This observation has prompted several authors (cf. [Deser *et al.* (1954); Deloff (1976); Mandelzweig (1977); Trueman (1961); Partensky *et al.* (1967); Lambert (1969)]) to introduce small shift approximation based on the idea that if atomic bound state pole has been shifted from the Coulomb position $k_n$ by $\delta k$, then $|\delta k/k_n|$ can be regarded as a small expansion parameter. One of the implications of SSA is the shift formula expressing the energy shift in a hadronic atom in terms of hadron-nucleus low-energy scattering parameters. This formula relating only experimentally measurable quantities without reference to the underlying nuclear potential has very important phenomenological applications. On the other hand, it is equally important to know how reliable SSA really is and we wish now to see into it.

In order to derive the shift formula we are going to obtain an approximate solution of the bound state equation (5.46). The function $g_\ell^+(k)$ which enters (5.46) is singular at $k = k_n$ and for $k$ such that $|k - k_n|$ is small in comparison with the level separation $|k_n - k_{n+1}|$, we can approximate $g_\ell^+(k)$

by a single pole term

$$g_\ell^+(k) \approx \frac{k_n\, R_{\ell N}}{k - k_n}, \tag{7.19}$$

where $R_{\ell N}$ is the residue which could be worked out from the definition of $g_\ell^+(k)$. Another alternative is to start with eq. (3.20) in which the function $g_\ell^-(k)$ regular at $k_n$ can be dropped. Writing $F_\ell^+(k)$ as $\dot{F}_\ell^+(k_n)(k - k_n)$, with the aid of (6.28), we arrive at the expression

$$R_{\ell N} = \left\{2|k_n|^2 \int_0^\infty \phi_\ell(k_n, r)^2\, \mathrm{d}r\right\}^{-1} \tag{7.20}$$

where the normalization integral, is

$$\int_0^\infty \phi_\ell(k_n, r)^2\, \mathrm{d}r = \frac{nN!\,[(2\ell+1)!!\,\ell!]^2}{4|k_n|^{2\ell+3}(\ell+n)!}. \tag{7.21}$$

In order to obtain an approximate solution of equation (5.46), it is assumed that the K-matrix is a slowly varying function of $k$ so that $\mathcal{K}_\ell(s,k)$ is replaced by $\mathcal{K}_\ell(s,k_n)$. Using the single pole approximation (7.19), the solution of equation (5.46) is

$$k \approx k_n[1 + \mathcal{K}_\ell(s, k_n)\, R_{\ell N}], \tag{7.22}$$

and this yields the celebrated level shift formula in the most general form

$$\delta E = -\epsilon - \mathrm{i}\tfrac{1}{2}\Gamma = E - E_n \approx -\frac{\mathcal{K}_\ell(s, k_n)}{2\mu \int_0^\infty \phi_\ell(k_n, r)^2\, \mathrm{d}r} \tag{7.23}$$

where $E_n = k_n^2/2\mu$. Approximating in (7.23) the K-matrix by the scattering length one obtains the Deser–Trueman formula. Repeating the same arguments which led to formula (6.14), we can write analogous expression for the full K-matrix which is also a meromorphic function of $s$. To be more explicit, we represent the K-matrix in the form

$$\mathcal{K}_\ell(s, k_n) = \frac{\int_0^\infty v(r)\, \phi_\ell(k_n, r)^2\, dr}{\int_0^\infty v(r)\, r\, \mathrm{d}r}\, \frac{s}{1 - s/\tilde{s}_{\ell 0}(k_n)} \prod_{N=1}^\infty \frac{1 - s/\tilde{z}_{\ell N}(k_n)}{1 - s/\tilde{s}_{\ell N}(k_n)} \tag{7.24}$$

where the poles and zeros in (7.24) depend now upon $k$ but when the effective range expansion is rapidly converging they are slowly varying functions. The dependence upon $k$ appears also in the integral in the numerator of

(7.24). Using (7.24) in (7.23), we can cast the shift formula to the form

$$\delta E = \int_0^\infty V(r)\, u_{\ell N}(r)^2\, \mathrm{d}r\, \frac{1}{1 - s/\tilde{s}_{\ell 0}(k_n)} \prod_{N=1}^{\infty} \frac{1 - s/\tilde{z}_{\ell N}(k_n)}{1 - s/\tilde{s}_{\ell N}(k_n)}, \tag{7.25}$$

where the first factor in (7.25) is recognized as the standard perturbative expression of the first order in $V$ whereas the remaining terms represent corrections non-linear in $V$. Usually, $\tilde{s}_{\ell 0}(k_n)$ is not far from the pole of the scattering length so that the pole terms in (7.25) are very important, and, except for the smallest $s$, the perturbative expression makes extremely poor approximation. The smallness of the shifts and widths is a necessary, but not a sufficient, condition for the first order perturbation theory to be applicable (cf. [Seki *et al.* (1971); Seki *et al.* (1987)]). It is apparent from (7.25) that the first order perturbation theory in $s$ is valid for weak potentials, i.e. when $|s| \ll \tilde{s}_{\ell 0}$ because only in this case the scattering length $A_\ell(s)$ could be approximated by a first order polynomial in $s$. The presence of poles induces rapid variation in $\delta E$ whenever $s$ comes close to any of the poles in (7.25). In consequence, keeping $Im\, s$ small and varying $Re\, s$ both the width and the level shift will show oscillatory behavior similar to that presented in Fig 6.2. This is the Krell effect.

Since SSA is based on the assumption that $|\delta k| \ll |k_n - k_{n+1}|$, by taking into account (7.22) we obtain the following upper bound

$$|\mathcal{K}_\ell(s, k_n)| \ll \frac{2|k_n|^2}{n+1} \int_0^\infty \phi_\ell(k_n, r)^2\, \mathrm{d}r. \tag{7.26}$$

When this condition is satisfied the energy shift can be computed reliably from formula (7.23). It should be noted that the K-matrix on the right hand side of (7.23) requires analytic continuation below threshold. In their pioneering work [Deser *et al.* (1954)] replaced the Coulomb K-matrix in (7.23) by the scattering length $a_\ell$ thus ignoring all Coulomb corrections not to mention the continuation below threshold. This crude approximation has been subsequently improved by several authors who argued that the Coulomb energies are so close to threshold that the K-matrix could be approximated by its threshold value setting $\mathcal{K}_\ell(s, k_n) \approx A_\ell(s)$. In principle, the scattering length approximation might be further improved by adding more terms in the effective range expansion to continue the K-matrix below threshold but, in fact, the K-matrix in (7.23) could be easily computed directly without any approximations. The method of calculation [Deloff (1981)] employs the Wronskian formula (5.26) and is based on the observation that for $k = k_n$ major simplifications take place. Firstly, the regular Coulomb solution $\phi_\ell(k_n, r)$ reduces to the standard hydrogen-like wave function, and, secondly, it turns out that the second linearly independent solution, namely the in-going wave solution $H_\ell^-(k_n, r)$ can also be obtained in an analytic form. Writing the bound state condition (5.46) as

$$g_\ell^+(k) = -\frac{w\left[\phi_\ell(s,k,R), \theta_\ell(k,R)\right]}{w\left[\phi_\ell(s,k,R), \phi_\ell(k,R)\right]} \tag{7.27}$$

and eliminating the function $\theta_\ell(k, R)$ in favor of $H_\ell^-(k, R)$ by means of (3.66) and then using (3.20), we have

$$\frac{2\mathrm{i}k}{F_\ell^+(k)F_\ell^-(k)} = -\frac{w\left[\phi_\ell(s,k,R), H_\ell^-(k,R)\right]}{w\left[\phi_\ell(s,k,R), \phi_\ell(k,R)\right]}, \tag{7.28}$$

which is still exact. To obtain approximate solution of (7.28) we set again $k = k_n + \delta k$ noting that it is the pole due to the zero of the Coulomb Jost function $F_\ell^+(k)$ at $k = k_n$ that gives the dominant contribution to the left hand side of (7.28). Using (6.28), we immediately arrive at an alternative form of the shift formula

$$\frac{\delta k}{k_n} = -\frac{w\left[\phi_\ell(s,k_n,R), \phi_\ell(k_n,R)\right]}{w\left[\phi_\ell(s,k_n,R), H_\ell^-(k_n,R)\right]}\,\frac{1}{2|k_n|^2\int_0^\infty \phi_\ell(k_n,r)^2\,\mathrm{d}r}, \tag{7.29}$$

and by comparing (7.29) with (7.23) the first factor on the right hand side of (7.29) can be recognized as $\mathcal{K}_\ell(s, k_n)$. The in-going wave solution $H_\ell^-$ which enters (7.29) can be obtained directly from the following integral representation given in [Abramowitz *et al.* (1965)]

$$H_\ell^-(k,r) = \frac{\mathrm{e}^{-\mathrm{i}kr}}{(kr)^\ell\,(2\ell+1)!\,C_\ell(\eta)}\int_0^\infty \mathrm{e}^{-t}\,t^{\ell-\,\mathrm{i}\eta}\,(t+2\mathrm{i}kr)^{\ell+\,\mathrm{i}\eta}\,\mathrm{d}\,t, \tag{7.30}$$

which is the Laplace transform of the Kummer function. For $k = k_n$ this integral can be explicitly evaluated, and, the result is

$$H_\ell^-(k_n,R) = \mathrm{e}^{\frac{1}{2}x}x^{-\ell}\frac{\chi_{\ell N}(x)}{(2\ell+1)!\,N!}; \tag{7.31a}$$

$$\chi_{\ell N}(x) = \left(\frac{\mathrm{d}}{\mathrm{d}x}\right)^N\left[\sum_{j=0}^{2\ell+N}(2\ell+N+j)!\,x^j - x^{\ell+n}\mathrm{e}^{-x}Ei(x)\right], \tag{7.31b}$$

where $x = 2|k_n|R$ and $Ei(x)$ is the exponential integral function defined in [Abramowitz *et al.* (1965)]. Using these expressions we can write down our ultimate formula amenable for computations of the energy shift. We have

$$\frac{\delta k}{k_n} \approx -\mathrm{e}^{-x} x^{2\ell+1} \frac{N! \left[\ell + 1 - \frac{1}{2}x + \mathcal{D}_\ell(s,x)\right] L_N^{2\ell+1}(x)}{n \left[-\ell + \frac{1}{2}x + \mathcal{D}_\ell(s,x)\right] \chi_{\ell N}(x)}, \tag{7.32}$$

where $L_N^{2\ell+1}(x)$ denotes Laguerre polynomial and $\mathcal{D}_\ell(s,x)$ is defined as

$$\mathcal{D}_\ell(s,x) = -R\frac{\phi'_\ell(s,k_n,R)}{\phi_\ell(s,k_n,R)} + x\frac{\mathrm{d}}{\mathrm{d}x}. \tag{7.33}$$

The wave functions which enter (7.33) usually could be obtained only by numerical integration of the wave equation. The advantage of formula (7.32) is evident for it provides the correct continuation below threshold and yet the amount of numerical effort is the same as in the scattering length approximation.

We shall now examine how reliable SSA really is under the most unfavorable circumstances, i.e. when $s$ is real so that the scattering length can grow very large. It will be convenient to introduce the following error function

$$e(s) = \left|1 - \frac{\delta E^{SSA}}{\delta E^{exact}}\right|, \tag{7.34}$$

where the "exact" energy shift is that determined by locating the bound state without approximations. In Figs. 7.13 and 7.14 we have plotted the error versus $s$ for the Saxon–Woods potential (6.20) for $\ell = 1, 2$ and to obtain $\mathcal{K}_\ell(s, k_n)$ in (7.23) we used: (i) continuation below threshold by means of (7.32), or (ii) scattering length approximation. In general, the performance of SSA is in both cases quite good: SSA works well for all $s$ up to the value at which the nuclear potential is capable of supporting a bound state of its own. At this value the energy shifts become very large and SSA breaks down completely. On the whole, the scattering length approximation is slightly less accurate and even for $s = 0$ the error does not go to zero. Indeed, the SSA shift formula in the lowest order in $s$ reduces properly to the first order perturbation result only if continuation below threshold has been made. With the scattering length approximation this is not the case and we have $e(0) \neq 0$. For complex $s$ the performance of SSA rapidly improves with increasing absorption. When $Im\, s$ is above certain limiting value, the condition (7.26) is satisfied for all $Re\, s$ and SSA always makes excellent approximation.

Summarizing, SSA is quite reliable but only for $s$ not exceeding the value at which the first zero-energy nuclear bound state appears. Thus, before ap-

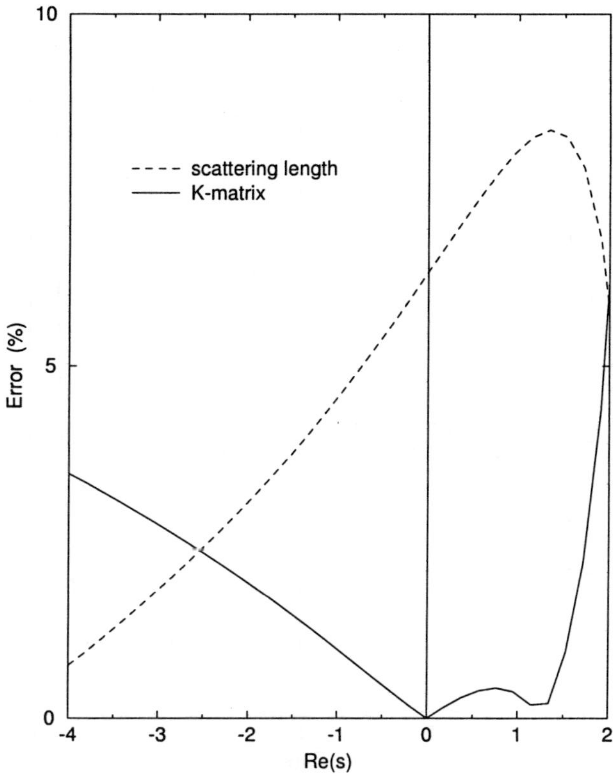


Fig. 7.13 Accuracy of small shift approximation as function of the potential strength for $\ell = 1$ ground state level. The presented curves have been obtained for the potential (6.20) and show the magnitude of the error in % as defined in (7.34). The continuous curve corresponds to the approximation using the full K-matrix extrapolated below threshold to the Coulomb binding energy and the dashed curve corresponds to the scattering length approximation.

plying the SSA shift formula in phenomenology it is vital checking whether (7.26) is satisfied. The continuation below threshold is recommended as it improves the accuracy of the shift formula.

Going beyond SSA appears to be prohibitively difficult because it requires taking into account more poles in $g_\ell^+(k)$, preferably all of them, which is then equivalent to solving the full problem. Nevertheless, the two pole approximation certainly deserves attention (cf. also ref. [Badalyan *et al.* (1982)]) as it leads to a bound state equation which is still simple enough to be solved analytically. This is instructive and helps to understand better the mechanism responsible for the complicated deformation of

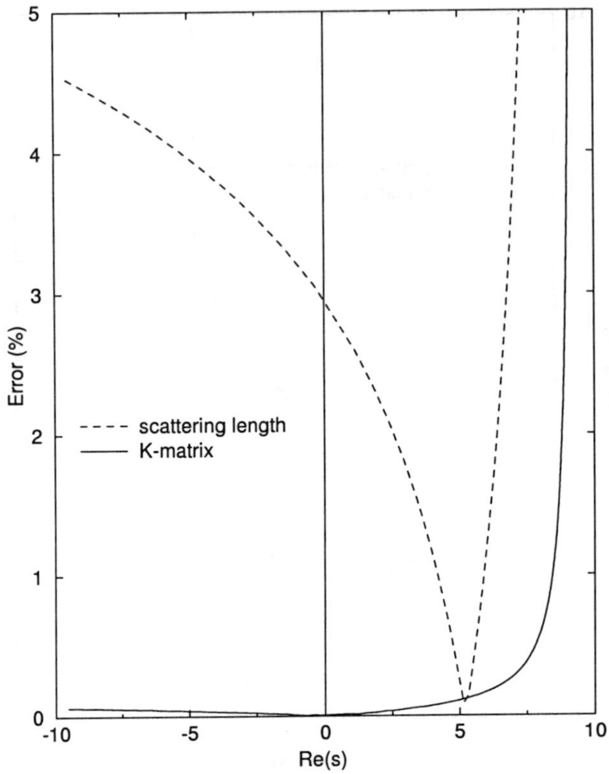


Fig. 7.14 Accuracy of small shift approximation as function of the potential strength for $\ell = 2$ ground state level. The presented curves have been obtained for the potential (6.20) and show the magnitude of the error in % as defined in (7.34) The continuous curve corresponds to the approximation using the full K-matrix extrapolated below threshold to the Coulomb binding energy and the dashed curve corresponds to the scattering length approximation.

pole trajectories known from numerical studies. Since our discussion has purely qualitative character, we are going to simplify the problem as much as possible. So, we replace $\mathcal{K}_\ell(s,k)$ by the scattering length $A_\ell(s)$ and we approximate $g_\ell^+(k)$ by a sum of pole terms, corresponding to the Coulombic ground state and the first excited state. With these simplifications, from (5.46) we obtain the approximate bound state equation

$$\frac{k_0 R_{\ell 0}}{k - k_0} + \frac{k_1 R_{\ell 1}}{k - k_1} = \frac{1}{A_\ell(s)}, \tag{7.35}$$

where $k_0$ and $k_1$ denote the ground and excited state momenta, respectively.

The residues $R_{\ell 0}$ and $R_{\ell 1}$ are real numbers and can be computed from (7.20) and (7.21). Equation (7.35) has two roots which we shall denote as $k_0(s)$ and $k_1(s)$ and they are arranged in such a way that for $s \to 0$ they resume the unperturbed Coulombic positions $k_0$ and $k_1$, respectively. The two solutions of equation (7.35), are

$$k_0(s) = k_0\Big\{1 + R_{\ell 0}\, A_\ell(s) + \frac{\ell+1}{2}\frac{\delta(s)}{\Delta(s) + \sqrt{\Delta(s)^2 + \delta(s)}}\Big\}; \quad (7.36)$$

$$k_1(s) = k_1\Big\{1 + R_{\ell 1}\, A_\ell(s) - \frac{\ell+2}{2}\frac{\delta(s)}{\Delta(s) + \sqrt{\Delta(s)^2 + \delta(s)}}\Big\}; \quad (7.37)$$

where

$$\delta(s) = \frac{4\, R_{\ell 0}\, R_{\ell 1}\, A_\ell(s)^2}{(\ell+1)(\ell+2)}; \quad (7.38)$$

$$\Delta(s) = \frac{1 + R_{\ell 0}\, A_\ell(s)}{\ell+1} - \frac{1 + R_{\ell 1}\, A_\ell(s)}{\ell+2}. \quad (7.39)$$

The three terms in the curly brackets in (7.36) and (7.37) can be interpreted, respectively, as purely Coulombic part, the SSA correction, and a two pole interference term which is non-linear in $A_\ell(s)$. For real $s$ the whole curly bracket expression is real so that the two solutions (7.36) and (7.37) are confined to the imaginary axis, as they should be. The trajectories corresponding to the two levels can cross and this happens when the expression under the square root in (7.36) vanishes, i.e. for $\Delta(s)^2 = -\delta(s)$. Although for real $s$ this equation has no solution because $\Delta(s)^2$ and $\delta(s)$ are then both real and positive but the situation changes when complex $s$ are admitted. Since $\Delta(s)$ and $\delta(s)$ depend upon $A_\ell(s)$ we can solve the above equation for $A_\ell(s)$ obtaining the following condition which determines the critical $s$ at which crossing takes place

$$\frac{1}{A_\ell(s)} = (\ell+1)(\ell+2)\Big\{\sqrt{\frac{R_{\ell 1}}{\ell+2}} - \mathrm{i}\sqrt{\frac{R_{\ell 1}}{\ell+2}}\Big\}^2 \quad (7.40)$$

and in order to obtain the critical strength one has to solve the above equation. The general scheme is now complete and the computational effort can be greatly reduced by using the product representation (6.14) for $A_\ell(s)$. In particular, inserting the truncated product formula in (7.40), we end up with a polynomial equation which can be solved numerically by standard methods.

To find out how this scheme works in practise, we took again the potential (6.20) with $\ell = 2$ and in Fig. 7.15 we present the results of our

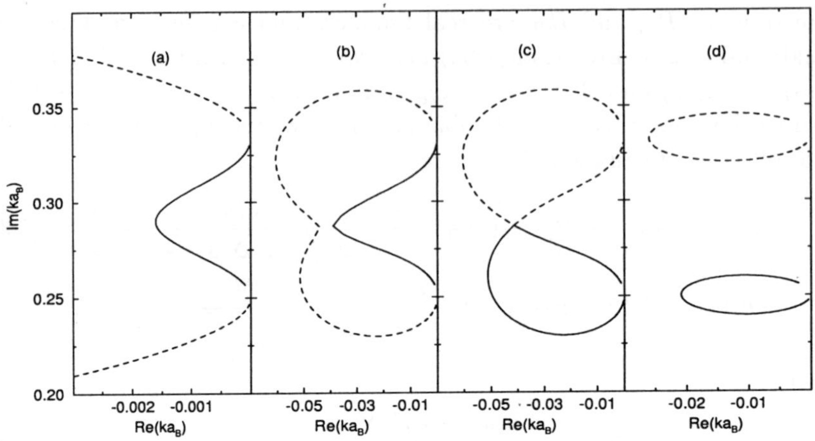


Fig. 7.15 Shape deformation of pole trajectories caused by increasing absorption. The curves have been obtained for nuclear potential (6.20) and $\ell = 2$ using two-pole approximation (7.36) and (7.37) in which we varied $Re(s)$ in the interval [8.0,10.0] with $Im(s)$ fixed. Continuous and dashed curves correspond to $N = 1$ and $N = 0$, respectively. The absorption is different in each of the four cases (a), (b), (c), (d), and the values of $Im(s)$, respectively, are 0.004, 0.0052, 0.00520943, 0.01.

computations. The imaginary part of $s$ has been here always regarded as a fixed parameter while $Re\, s$ plays the role of the independent variable. For each complex value of $s$ we compute $A_\ell(s)$ from the wave equation and then from (7.36) and (7.37) we obtain the two trajectories and all curves shown in Fig. 7.15 result from varying $Re\, s$ from 8.0 to 10.0. First we consider very small absorption case displayed in Fig. 7.15(a) where we can see that the $N = 1$ trajectory (solid curve) commences at the $N = 1$ pole and finishes at the ground state pole position. This is a manifestation of Zeldovich effect but owing to non-zero absorption the curve goes off the imaginary axis. The dashed curve corresponding to $N = 0$ starts at the Coulombic ground state position and rapidly goes away from the imaginary axis. This behavior is also correct for the ground state trajectory but as the distance from the ground state pole increases the basic two-pole approximation (7.35) becomes quite poor. The dashed curve makes a huge loop diving into lower half of the k-plane (not shown on the plot) eventually reappearing again close to the $N = 1$ pole (shown as a second branch of the dashed curve). This incorrect behavior indicates that the model has been pushed too far, outside the range of validity. When the absorption is increased the big loop shrinks whereas the $N = 1$ trajectory is deflected further off the imaginary axis. This can be seen in Fig. 7.15(b) where we observe a small concavity building up in the dashed curve which signals that crossing is imminent with further increase of the absorption. In Fig. 7.15(c) we show crossing of the two trajectories. The critical value of $s$ required for crossing has been determined very precisely from equation (7.40). At the crossing point the two curves loose their identity and instead of two touching loops we could have drawn the curves as in Fig. 7.15(b) except that now the concavity in the dashed curve melts with the kink in the continuous curve. Finally, with further increase of absorption the scattering lengths become strongly suppressed, just as those presented in Fig. 6.3, so that SSA makes excellent approximation and the non-linear term in (7.36)–(7.37) proportional to $\delta$ is negligible. As a result we obtain in Fig. 7.15(d) two well separated small loops. This leads to the Krell effect where the level shift and the width become oscillating functions of $Re\, s$. Although the crossing of the two trajectories captured in Fig. 7.15(c) does not look quite the same as in the exact case presented in Fig. 7.8 but the two pole approximation illustrates well the passage from the Zeldovich phase to the Krell phase. The former is non-linear whereas the latter is linear in $A_\ell(s)$ and the key issue is the size of the scattering length which is determined by the value of $Im\, s$. Thus the latter parameter decides which phase sets in.

# Chapter 8

# Gamow States and Completeness Problem

Quasi-stable or resonant states play a very important role in physics: in fact, most of the elementary particles belong to that category, as well as many nuclear species included in the periodic table of elements together with stable nuclei. In addition to that, we have also unstable atomic states. Unstable particles and nuclei decay through various interactions so that their lifetimes are finite but may vary by many orders of magnitude. Apart from a finite lifetime, these resonant states share all the remaining properties attributed to stable particles. In this situation it seems natural to look for a unifying description of both, bound states and resonances. This has been achieved in potential scattering theory where the bound states and resonances are associated with the zeroes of the Jost function and therefore appear as poles of the S-matrix and poles of the Green's function. The zero of the Jost function determines also the eigenenergy of a Gamow state and therefore the latter provides an alternative description of a resonant state. There is a one-to-one correspondence between the complex poles of the S-matrix and the complex eigenvalues obtained from imposing Siegert boundary condition. Historically, the energy eigenfunction with a complex eigenvalue was originally considered by Gamow [Gamow (1928)] with the purpose to understand $\alpha$ decay, but within the Hilbert space theory his treatment was heuristic as self-adjoint operators on a Hilbert space admit only real eigenvalues. A rigorous mathematical treatment of Gamow states required an extension of Hilbert space to the Rigged Hilbert Space [Bohm *et al.* (1989)] where Gamow vectors are eigenkets of a dual extension of the self-ajoint Hamiltonian. Since this extension is beyond the scope of this book, we shall consider the normalization of Gamow states and the associated completeness problem on a somewhat pedestrian level.

## 8.1 Normalization of Gamow States

From the results of the preceding chapter it has become apparent the close relation between bound states and the Gamow states. We have seen that both are obtained by solving the same equation and by varying the strength of the nuclear potential there is a continuous transition between Gamow states and bound states and the two types of states are located on a single pole trajectory. This formal similarity has attracted many people to the idea that bound states and Gamow states should be treated on equal footing. On the other hand, for a Gamow state we have $Im\, k < 0$ and in consequence the corresponding eigenfunction increases exponentially with distance what implies that the usual rules concerning normalization, orthogonality and completeness of the eigenfunctions require extension. Especially important normalization problem has been tackled by various approaches such as regularization of the integrals [Zel'dovich (1961); Berggren (1968); Berggren (1971)], analytic continuation from the upper half of the k-plane [Romo (1968)] and by using the residue of the resolvent for defining the norm [Hokkyo (1965); Garcia-Calderon *et al.* (1976)]. It is gratifying that all the different methods resulted in the same normalization convention.

In this chapter we are going to show that when the Coulomb potential is present the norm can be defined in a similar way as for the Coulomb-free case. We find particularly appealing the arguments that the norm of the wave function is given by the residue at the pole of the resolvent corresponding to the Gamow state and in what follows we study the behavior of the resolvent near one of such poles.

The resolvent, or the complete Green's function of the problem at issue, satisfies the wave equation

$$\left[\frac{\mathrm{d}^2}{\mathrm{d}r^2} - \frac{\ell(\ell+1)}{r^2} + \frac{2\mu\alpha Z}{r} + sv(r) + k^2\right] \langle r|\mathcal{G}_\ell^{\pm}(k)|r'\rangle = \delta(r-r'), \quad (8.1)$$

and can be expressed in terms of the wave functions

$$\langle r|\mathcal{G}_\ell^{\pm}(k)|r'\rangle = -\frac{\phi_\ell(s,k,r_<) f_\ell^{\pm}(s,k,r_>)}{F_\ell^{\pm}(s,k)}. \quad (8.2)$$

At any of the eigenvalues $k = k_n$ the Jost function $F_\ell^{+}(s,k)$ vanishes so that $\mathcal{G}_\ell^{+}(k)$ has a pole there and we wish to work out the residue of this pole. Close to the pole, we can write

$$\langle r|\mathcal{G}_\ell^\pm(k)|r'\rangle = -\frac{4ik_n^2\phi_\ell(s,k_n,r)\phi_\ell(s,k_n,r')}{\dot{F}_\ell^+(s,k_n)F_\ell^-(s,k_n)(k^2-k_n^2)} + \text{regular part}, \tag{8.3}$$

where we have used (5.4) setting $F_\ell^+(s,k_n) = 0$. Introducing normalized wave functions $\tilde{\phi}_\ell = \phi_\ell/N$, we can cast formula (8.3) to the familiar form

$$\langle r|\mathcal{G}_\ell^\pm(k)|r'\rangle = \frac{\tilde{\phi}_\ell(s,k_n,r)\,\tilde{\phi}_\ell(s,k_n,r')}{k^2-k_n^2} + \text{regular part}, \tag{8.4}$$

and by comparing (8.3) with (8.4) we conclude that the norm is given by the following expression

$$N_n^2(s) = -\frac{\dot{F}_\ell^+(s,k_n)F_\ell^-(s,k_n)}{4ik_n^2}, \tag{8.5}$$

and this normalization is applicable for all eigenvalues so that bound states and Gamow states are treated on equal footing. In particular, for bound states ($Im\, k_n > 0$) inserting (6.28) in (8.5), we recover the standard normalization

$$N_n^2(s) = \int_0^\infty \phi_\ell(s,k_n,r)^2\,\mathrm{d}r.$$

Since $k_n$ is fixed by the value of $s$ the normalization is a function of the single variable $s$. To obtain analogous expression for Gamow states it will prove useful to introduce some large radius $R$ beyond which the nuclear potential becomes negligibly small. The derivative of the Jost function can be expressed as a sum of two Wronskians

$$\begin{aligned}\dot{F}_\ell^+(s,k_n) =& w\left[\dot{f}_\ell^+(s,k_n,R), f_\ell^+(s,k_n,R)\right]\frac{F_\ell^-(s,k_n)}{2ik_n} + \\ &+ w\left[\phi_\ell(s,k_n,R), \dot{\phi}_\ell(s,k_n,R)\right]\frac{2ik_n}{F_\ell^-(s,k_n)};\end{aligned} \tag{8.6}$$

where the first Wronskian can be written as

$$w\left[\dot{f}_\ell^+(s,k_n,R), f_\ell^+(s,k_n,R)\right] = -f_\ell^+(s,k_n,R)^2\frac{\mathrm{d}}{\mathrm{d}k_n}\frac{f_\ell^{+\prime}(s,k_n,R)}{f_\ell^+(s,k_n,R)}, \tag{8.7}$$

and the second Wronskian , is

$$w\left[\phi_\ell(s,k_n,R), \dot{\phi}_\ell(s,k_n,R)\right] = -2k_n\int_0^R \phi_\ell(s,k_n,r)^2\,\mathrm{d}r. \tag{8.8}$$

Inserting (8.6) in (8.5), the normalization for Gamow states takes the following form

$$N_n^2(s) = \int_0^R \phi_\ell(s,k_n,r)^2\,\mathrm{d}r + \phi_\ell(s,k_n,R)^2 \frac{\mathrm{d}}{\mathrm{d}k_n^2} \frac{f_\ell^{+\prime}(s,k_n,R)}{f_\ell^{+}(s,k_n,R)}, \tag{8.9}$$

and it can be easily verified that the appearance of $R$ in this expression is spurious so that the norm does not really depend upon the choice of $R$. So far no approximation has been made and the above normalization is a natural extension of the results that have been obtained without the Coulomb potential. In practice, the logarithmic derivative in (8.9) may be approximated with arbitrary accuracy by the logarithmic derivative of purely Coulombic Jost solutions which could be conveniently evaluated by continued fraction summation.

It is physically rather satisfactory that with the above normalization rule the first order perturbation formula for energy can be written in an unified form valid for all eigenvalues including Gamow states. To demonstrate that, we add to the nuclear potential a small perturbation $\delta v(r)$ which for simplicity is assumed to vanish for $r > R$. Adding the first order correction to the Jost function and expanding the Jost function about the unperturbed eigenvalue $k_n$ in powers of $\delta k$, we obtain

$$\dot{F}_\ell^{+}(s,k_n)\delta k - \int_0^R \delta v(r)\, f_\ell^{+}(s,k_n,r)\phi_\ell(s,k_n,r)\,\mathrm{d}r = 0,$$

which yields the generalized perturbation formula

$$\delta k^2 = -\frac{\int_0^R \delta v(r)\,\phi_\ell(s,k_n,r)^2\,\mathrm{d}r}{N_n^2(s)},$$

with $N_n^2(s)$ given by formula (8.5).

## 8.2 Completeness Problem

When the potential is complex level crossing is not prohibited and when this happens not only the eigenenergies but also the corresponding wave functions become identical. As a result we have one function less and an important question arises whether the solutions of the wave equation still form a complete set. Actually, a similar query was raised many years ago by Peierls [Peierls (1948)] who was studying various discrete expansions and for that purpose considered real potential eigenvalue problem of Sturm–Liouville type but with a complex boundary condition imposed at large $r$. This led to complex eigenvalues and level crossing was possible. It is perhaps worthwhile to repeat here his simple arguments, so let $\psi_n$ and $\psi_{n+1}$ denote two normalized wave functions belonging to two different eigenvalues close to the crossing point $k_0$ which will be reached when $s \to s_0$. Instead of $\psi_n, \psi_{n+1}$ we may also use $(\psi_n + \psi_{n+1})$ and $(\psi_n - \psi_{n+1})$ as members of our complete set. When $s$ tends to $s_0$ the second of these functions tends to zero but so does the normalization constant and in consequence the second function does not drop out from the completeness expansion but will be replaced by the derivative with respect to $k$. Although the latter function is not a solution of the wave equation but is orthogonal to all solutions and is indispensable to sustain completeness.

In order to substantiate the intuitive arguments presented above in principle one has to resort to the rather advanced methods used in the spectral theory of non-selfadjoint operators. The main difficulty is connected with the existence of the so called spectral singularities, i.e. with the bound states located on the real axis what for complex potentials is possible so that the corresponding eigenfunctions are no longer square integrable. As a matter of fact, the real axis can be viewed as a division line on which a continuous transition takes place transforming a Gamow state into a bound state. For Gamow states, however, the just mentioned difficulty has been well known and several methods have been developed indicating how to normalize these seemingly divergent eigenfunctions by integrating over a finite volume and adding a surface term, by regularization techniques, or by analytic continuation. We can adopt these prescriptions in the derivation of completeness and, for concreteness, whenever any of the bound states mentioned hereafter happens to be on the real axis it should be always understood without further discussion that the corresponding normalizing integral contains a Gaussian regularizing factor $\exp(-\epsilon r^2), \epsilon \to 0$ ensuring convergence of the integrals. Having introduced this extension we can now actually prove completeness using a direct and fairly elementary method.

Let us consider first the more interesting case of Coulomb attraction. To demonstrate completeness when both Coulomb and a complex short ranged potentials are present let us first introduce, similarly as in (3.67), the physical wave function $\psi_\ell(s,k,r)$, defined as

$$\psi_\ell(s,k,r) = k[F_\ell^+(s,k)F_\ell^-(s,k)]^{-1/2}\phi_\ell(s,k,r), \tag{8.10}$$

which is regular at $r = 0$ and for $r \to \infty$ takes the form

$$\lim_{r\to\infty} \psi_\ell(s,k,r) = \sin(kr - \eta \log 2kr - \pi\ell/2 + \delta_\ell^c + \sigma_\ell). \tag{8.11}$$

For physical momenta the functions $\psi_\ell(s,k,r)$ obey the usual orthogonality condition

$$\frac{2}{\pi}\int_0^\infty \psi_\ell(s,p,r)\,\psi_\ell(s,p',r)\,\mathrm{d}r = \delta(p-p'),$$

where $p$ and $p'$ are real and positive quantities.

We consider the integral over the continuous spectrum

$$I = \frac{2}{\pi}\int_0^\infty \psi_\ell(s,k,r)\psi_\ell(s,k,r')dk, \tag{8.12}$$

and since $k$ is on the real axis we can use (5.4) to split $I$ into two integrals. We have

$$I = I_1 + I_2, \tag{8.13}$$

$$I_1 = -\frac{1}{2\pi}\int_{-\infty}^\infty [F_\ell^-(s,k)/F_\ell^+(s,k)]f_\ell^+(s,k,r)f_\ell^+(s,k,r')\mathrm{d}k, \tag{8.14}$$

$$I_2 = \frac{1}{4\pi}\int_{-\infty}^\infty [f_\ell^+(s,k,r)f_\ell^+(s,-k,r') + + f_\ell^+(s,-k,r)f_\ell^+(s,k,r')]\,\mathrm{d}k. \tag{8.15}$$

It should be observed that by bringing $I$ to the above form we end up with integrands which have no Coulomb cut in the upper half of the k-plane. So, in the integral (8.14) we shall close the integration path by adding the integral over the infinite radius semi-circle centered at the origin in the upper half of the complex k-plane. This is permissible because, the the big arc integral gives no contribution since the integrand of (8.14) vanishes on the arc. For potentials which fall-off faster than an exponential function the Jost solutions $f_\ell^+$ are analytic functions for $Im\, k > 0$ and the value of the integral (8.14) will be given by the residues at the poles of the $1/F_\ell^+(s,k)$ function. We assume that the Jost function $F_\ell^+(s,k)$ has infinite number of single zeros at $k = k_n, n = 1,2,...$ and, additionally, it might also have double zeros at $k = k_\nu, \nu = 1,2,....$ For the time being we exclude the possibility of triple, or higher zeros and we assume that $\ddot{F}_\ell^+(s,k_\nu) \neq 0$.

The residue theorem applied to the integral (8.14), gives

$$
\begin{aligned}
I_1 = -\mathrm{i}\sum_n \frac{F_\ell^-(s,k_n)}{\dot{F}_\ell^+(s,k_n)} f_\ell^+(s,k_n,r) f_\ell^+(s,k_n,r') - \\
-\mathrm{i}\sum_\nu \lim_{k\to k_\nu} \frac{\mathrm{d}}{\mathrm{d}k}\left\{ \frac{F_\ell^-(s,k) f_\ell^+(s,k,r) f_\ell^+(s,k,r')}{\frac{1}{2}\ddot{F}_\ell^+(s,k_\nu) + \frac{1}{6}\dddot{F}_\ell^+(s,k_\nu)(k-k_\nu) + \cdots} \right\};
\end{aligned}
\tag{8.16}
$$

where dot denotes the derivative with respect to $k$.

In the second integral (8.15), we shall also close the contour of integration by adding and subtracting the integral along the big arc in the upper half of the k-plane centered at the origin. Since there are no poles in the integrand in (8.15) the closed contour integral vanishes and we are left with the integral along the big arc which is easily evaluated by using the asymptotic form of $f_\ell^+(s,k,r)$. Denoting by $K$ the radius of the big arc, we have

$$
I_2 = \lim_{K\to\infty} \frac{1}{2\pi} \int_{-1}^{1} \cos[K(r-r')z]\, dz = \delta(r-r'). \tag{8.17}
$$

As seen from the form of the right hand side of (8.16) the integral $I_1$ represents the contribution from the discrete spectrum and formula (8.16) can be brought to a more familiar form by eliminating the Jost solutions in favor of the regular solutions representing bound states. At the bound states, we have

$$
f_\ell^+(s,k_n,r) = \frac{2\mathrm{i}k_n}{F_\ell^-(s,k_n)} \phi_\ell(s,k_n,r); \tag{8.18}
$$

and at the double poles where $\dot{F}_\ell^+(s,k_\nu) = 0$, we obtain

$$
\begin{aligned}
\dot{f}_\ell^+(s,k_\nu,r) = & \frac{2\mathrm{i}k_\nu}{F_\ell^-(s,k_\nu)} \Big\{ \dot{\phi}_\ell(s,k_\nu,r) + \\
& + \left[ \frac{1}{k_\nu} - \frac{\dot{F}_\ell^-(s,k_\nu)}{F_\ell^-(s,k_n)} \right] \phi_\ell(s,k_n,r) \Big\}.
\end{aligned}
\tag{8.19}
$$

Differentiating the Wronskian (5.16), we can calculate (cf. [Nikolayev *et al.* (1973)] ) the derivatives of the Jost functions and the result is

$$\dot{F}_\ell^+(s,k_n) = -[4\mathrm{i}k_n^2/F_\ell^-(s,k_n)]\,N_n^2; \tag{8.20}$$

$$\ddot{F}_\ell^+(s,k_\nu) = -[4\mathrm{i}k_\nu^2/F_\ell^-(s,k_\nu)]\,\dot{N}_\nu^2; \tag{8.21}$$

$$\dddot{F}_\ell^+(s,k_\nu) = -[12\mathrm{i}k_\nu^2/F_\ell^-(s,k_\nu)]\Big\{2M_\nu^2 + + \left[1/k_\nu - \dot{F}_\ell^-(s,k_\nu)/F_\ell^-(s,k_n)\right]\dot{N}_\nu^2\Big\}; \tag{8.22}$$

where

$$N_n^2 = \int_0^\infty \phi_\ell(s,k_n,r)^2\,\mathrm{d}r; \tag{8.23}$$

$$\dot{N}_\nu^2 = 2\int_0^\infty \phi_\ell(s,k_\nu,r)\,\dot{\phi}_\ell(s,k_\nu,r)\,\mathrm{d}r; \tag{8.24}$$

$$M_\nu^2 = \int_0^\infty \dot{\phi}_\ell(s,k_\nu,r)^2\,\mathrm{d}r. \tag{8.25}$$

Inserting the above expressions in (8.16), after simplifications we arrive at the extended completeness relation

$$\frac{2}{\pi}\int_0^\infty \psi_\ell(s,k,r)\,\psi_\ell(s,k,r')dk + \sum_n \phi_\ell(s,k_n,r)\,\phi_\ell(s,k_n,r')/N_n^2 + + \sum_\nu \Big[\dot{\phi}_\ell(s,k_\nu,r)\,\phi_\ell(s,k_\nu,r') + \phi_\ell(s,k_\nu,r)\,\dot{\phi}_\ell(s,k_\nu,r') - - (M_\nu^2/\tfrac{1}{2}\dot{N}_\nu^2)\,\phi_\ell(s,k_\nu,r)\phi_\ell(s,k_\nu,r')\Big]/\tfrac{1}{2}\dot{N}_\nu^2 = \delta(r-r'), \tag{8.26}$$

As seen from (8.26) when level crossing takes place in the completeness relation the derivative of the eigenfunction is indeed present in accord with the Peierls argument given at the beginning of this chapter. When Coulomb potential is repulsive bound states can appear solely on account of the strong interaction so they might or might not exist at all but in any case their number will be finite. Level crossing is perfectly possible at certain critical values of the strong interaction strength parameter so formula (8.26) remains valid with the restriction that the set of bound states is finite and could be empty.

The generalization of the above derivation to the case when not just two but an arbitrary number (say $n$, $n > 2$) of levels can cross, is straightforward. This is the case when for some $k = k_0, n$ of the roots of the equation $F_\ell^+(s,k) = 0$ coincide and not only the Jost function but also its first $(n-1)$ derivatives vanish at $k_0$ while the n-th derivative is different from zero at that point. Although $n$ of the eigenfunctions become identical but we can still obtain a complete set of functions if we supplement the single eigenfunction corresponding to $k_0$ by the first $(n-1)$ derivatives of the eigenfunction at $k_0$. The generalization of eq. (8.26) follows immediately by applying residue theorem to a pole of order $n$ when calculating the integral (8.14).

# Chapter 9

# X-Ray Transition Rate

The problem we address ourself in this chapter is an investigation of the effect which the nuclear potential might have on the total X-ray transition rate. Since the nuclear potential has a short range as compared with the Bohr radius these effects are negligible in all radiative transitions except for the very last one leading to the level from which only nuclear absorption takes place. Most authors have ignored all nuclear effects including this last transition, so much so that in the early papers even the quite obvious center of mass factor has been overlooked. The problem is not trivial and has practical implications. Since the last level is unstable and has a finite width the energy $\omega$ of the emitted photon is not discrete but will have certain distribution. Of course, there is no violation of energy conservation for the absorptive part of the nuclear potential acts as a sink taking away some portion of the available energy. It is important to know the functional form of the $\omega$ distribution as this function is being used in all fitting procedures to the experimental counting rate from which the widths and shifts of the last level are inferred. Adoption of a different form of the $\omega$ distribution in general leads to different values of the adjustable parameters changing the result of the measurement. The integration over the $\omega$ distribution provides us with the value of the total X-ray transition rate. This quantity is needed to determine the branching ratio of the two competing processes which can occur from the one-before-last level, i.e. the X-ray transition, or the nuclear absorption. The value of this branching ratio is used to deduce the width of the one-before-last level and inaccurate input of the X-ray transition rate can be a source of a systematic error on the indirectly measured value of the width. Although there have been some attempts to examine the nuclear effects on the X-ray transition rate the results obtained are incomplete. As a matter of fact, there is a number of possible corrections to the X-ray transition rate and to be able to trace down their origin and to assess the approximations used we are going to start from scratch.

Let us then consider a radiative transition process where in the initial state the hadronic atom is at rest in the state $(n', \ell')$ and in the final state we have a photon and the recoiling atom deposited in the state $(n, \ell)$. We assume that this is the last X-ray transition so that the nuclear interaction has negligible effect in the initial state but is felt in the final state so that the binding energy will be shifted from its Coulomb position and the atomic level acquires a width $\Gamma$. The width of the upper level is of the same order as the radiative transition rate to the lower level and both these quantities are much smaller than $\Gamma$. In the calculation of the X-ray transition rate we introduce time scale which is very large as compared to the lifetime $1/\Gamma$ of the lower level.

The X-ray transition is induced by a time dependent interaction $V(t)$ which couples the atomic system with the electromagnetic field. The time dependence is assumed to be exponential

$$V(t) = H' \mathrm{e}^{\mathrm{i}\omega t}, \tag{9.1}$$

where $\omega$ is the frequency of the field oscillation and the operator $H'$ will be specified later on. We assume that $V(t)$ is a small perturbation and we wish to calculate the first order correction in $H'$. All retardation effects will be disregarded. To calculate the transition probability we use Dirac perturbation theory in which we have to make allowance for the fact that we have to do with unstable states. As usual, the wave function $|\Psi(t)\rangle$ is expanded in a complete set of energy eigenstates of the unperturbed Hamiltonian $|\psi_\alpha\rangle$. This Hamiltonian is assumed to have complex eigenvalues $E_\alpha - \mathrm{i}\Gamma_\alpha/2$, and, we write

$$|\Psi(t)\rangle = \sum_\alpha c_\alpha(t) \mathrm{e}^{-\mathrm{i}tE_\alpha - \Gamma_\alpha t/2} |\psi_\alpha\rangle \,. \tag{9.2}$$

Inserting (9.2) in the time dependent Schrödinger equation containing the unperturbed Hamiltonian and the potential (9.1), we obtain a system of first order differential equations for the expansion coefficients $c_\alpha(t)$ which can be easily solved if we assume that at time $t = 0$ the system is in the state $|i\rangle$, i.e. we set in the initial conditions $c_i(0) = 1$ and $c_j(0) = 0$ if $j \neq i$.

The probability that after time $t$ the system will be in the state $|f\rangle$, is

$$|c_f(t)|^2 = |\langle f|H'|i\rangle|^2 \frac{[1 - \mathrm{e}^{-\Gamma t/2}\cos(\omega - \Delta E)t]^2 + \mathrm{e}^{-\Gamma t}\sin^2(\omega - \Delta E)t}{(\omega - \Delta E)^2 + (\frac{1}{2}\Gamma)^2}, \tag{9.3}$$

where $\Delta E = E_i - E_f$ and we have set $\Gamma_i = 0$; $\Gamma_f = \Gamma$. The transition probability averaged over time $T$ long in comparison with $1/\Gamma$ is independent of $T$ and is given as

$$\langle |c_f|^2 \rangle = \frac{1}{T}\int_0^T |c_f(t)|^2 \mathrm{d}t = \frac{|\langle f|H'|i\rangle|^2}{(\omega - \Delta E)^2 + (\frac{1}{2}\Gamma)^2}. \tag{9.4}$$

Taking as the unit of time the lifetime $1/\Gamma$ of the state $f$, we obtain from (9.4) the transition probability $w_{i\to f}$ per unit of time

$$w_{i\to f} = \Gamma \langle |c_f|^2 \rangle = \frac{\Gamma|\langle f|H'|i\rangle|^2}{(\omega - \Delta E)^2 + (\frac{1}{2}\Gamma)^2}. \tag{9.5}$$

For $\Gamma \to 0$ the above formula yields the celebrated "golden rule"

$$\lim_{\Gamma\to 0} w_{i\to f} = 2\pi|\langle f|H'|i\rangle|^2 \delta(\omega - \Delta E),$$

where the presence of the delta function ensures energy conservation so that the energy of the photon is fixed and equal to $\Delta E$. This is not the case for a transition to unstable state $f$ since some part of the energy is being absorbed by the nuclear interaction. The transition probability is a function of $\omega$ and the total X-ray transition rate is obtained by integrating over $\omega$.

In order to determine the transition amplitude $\langle f|H'|i\rangle$ we consider the interaction Hamiltonian $H_{int}$ whose form has been borrowed from classical treatment of electromagnetism

$$H_{int} = -\int \mathbf{j}(\mathbf{r}) \cdot \mathbf{A}(\mathbf{r})\, d^3r, \tag{9.6}$$

where $\mathbf{j}(\mathbf{r})$ is the density of the current represented by the hadronic atom and $\mathbf{A}(\mathbf{r})$ is vector potential in the radiation gauge

$$div\ \mathbf{A}(\mathbf{r}) = 0.$$

The quantization of the radiation field leads to the familiar expansion of the vector potential

$$\mathbf{A}(\mathbf{r}) = \sum_{\mathbf{k},\lambda} \sqrt{\frac{2\pi}{\omega L^3}} \left[ b_{\mathbf{k},\lambda} \mathbf{e}_{\mathbf{k},\lambda} \mathrm{e}^{\mathrm{i}\mathbf{k}\cdot\mathbf{r} - \mathrm{i}\omega t} + b^{\dagger}_{\mathbf{k},\lambda} \mathbf{e}^{\dagger}_{\mathbf{k},\lambda} \mathrm{e}^{-\mathrm{i}\mathbf{k}\cdot\mathbf{r} + \mathrm{i}\omega t} \right], \quad (9.7)$$

where $b_{\mathbf{k},\lambda}, b^{\dagger}_{\mathbf{k},\lambda}$ are, respectively, photon annihilation and creation operators and $e_{\mathbf{k},\lambda}$ is photon polarization vector. In (9.7) we assumed the periodic boundary conditions for $\mathbf{A}$ enclosed in a box taken to be a cube of side $L$.

Both $\mathbf{j}$ and $\mathbf{A}$ are second quantization operators so that $H_{int}$ provides the required matrix element for the transition from a state consisting of an atom in its initial state $|i\rangle$ and zero photons to a state consisting of one photon with frequency $\omega$ and the corresponding atomic final state $|f\rangle$. In the first approximation the atomic current can be taken as a sum of the currents connected with the motion of the charged particles making up the hadronic atom where each of them is assumed to be a point like object. In the non-relativistic limit, we have

$$\begin{aligned} \mathbf{j}(\mathbf{r}) = {} & \left\langle \Psi(\mathbf{x}_N, \mathbf{x}_H) \middle| Ze\tfrac{1}{2}\{\mathbf{v}_N, \delta(\mathbf{r} - \mathbf{x}_N)\} \middle| \Psi(\mathbf{x}_N, \mathbf{x}_H) \right\rangle - \\ & - \left\langle \Psi(\mathbf{x}_N, \mathbf{x}_H) \middle| e\tfrac{1}{2}\{\mathbf{v}_H, \delta(\mathbf{r} - \mathbf{x}_H)\} \middle| \Psi(\mathbf{x}_N, \mathbf{x}_H) \right\rangle, \end{aligned} \quad (9.8)$$

where $\mathbf{x}_N, \mathbf{x}_H$ are the coordinates of the nucleus and the hadron, respectively, $\Psi$ is hadron-nucleus wave function and the braces denote anticommutators. As a matter of fact, formula (9.8) gives merely the convection current arising from the motion of charges and should be supplemented by the magnetization current. However, since we are interested here in an electric dipole transition only we will ignore in the following the magnetization current. The quantities $\mathbf{v}_N, \mathbf{v}_H$ are the velocity operators, defined in the usual way

$$\mathbf{v}_j = \mathrm{i}[H, \mathbf{x}_j], \quad j = N, H \quad (9.9)$$

where $H$ denotes the full Hamiltonian which accounts for the presence of the electromagnetic field.i This Hamiltonian is obtained by applying the minimal substitution rule $\mathbf{p}_N \to (\mathbf{p}_N - Ze\mathbf{A})$, $\mathbf{p}_H \to (\mathbf{p}_H + e\mathbf{A})$ and contains additional dependence upon the vector potential $\mathbf{A}$.

It is advantageous to separate the center of mass motion in the expression for the current (9.8) by introducing the center of mass coordinate $\mathbf{X}$ and the hadron-nucleus separation vector $\mathbf{x}$

$$\mathbf{x}_N = \mathbf{X} + \frac{\mu}{M}\mathbf{x}; \ \mathbf{x}_H = \mathbf{X} - \frac{\mu}{m}\mathbf{x}, \quad (9.10)$$

where $m$, $M$ and $\mu$ are the masses of the hadron, the nucleus, and the reduced mass, respectively. Denoting the center of mass momentum as $\mathbf{P}$, we expand the two-particle wave function in a complete set of energy eigenfunctions $u_\alpha(\mathbf{x})$ describing the internal motion. We have

$$\Psi(\mathbf{x}_N, \mathbf{x}_H) = L^{-3/2} \mathrm{e}^{\mathrm{i}\mathbf{P}\cdot\mathbf{X}} \textstyle\sum_\alpha c_\alpha u_\alpha(\mathbf{x}), \tag{9.11}$$

where $c_\alpha^\dagger$ is the operator which creates the internal state $\alpha$. Spontaneous X-ray emission takes place when the internal atomic state $i$ is annihilated and at the same time another state $f$ is created together with a photon of momentum $\mathbf{k}$ and polarization $\mathbf{e}_{\mathbf{k},\lambda}$. This process requires a triple operator product $b_{\mathbf{k},\lambda}^\dagger c_f^\dagger c_i$ in the interaction Hamiltonian and the quantity of interest for us is the coefficient multiplying that operator because it provides the transition amplitude. This coefficient could be worked out by inserting the expansions (9.11) and (9.7) in the interaction Hamiltonian. We have

$$\langle 1_\mathbf{k}; f|H_{int}|0_\mathbf{k}; i\rangle = \langle f|H'|i\rangle\, \mathrm{e}^{\mathrm{i}\omega t},$$

and the transition amplitude takes the form

$$\begin{gathered}\langle f|H'|i\rangle = \mathrm{i}\frac{eM}{M+m}\sqrt{\frac{2\pi}{\omega L^3}\frac{2\pi}{L^3}}\,\delta(\mathbf{P}+\mathbf{k})\,\times \\ \mathbf{e}_{\mathbf{k},\lambda}^\dagger \cdot \langle u_f(\mathbf{x})|\,[h,\mathbf{x}]\left\{\exp(\mathrm{i}\frac{\mu}{m}\mathbf{k}\cdot\mathbf{x}) + \frac{Zm}{M}\exp(-\mathrm{i}\frac{\mu}{M}\mathbf{k}\cdot\mathbf{x})\right\}|u_i(\mathbf{x})\rangle\end{gathered} \tag{9.12}$$

where we have assumed that the hadronic atom is initially at rest and after emission recoils with momentum $\mathbf{P}$. To obtain the above expression the two-body Hamiltonian is written as $T_X + h$ where $h$ denotes the internal Hamiltonian and $T_X$ is the kinetic energy of the center of mass motion. The commutator $[T_X, X]$ yields a term proportional to $\mathbf{P}+\mathbf{k}$ but because of the presence of the $\delta(\mathbf{P}+\mathbf{k})$ factor this term drops out and the center of mass motion gives no contribution. Formula (9.12) can be further simplified by introducing the long wavelength approximation which is based on the observation that the wavelength of the emitted photon is much bigger than the linear dimension of the hadronic atom. Although in the hadronic atom case the wavelength is shorter than that in an ordinary atom but the size of the hadronic atom is smaller by the same amount as compared with an ordinary atom so that $\lambda_{photon} \gg r_{atom}$ holds in both cases. This means that the exponentials in (9.12) vary so slowly that we can put $\mathbf{k}\cdot\mathbf{x} \approx 0$ and formula (9.12) takes the form

$$\langle f|H'|i\rangle = \mathrm{i}e\frac{M+Zm}{M+m}\sqrt{\frac{2\pi}{\omega L^3}}\left(\frac{2\pi}{L}\right)^3 (B_f - B_i + \mathrm{i}\tfrac{1}{2}\Gamma)\ \times$$
$$\times\ \delta(\mathbf{P}+\mathbf{k})\left\langle u_f(\mathbf{x})|\mathbf{e}^{\dagger}_{\mathbf{k},\lambda}\cdot\mathbf{x}|u_i(\mathbf{x})\right\rangle, \qquad (9.13)$$

where $B_{i,f}$ are the binding energies. The X-ray transition rate is obtained by multiplying the transition probability (9.5) by the density of states $(L/2\pi)^6 d^3k\, d^3P$. By carrying out all integrations and summing over photon polarizations, we obtain

$$\frac{d\Gamma_{i\to f}}{d\omega} = \frac{2\alpha}{3\pi}\,\frac{\Gamma\omega[(B_i-B_f)^2+\Gamma^2/4]}{(\omega-\Delta E)^2+\Gamma^2/4}\,|\,\langle u_f(\mathbf{x})|\mathbf{x}|u_i(\mathbf{x})\rangle\,|^2\, c_{cm}, \qquad (9.14)$$

where $\Delta E = B_i - B_f - \frac{1}{2}\omega^2/(M+m)$ contains the recoil energy and $c_{cm}$ is center of mass correction

$$c_{cm} = \left[\frac{M+Zm}{M+m}\right]^2.$$

As mentioned before, the upper level is almost Coulombic but the nuclear interaction shifts the energy of the lower level by $\epsilon$ relatively to the purely Coulomb position and we can write $B_i - B_f = \omega_0 + \epsilon$ where $\omega_0$ is the difference of the binding energies in absence of nuclear interaction. Formula (9.14) exhibits the dependence upon $\omega$ of the partial rate and gives the "line shape" which is not quite Lorentzian (note that $|\,\langle \mathbf{x}\rangle\,|^2$ is independent of $\omega$). Contrary to some assertions in the literature this shape is not $\omega^3$ times a Lorentz factor. There is also no problem with the integration over $\omega$ as, owing to the recoil correction, the integral is always convergent and the range of integration could be extended to infinity. On the other hand, the allowed values of $\omega$ for physical reasons should not be pushed to the limit $\omega \to \infty$ because this would be in disaccord with the long wavelength approximation. We think that it is more appropriate to integrate over $\omega$ in the range $[0, 2\omega_0]$ as in this case the peak comes roughly in the middle of the range, neglecting the small recoil correction. Using this prescription, we obtain the total X-ray transition rate

$$\Gamma_{i\to f} = (4\alpha/3)\,\omega_0^3\,|\,\langle u_f(\mathbf{x})|\mathbf{x}|u_i(\mathbf{x})\rangle\,|^2\, c_{cm}c_L, \qquad (9.15)$$

where $c_L$ is a correction due to the line shape and we have

$$c_L = [(1+\frac{\epsilon}{\omega_0})^2 + (\frac{\Gamma}{2\omega_0})^2](1+\frac{\epsilon}{\omega_0})\frac{2}{\pi}\arctan\left(\frac{1+\epsilon/\omega_0}{\Gamma/2\omega_0}\right). \qquad (9.16)$$

The nuclear interaction distorts also the wave functions and additional correction comes from the $\langle \mathbf{x} \rangle$ matrix element in (9.15). If the transition is between $(n', \ell')$ and $(n, \ell)$ states (with the dipole selection rule $\Delta\ell = \pm 1$) the formula for the transition rate can be written as a product of the standard atomic transition rate times three corrections. We have

$$\Gamma_{n'\ell' \to n\ell} = \frac{4\alpha}{3}\,\omega_0^3\,\frac{\max(\ell', \ell)}{2\ell' + 1}\left|\int_0^\infty u_{n\ell}(r)\, r\, u_{n'\ell'}(r)\,\mathrm{d}r\right|^2 c_{cm}\, c_L\, c_I, \quad (9.17)$$

where the correction $c_I$ resulting from distortion of the radial integral can be written as

$$c_I = \left|\frac{\displaystyle\int_0^\infty u_{n\ell}(s,r)\, r\, u_{n'\ell'}(s,r)\,\mathrm{d}r}{\displaystyle\int_0^\infty u_{n\ell}(r)\, r\, u_{n'\ell'}(r)\,\mathrm{d}r}\right|^2 . \quad (9.18)$$

The presence of the nuclear interaction strength $s$ in the wave functions occurring in the numerator of (9.18) is meant to indicate that they are the eigenfunctions of the full Hamiltonian. When the nuclear potential is known, then $c_I$ can be computed numerically. Using (10.1) in (9.18), we obtain a lengthy expression

$$\begin{aligned} c_I = & \left|\int_0^R \phi_\ell(s, k_n, r)\, r\, u_{n'\ell'}(r)\,\mathrm{d}r + \right. \\ & \left. + \frac{\phi_\ell(s, k_n, R)}{f_\ell^+(k_n, R)}\int_R^\infty f_\ell^+(k_n, r)\, r\, u_{n'\ell'}(r)\,\mathrm{d}r\right|^2 \\ \times & \left|\int_0^R \phi_\ell(s, k_n, r)^2\,\mathrm{d}r + \frac{\phi_\ell(s, k_n, R)^2}{f_\ell^+(k_n, R)^2}\int_R^\infty f_\ell^+(k_n, r)^2\,\mathrm{d}r\right|^{-1} \\ \times & \left|\int_0^\infty u_{n\ell}(r)\, r\, u_{n'\ell'}(r)\,\mathrm{d}r\right|^{-2}, \end{aligned}$$

where the upper level has been assumed to be purely Coulombic. The momentum $k_n$ is the complex eigenmomentum of the full Hamiltonian. To estimate the size of the correction $c_I$ we consider the $2p \to 1s$ transition where the nuclear effects are expected to be the largest. To simplify matters we assume zero range forces neglecting the integrals in the $[0, R]$ range what is next partly compensated by pushing the lower limit in the remaining integrals down to zero. The resulting expression for $c_I$ is

$$c_I \approx \frac{\left|\int_0^\infty f_0^+(k_1,r)\, r\, u_{2p}(r)\, \mathrm{d}r\right|^2}{\left|\int_0^\infty f_0^+(k_1,r)^2\, \mathrm{d}r\right| \left|\int_0^\infty u_{1s}(r)\, r\, u_{2p}(r)\, \mathrm{d}r\right|^2}, \tag{9.19}$$

and the dependence upon the nuclear potential enters only via $k_1$ which is a function of $\epsilon$ and $\Gamma$. In formula (9.19) all integrals can be performed analytically and the ultimate expression for $c_I$ takes the form

$$c_I = \left|\frac{3^{10}}{(2-\mathrm{i}\eta_1)^{10}} \frac{36}{\Gamma(5+\mathrm{i}\eta_1)^2}\right| \times$$
$$\times \left|\frac{(-\mathrm{i}\eta_1)^7 {}_2F_1(5, 1+\mathrm{i}\eta_1; 5+\mathrm{i}\eta_1; z)^2}{\Gamma(1-\mathrm{i}\eta_1)^2 - [(-\mathrm{i}\eta_1)^2\psi'(-\mathrm{i}\eta_1) - 1 + 2\mathrm{i}\eta_1]/\Gamma(1+\mathrm{i}\eta_1)^2}\right|,$$

where $\psi(z)$ is the logarithmic derivative of the Gamma function and prime denotes derivative with respect to the argument. The argument in the hypergeometric function is $z = (\mathrm{i}\eta_1+2)/(\mathrm{i}\eta_1-2)$ and the parameter $\eta_1$ can be expressed in terms of $\epsilon/\omega_0$ and $\Gamma/2\omega_0$

$$\eta_1 = -\frac{\mu\alpha Z}{k_1} = \mathrm{i}\left\{1 + \frac{3}{4}\frac{\epsilon + \frac{1}{2}\mathrm{i}\Gamma}{\omega_0}\right\}^{-1}.$$

We have computed $c_L$ and $c_I$ varying the two parameters $\epsilon/\omega_0$ and $\Gamma/2\omega_0$ and the results are presented in Table 9.1. It is interesting to note that the correction from $c_L$ goes in the opposite direction than that from $c_I$ and as a result of strong cancelation the total effect is reduced typically by a factor of two or more. It is seen from Table 9.1 that we do not have to worry about the nuclear corrections unless the shifts or/and widths are exceptionally large, their values being at least 10% of the transition energy. There are not many such cases known, but a prominent example here is the antiprotonic hydrogen where according to a recent measurement [Heitlinger *et al.* (1992); Gotta *et al.* (1999)] the magnitude of the lowest, spin averaged s-level shift and width is of about $1\,keV$ whereas the $2p \to 1s$ total electro-magnetic transition energy is $9.4\,keV$.

Summarizing, the nuclear interaction gives rise to line shape which to a good approximation takes the functional form $\omega$ times a Lorentz factor and a similar conclusion has been reached by [Ericson *et al.* (1977); Batty (1989)]. The line-shape is slightly shifted towards higher energies but the correction is affecting mainly the tail region. The nuclear effects on the total X-ray transition rate are in most cases negligible.

Table 9.1 Corrections to the $2p \to 1s$ X-ray transition rate due to nuclear interaction. The quantity $c$ is the product of $c_L$ and $c_I$ defined in the text.

$\frac{1}{2}\Gamma/\omega_0 = 10^{-4}$

| $\epsilon/\omega_0$ | $c$ | $\epsilon/\omega_0$ | $c$ |
|---|---|---|---|
| $-10^{-4}$ | 1.00 | $10^{-4}$ | 1.00 |
| $-10^{-3}$ | 1.00 | $10^{-3}$ | 1.00 |
| $-10^{-2}$ | 0.99 | $10^{-2}$ | 1.01 |
| $-10^{-1}$ | 0.92 | $10^{-1}$ | 1.07 |

$\frac{1}{2}\Gamma/\omega_0 = 10^{-3}$

| $\epsilon/\omega_0$ | $c$ | $\epsilon/\omega_0$ | $c$ |
|---|---|---|---|
| $-10^{-4}$ | 1.00 | $10^{-4}$ | 1.00 |
| $-10^{-3}$ | 1.00 | $10^{-3}$ | 1.00 |
| $-10^{-2}$ | 0.99 | $10^{-2}$ | 1.01 |
| $-10^{-1}$ | 0.92 | $10^{-1}$ | 1.07 |

$\frac{1}{2}\Gamma/\omega_0 = 10^{-2}$

| $\epsilon/\omega_0$ | $c$ | $\epsilon/\omega_0$ | $c$ |
|---|---|---|---|
| $-10^{-4}$ | 0.99 | $10^{-4}$ | 0.99 |
| $-10^{-3}$ | 0.99 | $10^{-3}$ | 0.99 |
| $-10^{-2}$ | 0.99 | $10^{-2}$ | 1.00 |
| $-10^{-1}$ | 0.91 | $10^{-1}$ | 1.06 |

$\frac{1}{2}\Gamma/\omega_0 = 10^{-1}$

| $\epsilon/\omega_0$ | $c$ | $\epsilon/\omega_0$ | $c$ |
|---|---|---|---|
| $-10^{-4}$ | 0.94 | $10^{-4}$ | 0.94 |
| $-10^{-3}$ | 0.94 | $10^{-3}$ | 0.94 |
| $-10^{-2}$ | 0.93 | $10^{-2}$ | 0.95 |
| $-10^{-1}$ | 0.86 | $10^{-1}$ | 1.01 |

# Chapter 10

# Computational Methods

From the foregoing discussion it has become apparent that the problem of finding the eigenvalues of a non-selfadjoint Hamiltonian presents a highly demanding task as far as computational methods are concerned. For one thing, the methods must be quite general so that the search for the eigenvalues could be conducted all over the complex k-plane, and, for the other, the algorithms are expected to be highly accurate so that such subtle effects could be examined as crossing of two pole trajectories. Finally, in the actual practice the effective potential usually contains adjustable parameters whose values are determined from a fit to a large number of data in which the eigenvalues have to be repeatedly evaluated many times. Therefore, to accomplish this task it is crucial to have fast and effective numerical procedures. As a matter of fact, the methods described in the literature are not fully adequate to meet the above criteria. For local potentials two alternative methods have been used, one by [Blatt (1967)], the other due to [Krell *et al.* (1968)]. Basically, both methods attempt to solve the equation $F_\ell^+(s,k) = 0$, the former by using the Newton–Raphson algorithm whereas the latter by minimizing the square of the modulus of the Jost function. The performance of these methods in the upper part of the k-plane is quite satisfactory but neither of them is applicable for Gamow states. The other disadvantage is the requirement that the wave equation must be integrated very far out until the wave function assumes the asymptotic form. This increases execution times and may lead to accumulation of rounding errors deteriorating the ultimate accuracy. We shall show in a moment how to avoid that. For non-local potentials momentum space methods are usually employed and the main difficulty here derives from the fact that the Coulomb potential in momentum space is singular on-shell. This singularity, as shown in by [Kwon *et al.* (1978)], can be removed by subtraction so that the resulting homogeneous Fredholm integral equation can be solved by standard methods. The subtraction technique, however, works only for bound state problems ($Im\, k > 0$), and, in particular, it cannot be used for Gamow states. An alternative approach developed by [Deloff *et al.* (1980)] employs Coulomb Green's function so that Coulomb potential is no longer present in the wave equation and the latter could be conveniently solved using Sturmian expansion.

In the following we are going to discuss at some length one by one a number of practical methods for finding the eigenvalues of a non-selfadjoined Hamiltonian. The common feature shared by all of them is that at the end one has to solve numerically the equation $F_\ell^+(s,k) = 0$ regarding $k$ as the independent variable and $s$ as a given parameter. Such an equation is usually solved by one of the iterative methods in which the user supplies one or more starting values for the root. We do know that the equation to be solved has infinite number of roots and some of them may lie close to each other so that a problem arises how to make sure that the iterative procedure has converged to the root we wanted. It is our experience that for atomic states the best strategy is to set at the start $s = 0$ which should force the root to be at the purely Coulombic position whose analytic form is well known. It is advantageous then to break the path leading from zero to the actual value of $s$ into a number of intermediate steps forming a sequence of strength values $s_0, s_1, ..., s_n$ with $s_0 = 0$ and $s_n = s$. For each $s_j$ there will be a corresponding root $k_j$ and one can use the root $k_{j-1}$ as the approximation to the $k_j$ root in the j-th root search. Although instead of locating a single root we have now to do this job $n$ times but the convergence in each step is very fast so that usually $n$ searches take less iterations than a single search with a fixed value of $s$. To locate the quasi-nuclear states and the Gamow states in principle a similar method can be used. The only difference is that the reference point is not the purely Coulombic state, but the zero-energy bound state. For this purpose we need to know first the root of the equation (6.26) and the path between this critical value of $s$ and the actual $s$ value can be divided into a number of steps. In the search for Gamow states extra care is required because of the presence of the Coulomb branch cut.

## 10.1 The Matching Method

In the matching method the energy eigenvalues are obtained by locating the zeros of the Jost function $F_\ell^+(s,k)$. For any value of $k$ it is possible to find a solution of (2.3) regular at $r=0$ and satisfying Siegert boundary condition. This solution can be written as

$$u_\ell(s,k,r) = \begin{cases} \phi_\ell(s,k,r) & \text{if } r \le R \\ f_\ell^+(s,k,r)\dfrac{\phi_\ell(s,k,R)}{f_\ell^+(s,k,R)} & \text{if } r \ge R \end{cases} \tag{10.1}$$

where in (10.1) the matching radius $R$ in principle could be any positive number. Although by construction the function $u_\ell(s,k,r)$ is continuous but in general the matching will not be smooth because at $r=R$ the derivative $u_\ell'(s,k,r)$ suffers a discontinuity. It is only possible for some special values of $k$ that both the function $u_\ell(s,k,r)$ and also its derivative can be made continuous at $r=R$. These particular $k$ values correspond to eigenvalues and the function $u_\ell(s,k,r)$ becomes then an eigenfunction. The requirement of continuity of the derivative of $u_\ell(s,k,r)$ takes the form of an utmost important equation

$$\Delta(s,k) \equiv \phi_\ell'(s,k,R) - \phi_\ell(s,k,R)\frac{f_\ell^{+\prime}(s,k,R)}{f_\ell^+(s,k,R)} = 0, \tag{10.2}$$

which is of course equivalent to the vanishing of the Jost function $F_\ell^+(s,k)$. Since the nuclear potential has a short range by taking $R$ large enough we can approximate with arbitrary accuracy the logarithmic derivative in (10.2) $f_\ell^{+\prime}(s,k,R)/f_\ell^+(s,k,R)$ by the logarithmic derivative of the purely Coulombic wave functions $f_\ell^{+\prime}(k,R)/f_\ell^+(k,R)$ making (10.2) ideally suited for locating eigenvalues for the following reasons. Firstly, the function $\Delta(s,k)$ can be easily computed as the regular solution $\phi_\ell(s,k,r)$ and its derivative are obtained by solving the initial value problem for the underlying differential equation while the logarithmic derivative of the Coulomb Jost solutions can be obtained by continued fraction summation. Secondly, to be able to locate a zero of a complex function it is essential to know the distribution of its singularities. The important point is that the function (10.2) is free from singularities in the upper half of the complex k-plane. Since the boundary conditions are independent of $s$ the function $\Delta(s,k)$ will be analytic in the variable $s$ and in addition to that it is also analytic and single-valued function everywhere throughout the complex k-plane away from the branch cut along the negative imaginary semi-axis. Thirdly, since it is irrelevant in which part of the k-plane the zero is positioned all eigenvalues could be determined and the method is valid for bound states and Gamow states as well. Fourthly, the method is rather versatile in that it applies to attractive, or repulsive Coulomb potential and also to the chargeless case.

For fixed $s$ equation (10.2) can be solved in the complex k-plane by any of the iterative methods but we have found particularly effective the Mueller–Traub method (cf. [Press *et al.* (1992)]). The advantage of this method is that all that it takes is a calculational scheme returning the function $\Delta(s,k)$ for any values of the arguments and in particular it does not require the knowledge of the derivative. The idea of the method is to approximate locally the function by a second order interpolating polynomial whose coefficients are determined by supplying the values of the function $\Delta(s,k)$ at three different $k$ values. Actually, we have been using a variant of this method based on a rational rather than polynomial approximation which is equivalent to the [2,1] Padé approximant and has better global convergence properties. The function is locally represented in the form

$$\Delta(s,k) = \frac{c + b\,k + a\,k^2}{1 + d\,k},$$

but at start we set $d = 0$ and from the knowledge of the value of the function at three initial guesses $k_1, k_2, k_3$ we determine the three coefficients of the numerator. The roots of the polynomial are then easily established thus providing the approximate value of the zero of $\Delta(s,k)$. Subsequently, this estimated position of the zero $k_4$ and the value of the function at this point are used to pin down $a, b, c, d$. In the next iteration we obtain a second approximation to the zero and a new point which could be used for the interpolation. Among the five points we reject the point for which the absolute value of the function assumes the largest value and the iteration cycle could be continued until convergence is attained. A good check of the performance is to calculate the Coulomb spectrum setting the nuclear potential equal to zero. With a good differential equation solver it is not difficult to compute Coulomb energies to machine accuracy.

As an alternative to the above scheme one may also consider using the celebrated Newton–Raphson method. The distinctive feature of the latter is that it requires the evaluation of both the function and the derivative at arbitrary value of the argument. Thus equation (10.2) is solved by trying to refine the first guess $k_0$ to the zero and this is achieved by expanding the function $\Delta(s,k)$ about $k_0$ in a Taylor series. Assuming that terms beyond linear in $(k - k_0)$ are unimportant, we obtain

$$k \approx k_0 - \frac{\Delta(s, k_0)}{\dot{\Delta}(s, k_0)} \tag{10.3}$$

where dot denotes the derivative with respect to $k$. Differentiating formula (10.2), one has

$$\dot{\Delta}(s, k_0) = \frac{w\left[\phi_\ell(s, k_0, R), \dot{\phi}_\ell(s, k_0, R)\right] + \Delta(s, k_0)\dot{\phi}_\ell(s, k_0, R)}{\phi_\ell(s, k_0, R)} - \phi_\ell(s, k_0, R)\frac{\mathrm{d}}{\mathrm{d}k_0}\frac{f_\ell^{+\prime}(s, k_0, R)}{f_\ell^{+}(s, k_0, R)} \tag{10.4}$$

where the Wronskian $w\left[\phi_\ell, \dot{\phi}_\ell\right]$ can be simplified to the form

$$w\left[\phi_\ell(s, k_0, R), \dot{\phi}_\ell(s, k_0, R)\right] = -2k_0 \int_0^R \phi_\ell(s, k_0, r)^2 \, \mathrm{d}r \tag{10.5}$$

and for bound states also the last term in (10.4) can be written as an integral

$$\frac{\mathrm{d}}{\mathrm{d}k_0}\frac{f_\ell^{+\prime}(s, k_0, R)}{f_\ell^{+}(s, k_0, R)} = \frac{2k_0}{f_\ell^{+}(s, k_0, R)^2} \int_R^\infty f_\ell^{+}(s, k_0, r)^2 \, \mathrm{d}r. \tag{10.6}$$

Neglecting the $\Delta$ term in the expression (10.4) for $\dot{\Delta}$ and inserting (10.5) and (10.6) in (10.4), from (10.3) we have

$$k^2 \approx k_0^2 + \tilde{u}_\ell(s, k_0, R)\left[\tilde{u}'_\ell(s, k_0, R - \varepsilon) - \tilde{u}'_\ell(s, k_0, R + \varepsilon)\right] \tag{10.7}$$

where the presence of $\varepsilon \to 0$ indicates how to calculate the derivative from (10.1). In (10.7) we introduced normalized wave function $\tilde{u}_\ell = u_\ell/N$ and the normalization $N$ is defined in (8.5). Formula (10.7) is valid everywhere throughout the k-plane without restrictions away from the branch cut and may be plugged into an iterative scheme. In this form the matching method generalizes the well known Blatt's method [Blatt (1967)].

The Jost function can have also multiple roots and the multiplicity of the root $k_n$ can be established by calculating the derivatives $\dot{F}_\ell^{+}(s, k_n), \ddot{F}_\ell^{+}(s, k_n)$ and so on, and it is understood that a zero has multiplicity $m$, if the first $(m - 1)$ derivatives vanish while $m$-th derivative is non-zero. In practice, we are interested in the location of double poles as this gives us the crossing points of two pole trajectories. At a double zero the normalization of the wave function is bound to vanish, so if the normalization is computed simultaneously with the corresponding eigenvalue the multiple roots can be easily recognized. To calculate the normalization we find very useful formula (8.9) which is valid for all eigenvalues and both the integral and the derivative can be computed numerically. Of particular interest might be a search specifically for a double zero, and, to accomplish this task the equation

$$N_n^2(s_c) = 0$$

must be solved for $s_c$, e.g. by the Newton–Raphson method. Supplying two initial guesses $s_1$ and $s_2$, the estimated position of the zero is

$$s_c = s_2 - (s_2 - s_1)\frac{N_n^2(s_2)}{N_n^2(s_2) - N_n^2(s_1)}$$

and the above expression is used iteratively until convergence is attained. The eigenvalue corresponding to the ultimate value of $s_c$ is a double root of the Jost function.

The matching method can be also generalized to accommodate a multichannel situation. In this case, assuming $n$ channels, we have a wave function with $n$ components which satisfies $n$ dimensional wave equation. By solving a system of $n$ differential equations $n$ times changing perpetually the initial boundary conditions we can generate $n$ particular linearly independent solutions. To simplify our notation we suppress the orbital momentum $\ell$ labeling the different components of the wave function by a suffix $i$ writing the wave function as $u_i^{(j)}(k,r)$ where the superfix $j$ enumerates the different particular solutions and $k$ is now the momentum corresponding to the channel with the lowest threshold. By taking a linear superposition of these particular solutions we seek a wave function which for large separations will have only Coulomb distorted outgoing waves in all channels, i.e. we set

$$\sum_{j=1}^{n} C_j u_i^{(j)}(k,r) = A_i f_i^+(k,r), \text{ for } r > R \tag{10.8}$$

where $A_i$ are some arbitrary amplitudes, $C_i$ are coefficients to be determined and $f_i^+(k,r)$ denotes the Coulomb Jost solution for the i-th channel. The coefficients $C_i$ can be always determined from the $n$ matching conditions (10.8) for any values of the amplitudes $A_i$. However, the requirement of smooth matching imposes a constraint for the derivative of the wave function

$$\sum_{j=1}^{n} C_j u_i^{(j)\prime}(k,R) = A_i f_i^{+\prime}(k,R), \tag{10.9}$$

and by eliminating the amplitudes $A_i$ between (10.8) and (10.9), we obtain a set of homogeneous linear algebraic equations for the coefficients

$$\sum_{j=1}^{n} C_j \left[ u_i^{(j)\prime}(k,R) - u_i^{(j)}(k,R)\,\frac{f_i^{+\prime}(k,R)}{f_i^{+}(k,R)} \right] = 0. \tag{10.10}$$

The condition which ensures that this system has a non-trivial solution can be written as

$$\Delta(k) \equiv \det \left[ u_i^{(j)\prime}(k,R) - u_i^{(j)}(k,R)\,\frac{f_i^{+\prime}(k,R)}{f_i^{+}(k,R)} \right] = 0. \tag{10.11}$$

and this formula is a multi-channel analog of (10.2). Apart from purely algebraic complications, (10.11) brings one important change in the analytic structure of the function $\Delta(k)$ which now exhibits $(n-1)$ branch points connected with the presence of the higher lying channels.

## 10.2 Variational Methods

Variational principles have proved to be one of the most effective approximation methods in all of mathematical physics and it is not surprising that they can be also profitably applied to the two-potential problem considered in this book. Since we are interested in locating poles of the T-matrix the best starting point is the simplest Schwinger variational principle for the T-matrix. To begin with, we consider single potential scattering and let $T, V, G_0$ denote, respectively, the transition matrix, the interaction and the free resolvent operator. Following Schwinger [Lippmann *et al.* (1950)], one can obtain a variational estimate of the T-matrix in the form of a functional

$$\langle \mathbf{p}| T |\mathbf{p}'\rangle = \left\langle \psi_{\mathbf{p}}^{(-)} \right| V |\mathbf{p}'\rangle + \langle \mathbf{p}| V \left| \psi_{\mathbf{p}'}^{(+)} \right\rangle - \left\langle \psi_{\mathbf{p}}^{(-)} \right| V - VG_0V \left| \psi_{\mathbf{p}'}^{(+)} \right\rangle, \tag{10.12}$$

which is stationary with respect to small variations of the trial functions $\psi_{\mathbf{p}}^{(\pm)}\rangle$ about their exact values. Imposing the requirement for the above expression to be stationary leads immediately to the Lippmann–Schwinger equations for the wave functions

$$\left|\psi_{\mathbf{p}}^{(+)}\right\rangle = |\mathbf{p}\rangle + G_0 V \left|\psi_{\mathbf{p}}^{(+)}\right\rangle, \tag{10.13}$$

$$\left\langle\psi_{\mathbf{p}}^{(-)}\right| = \langle\mathbf{p}| + \left\langle\psi_{\mathbf{p}}^{(-)}\right| V G_0. \tag{10.14}$$

There are many other forms possible, e.g. one such form is obtained setting $|\psi_{\mathbf{p}}^{(+)}\rangle = A|\chi_{\mathbf{p}}^{(+)}\rangle$ and $\langle\psi_{\mathbf{p}}^{(-)}| = \langle\chi_{\mathbf{p}}^{(-)}|B$ where the coefficients $A$ and $B$ could be regarded as variational parameters to be determined from the conditions

$$\frac{\partial}{\partial A}\langle\mathbf{p}|\,T\,|\mathbf{p}'\rangle = \frac{\partial}{\partial B}\langle\mathbf{p}|\,T\,|\mathbf{p}'\rangle = 0.$$

This procedure yields another Schwinger-type functional

$$\langle\mathbf{p}|T|\mathbf{p}'\rangle = \frac{\langle\chi_{\mathbf{p}}^{(-)}|V|\mathbf{p}'\rangle\langle\mathbf{p}|V|\chi_{\mathbf{p}'}^{(+)}\rangle}{\langle\chi_{\mathbf{p}}^{(-)}|V - VG_0V|\chi_{\mathbf{p}'}^{(+)}\rangle}. \tag{10.15}$$

The presence of the resolvent in the Schwinger-type functionals leads to additional complications and is not a welcome feature. Therefore, in practical applications the Kohn-type [Kohn (1948); Hulthen (1948)] variational principle of a much simpler form has been given preference (applications of Schwinger principle to electron-molecule collisions and molecular photoionization have been reviewed by [Lucchese *et al.* (1986)]). The importance of Schwinger variational principle only recently has been fully appreciated when a connection has been established between variational approach and the degenerate kernel schemes leading to separable expansions successfully used for solving the scattering equations. It has been established [Adhikari *et al.* (1987)] that an underlying variational principle is responsible for the good convergence properties obtained with the method of separable expansions. Indeed, in Schwinger variational principle the trial wave function could be sought as a linear superposition of some complete set of basis functions where the coefficients would be regarded as variational parameters. When the derivatives of the Schwinger functional with respect to these coefficients are forced to vanish we obtain a set of linear equations for the expansion coefficients and these equations are identical with those which would follow in result of a degenerate kernel approximation using the same basis. It is therefore fair to say that Schwinger principle has been widely used but this fact has seldom been acknowledged and even realized. The functional (10.15) has a rational form suggesting that the poles of the T-matrix are connected with the zeros of the expectation value of the operator $V - VG_0V$. As a matter of fact, one can also write a Schwinger variational principle for bound states and resonances in the form

$$F[\chi] = \langle\chi| V - VG_0V |\chi\rangle , \tag{10.16}$$

so that varying the trial function $\chi$ about the exact value yields a homogeneous Lippmann–Schwinger equation appropriate for bound states

$$|\chi\rangle = G_0 V |\chi\rangle . \tag{10.17}$$

To proceed further we introduce a complete set of functions $f_i$ not necessarily orthogonal and we represent the trial function as a superposition of the basis functions

$$|\chi\rangle = \sum_n C_n |f_n\rangle . \tag{10.18}$$

Using (10.18) in (10.17), we obtain a set of homogeneous equations for the coefficients $C_n$

$$\sum_n \langle f_m| 1 - G_0 V |f_n\rangle C_n = 0. \tag{10.19}$$

A non-trivial solution of this set exists if, and only if, the following determinant vanishes

$$\det |\mathbf{1} - \mathbf{G}_0 \mathbf{V}| = 0, \tag{10.20}$$

where all matrices (denoted in (10.20) by boldface symbols) are evaluated in the $f_n$ basis. Using matrix notation, we have

$$G_{ij} = \langle f_i| G_0 |f_j\rangle ,$$
$$V_{ij} = \langle f_i| V |f_j\rangle ,$$

The specific values of the energy at which the determinant (10.20) vanishes are identified with the eigenvalues of the system.

The generalization of the above scheme to the two-potential case can be achieved with the following substitutions: the nuclear potential $-s\,v(r)$ is identified with $V$, the free resolvent $G_0$ must be replaced by the Coulomb Green's function and the plane waves in the scattering problem become Coulomb physical solutions. Since the determinant (10.20) involves Coulomb Green's function it is essential to be able to find a suitable basis in which matrix elements calculations could be efficiently carried out. The basis which might be capable of bringing major simplifications should be somehow related to the Coulomb problem. When we add the restriction that we want a basis with only discrete spectrum there is just one choice: the Sturmians.

Sturmians, or Sturm–Liouville functions are defined as eigenfunctions of the kernel $K = G_0V$ of the Lippmann–Schwinger equation (cf. [Weinberg (1963); Rotenberg (1970); Y Chen *et al.* (1972)]). It has been shown that these eigenfunctions form a complete discrete set without a continuum, regardless of the nature of the interaction potential $V$ as long as the kernel is a completely continuous operator. In general, Sturmians are itself not eigenfunctions of a physical system but just because they form a discrete set they provide a convenient basis for various expansions. A particularly useful choice is to set $V$ to be the Coulomb potential for in this case Sturmians are immediately derived from the hydrogenic solutions of the Schrödinger equation. Here, we have also made this choice and by Sturmians we shall mean hereafter the Coulomb Sturm–Liouville eigenfunctions.

In coordinate space the Sturmians $S_{\ell N}(b,r)$ enumerated by the nodal quantum number $N = 0, 1, 2, \dots$ are obtained as solution of the differential equation

$$\left[\frac{\mathrm{d}^2}{\mathrm{d}r^2} - \frac{\ell(\ell+1)}{r^2} + \frac{2b(N+\ell+1)}{r} - b^2\right] S_{\ell N}(b,r) = 0, \tag{10.21}$$

with the following boundary conditions imposed, respectively, at zero, and at infinity

$$\lim_{r\to 0} S_{\ell N}(b,r) \sim r^{\ell+1}; \tag{10.22}$$

$$\lim_{r\to\infty} S_{\ell N}(b,r) \sim \mathrm{e}^{-br}, \tag{10.23}$$

where $b$ is a free parameter ($Re\, b > 0$). The eigenfunctions are given in a closed form

$$S_{\ell N}(b,r) = \sqrt{\frac{N!}{(2\ell+1+N)!}}(2br)^{\ell+1}\mathrm{e}^{-br}L_N^{2\ell+1}(2br), \tag{10.24}$$

where $L_N^{2\ell+1}$ denotes Laguerre polynomial. These functions are orthogonal and form a complete set with respect to the weight function $1/r$. One would like to get rid of the weighting function and the simplest way out would be to redefine the Sturmians by multiplying them by the factor $1/\sqrt{r}$. However, a better idea is to introduce a complementary set of functions, defined as follows

$$s_{\ell N}(b,r) = \frac{S_{\ell N}(b,r)}{r},$$

so that $s_{\ell N}(b,r)$ together with $S_{\ell N}(b,r)$ satisfy orthogonality and completeness relations

$$\int_0^\infty S_{\ell N}(b,r) s_{\ell N'}(b,r)\,\mathrm{d}r = \delta_{NN'}; \tag{10.25}$$

$$\sum_{N=0}^\infty S_{\ell N}(b,x)\, s_{\ell N}(b,y) = \delta(x-y). \tag{10.26}$$

appropriate for a biorthogonal set. In momentum space the Sturmians $\tilde{S}_{\ell N}(b,p)$ and $\tilde{s}_{\ell N}(b,p)$ are obtained in the usual way by means of a Fourier transform, i.e.

$$\tilde{S}_{\ell N}(b,p) = (2/\pi)^{1/2} \int_0^\infty r j_\ell(pr) S_{\ell N}(b,r)\,\mathrm{d}r, \tag{10.27}$$

and similar expression can be written for $\tilde{s}_{\ell N}(b,p)$. When the Fourier transform (10.27) is effected, we obtain

$$\tilde{s}_{\ell N}(b,p) = \sqrt{\frac{2}{\pi}\frac{N!}{(2\ell+1+N)!}}\,\frac{2^\ell\,\ell!}{p}\,(\sin\phi)^{\ell+1} C_N^{\ell+1}(\cos\phi), \tag{10.28}$$

where $C_N^{\ell+1}$ denotes Gegenbauer polynomial and $\phi = 2\arctan(p/b)$. The functions $\tilde{S}_{\ell N}(b,p)$ and $\tilde{s}_{\ell N}(b,p)$ differ only by a factor, and the relation between them is

$$\tilde{S}_{\ell N}(b,p) = \frac{N+\ell+1}{b}(1+\cos\phi)\tilde{s}_{\ell N}(b,p). \tag{10.29}$$

The orthogonality and completeness relations in momentum space, are

$$\int_0^\infty \tilde{S}_{\ell N}(b,p)\tilde{s}_{\ell N'}(b,p) p^2\,\mathrm{d}p = \delta_{NN'};$$

$$\sum_{N=0}^\infty \tilde{S}_{\ell N}(b,p)\,\tilde{s}_{\ell N}(b,p') = \frac{\delta(p-p')}{pp'}.$$

The differential equation (10.21) takes the form of an integral equation in momentum space but by making use of (10.29), it can be written as an integral transform connecting the functions $\tilde{S}_{\ell N}(b,p)$ and $\tilde{s}_{\ell N}(b,p)$

$$p\,\tilde{s}_{\ell N}(b,p) = \frac{1}{\pi}\int_0^\infty Q_\ell\left(\frac{p^2+p'^2}{2pp'}\right)\tilde{S}_{\ell N}(b,p')p'\,\mathrm{d}p',$$

where $Q_\ell$ denotes Legendre polynomial of the second kind.

So far the parameter $b$ has been unspecified and in particular it could be set to be $b = -ik$. The reason for this choice is that the Coulomb Green's function (cf. (5.11)) $G_\ell^+(k)$ for $Im\, k > 0$ could be then very simply represented in the form of a Sturmian expansion. In coordinate representation, we have

$$\langle r|G_\ell^+(k)|r'\rangle = \frac{1}{2ik}\sum_{N=0}^{\infty}\frac{S_{\ell N}(-ik,r)\; S_{\ell N}(-ik,r')}{N+\ell+1+i\eta}. \qquad (10.30)$$

To actually prove formula (10.30) we note that the right hand side of (10.30) is manifestly symmetric in both variables $r$ and $r'$ and satisfies the boundary conditions required for the Green's function. To check that the equation (5.12) is indeed satisfied we apply the operator occurring on the left hand side of eq (5.12) to formula (10.30). By making use of (10.21) the pole terms are cancelled and then by invoking the completeness relation (10.26) we end up with a delta function which completes the demonstration. When $k$ is equal to $k_n$ the Sturmians are proportional to the hydrogenic wave functions

$$\lim_{k\to k_n} S_{\ell N}(-ik,r) = \sqrt{a_B}(N+\ell+1)\, u_{\ell N}(r)$$

and it is easily verified that the residues of the poles in (10.30) and in (5.13) are identical.

It is amusing that by using the Sturmian representation one can derive the perturbative expansion of the Green's function in the fine structure constant. Each term of the series can be obtained in a simple form and it is possible to perform effective summation of the series. The Coulomb Green's function satisfies the Lippmann–Schwinger equation $G = G_0 + G_0\, V_C\, G$ where $G_0$ and $V_C$ denote, respectively, the free resolvent and the Coulomb potential ($V_C(r) = 2k\eta/r$). Iterating this equation we obtain the perturbative expansion

$$G = G_0 + G_0\, V_C\, G_0 + G_0\, V_C\, G_0\, V_C\, G_0 + \cdots\,, \qquad (10.31)$$

and the Sturmian representation of $G_0$ follows immediately from (10.30) when we set $\eta = 0$. Since the Coulomb potential happens to be the weighting function in the Sturmian orthogonality relation (10.25) all intermediate state integrations become trivial, and the perturbative expansion can be written in the form

$$\langle r|G_\ell^+(k)|r'\rangle = \frac{1}{2ik}\left\{\sum_{N=0}^{\infty}\frac{S_{\ell N}(-ik,r)S_{\ell N}(-ik,r')}{N+\ell+1}+\right.$$
$$+(-i\eta)\sum_{N=0}^{\infty}\frac{S_{\ell N}(-ik,r)S_{\ell N}(-ik,r')}{(N+\ell+1)^2}+$$
$$\left.+(-i\eta)^2\sum_{N=0}^{\infty}\frac{S_{\ell N}(-ik,r)S_{\ell N}(-ik,r')}{(N+\ell+1)^3}+\cdots\right\},$$

where the first term represents the free resolvent . This geometric power series in $(-i\eta)$ can be summed explicitly which results in formula (10.30).

Formula (10.30) is quite remarkable in that the Green's function is given as a sum of the poles at the Coulomb bound states. The residues factorize what results in a separable representation very useful in practical applications. In particular, with the aid of expansion (10.30) one can calculate the expectation value of the Coulomb Green's function sandwiched between any pair of Sturmians. This quantity will be convenient for later use and we wish to introduce the Coulomb Green's function in the Sturmian representation defined by the formula

$$\langle s_{\ell N}(a)|G_\ell^+(k)|s_{\ell M}(b)\rangle = \int_0^\infty s_{\ell N}(a,r)\,\langle r|G_\ell^+(k)|r'\rangle\, s_{\ell M}(b,r')\,\mathrm{d}r\,\mathrm{d}r'. \tag{10.32}$$

The Green's function in Sturmian representation can be easily calculated by using the Sturmian expansion (10.30). Inserting (10.30) in (10.32), the Green's function is given in the form of the series

$$\langle s_{\ell N}(a)|G_\ell^+(k)|s_{\ell M}(b)\rangle = \frac{1}{2ik}\sum_{J=0}^{\infty}\frac{\langle s_{\ell N}(a)|S_{\ell J}(-ik)\rangle\,\langle s_{\ell J}(-ik)|S_{\ell M}(b)\rangle}{J+\ell+1+i\eta}, \tag{10.33}$$

where the symbol $\langle s_{\ell N}(a)|S_{\ell M}(b)\rangle$ denotes the overlap integral of two Sturmians

$$\langle s_{\ell N}(a)|S_{\ell M}(b)\rangle = \int_0^\infty s_{\ell N}(a,r)S_{\ell M}(b,r)\,\mathrm{d}r, \tag{10.34}$$

which can be explicitly evaluated, and the answer is

$$\langle s_{\ell N}(a)|S_{\ell M}(b)\rangle = \frac{(-1)^M\,(2\ell+1+M+N)!}{\sqrt{M!\,(2\ell+1+M)!\,N!\,(2\ell+1+N)!}}\times$$
$$\times\,(1-x^2)^{\ell+1}\,x^{N+M}\,{}_2F_1(-M,-N;-2\ell-1-M-N;1/x^2), \tag{10.35}$$

where $x = (b-a)/(b+a)$ and ${}_2F_1$ denotes hypergeometric function. It should be noted that since the first two arguments of the latter are negative integers the hypergeometric series stops after $\min(M, N)$ terms and the resulting expression reduces to a polynomial. When the parameters $a$ and $b$ are equal formula (10.35) yields the orthogonality relation (10.25).

In principle, formula (10.33) already gives the Green's function in Sturmian representation but the series form renders (10.33) impractical for calculating the full Green's matrix. Fortunately, this is not required because, as we shall see later on, all matrix elements in (10.33) can be generated by recurrence knowing just a single matrix element $\langle s_{\ell 0}(a)|G_\ell^+(k)|s_{\ell 0}(b)\rangle$ . To evaluate the latter we make use of the expansion (10.33) which in this case can be immediately converted to the hypergeometric series, and, we obtain

$$\begin{aligned}\langle s_{\ell 0}(a)|G_\ell^+(k)|s_{\ell 0}(b)\rangle &= \left[\frac{4ba}{(b+a)^2}\right]^{\ell+1} \times \\ &\times \frac{(1-z)^{2\ell+2}{}_2F_1(2\ell+2,\, \nu;\, \nu+1;\, z)}{2\mathrm{i}k\nu},\end{aligned} \tag{10.36}$$

with $\nu = \ell + 1 + \mathrm{i}\eta$ and $z = [(a+\mathrm{i}k)/(a-\mathrm{i}k)]\,[(b+\mathrm{i}k)/(b-\mathrm{i}k)]$. Although formula (10.36) has been derived under the assumption that $Im\, k > 0$ but it could be analytically continued to the lower part of the k-plane by using the well known properties of the hypergeometric function. Setting $a = b$ in (10.36), and using the familiar relation

$${}_2F_1(a,\, b;\, c;\, x) = (1-x)^{c-a-b}{}_2F_1(c-a,\, c-b;\, c;\, x),$$

we arrive at the ultimate formula

$$\langle s_{\ell 0}(b)|G_\ell^+(k)|s_{\ell 0}(b)\rangle = -\frac{2\,b\,{}_2F_1(1,\, -\ell+\mathrm{i}\eta;\, \ell+2+\mathrm{i}\eta;\, x)}{(\ell+1+\mathrm{i}\eta)(b-\mathrm{i}k)^2}, \tag{10.37}$$

where $x = (b+\mathrm{i}k)^2/(b-\mathrm{i}k)^2$. The hypergeometric function in (10.37) can be efficiently computed by continued fraction summation method.

To derive recurrence relations for the matrix elements (10.33) it is convenient to consider the inverse of the Green's operator for the Coulomb problem

$$G_\ell(k)^{-1} = k^2 + \frac{\mathrm{d}^2}{\mathrm{d}r^2} - \frac{\ell(\ell+1)}{r^2} + \frac{2\mu\alpha Z}{r} \tag{10.38}$$

and the expectation value of this operator in the Sturmian representation will be identified with the inverse Green's function matrix

$$\langle S_{\ell M}(b)|G_\ell^+(k)^{-1}|S_{\ell N}(b)\rangle = \int_0^\infty S_{\ell M}(b,r)G_\ell(k)^{-1}S_{\ell N}(b,r)\,\mathrm{d}r. \quad (10.39)$$

Using (10.21) and the well known recurrence relation for the Laguerre polynomials, the integral in (10.39) can be easily evaluated and the resulting inverse Green's matrix has a very simple tridiagonal (Jacobi) structure. We obtain

$$\langle S_{\ell M}(b)|G_\ell^+(k)^{-1}|S_{\ell N}(b)\rangle = u_{N-1}\,\delta_{M\,N-1} + d_N\,\delta_{M\,N} + u_N\,\delta_{M\,N+1}, \quad (10.40)$$

where $d_N$ and $u_N$ denote the diagonal and above diagonal element, respectively, and are given by the expressions

$$\begin{aligned} d_M &= \frac{k^2-b^2}{b}(M+\ell+1) + 2\mu\alpha Z; \\ u_M &= -\frac{k^2+b^2}{2b}\sqrt{(M+1)(M+2\ell+2)}, \end{aligned}$$

where $M = 0, 1, 2...$ Since in all practical applications one is forced to deal with matrices of finite sizes, in the following we shall introduce the discrete basis that has been truncated at some maximal value $N$ of the nodal quantum number, and, accordingly, $N$ hereafter will be a fixed number representing the size of the discrete space. To proceed further, we shall introduce a shorthand notation for the Green's matrix and its inverse (10.39) suppressing the orbital momentum $\ell$ and abbreviating

$$\begin{aligned} G_{ij} &= \langle s_{\ell i}(b)|G_\ell^+(k)|s_{\ell j}(b)\rangle\,; \\ G_{ij}^{-1} &= \langle S_{\ell i}(b)|G_\ell^+(k)^{-1}|S_{\ell j}(b)\rangle\,. \end{aligned}$$

The familiar resolvent equation whose operator form is $G^{-1}G = 1$, in matrix form becomes largely simplified owing to the Jacobi structure of the inverse Green's matrix. Using (10.40), we have

$$d_0\,G_{00} + u_0\,G_{01} = 1, \quad (10.41)$$

and

$$u_{m-1}\,G_{k\,m-1} + d_m\,G_{k\,m} + u_m\,G_{k\,m+1} = \delta_{m\,k}, \quad (10.42)$$

for $m > 0$ and $k = m, m+1, \ldots.$ It is apparent from the above expressions that the neighboring elements of the Green's matrix are connected by a three-term recurrence relation. To determine all the elements of the symmetric matrix $G_{ij}$ we need only $G_{00}$ to start the recurrence sequence. Since $G_{00}$ has already been calculated in (10.37), the remaining $G_{ij}$ can be obtained from (10.41) and (10.42). This works fine only for negative energies, i.e. for $k$ in the upper half of the complex k-plane but when $Im\, k \to 0$, the three-term recurrence becomes unstable and eventually converges to a dominant solution which is not the one we want. On the other hand, the Green's function is an analytic function of $k$ , so if there is a region of the complex plane where the function can be determined, this value could be used for analytic continuation to other parts of the complex plane. Accordingly, a foolproof calculational scheme should exist valid throughout the whole complex plane. This is indeed the case and we are going to present now the ultimate solution of this problem.

It turns out that analytic expression (10.37) is only a particular case of a general formula that gives the Sturmian representation Coulomb Green's function for arbitrary values of the quantum numbers. Actually, it has been shown (cf. [Heller (1975)] ) that the familiar product form of the coordinate representation Coulomb Green's function

$$\langle r'|G_\ell^+(k)|r\rangle = -(1/k)\, F_\ell(k, r_<)\, H_\ell^+(k, r_>), \tag{10.43}$$

has a completely analogous counterpart in Sturmian representation, where the switching prescription "greater than" and "smaller than" refers now to the discrete variables, i.e. to the nodal quantum numbers and the relevant formula, is

$$\langle s_{\ell n'}(b)|G_\ell^+(k)|s_{\ell n}(b)\rangle = -(1/k)\, \mathcal{F}_\ell(k, n_<)\, \mathcal{H}_\ell^+(k, n_>). \tag{10.44}$$

In (10.44) $n_< = \min(n, n'), n_> = \max(n, n')$ and the functions $\mathcal{F}_\ell$ and $\mathcal{H}_\ell^+$, should be regarded as Sturmian representation of the regular and irregular Coulomb wave function, respectively, i.e.

$$\mathcal{F}_\ell(k, n) = \int_0^\infty F_\ell(k, r)\, s_{\ell n}(b, r)\, \mathrm{d}r, \tag{10.45}$$

$$\mathcal{H}_\ell^+(k, n) = \int_0^\infty H_\ell^+(k, r)\, s_{\ell n}(b, r)\, \mathrm{d}r. \tag{10.46}$$

where the above integrations can be performed either directly, or by using more sophisticated methods [Yamani *et al* (1975); Ojha (1986)]. The important point is that as a result the functions $\mathcal{F}_\ell$ and $\mathcal{H}_\ell^+$, are given by analytic expressions suitable for efficient computations. The explicit formulas, are

$$\mathcal{F}_\ell(k,n) = \sqrt{\frac{(2\ell+1+n)!}{n!}}C_\ell(\eta)(-1)^n(\sin\phi)^{\ell+1}\mathrm{e}^{(\eta-\,\mathrm{i}n)\phi} \times$$
$$\times\, {}_2F_1(-n,\, \ell+1-\mathrm{i}\eta;\, 2\ell+2;\, 1-\mathrm{e}^{2\mathrm{i}\phi}), \tag{10.47}$$

for the regular function, and, respectively,

$$\mathcal{H}^+_\ell(k,n) = \frac{\sqrt{(2\ell+1+n)!\,n!}}{(2\ell+1)!}\,\frac{(-1)^n\mathrm{e}^{-(\eta-\,\mathrm{i}n)\phi}}{C_\ell(\eta)\,(\sin\phi)^\ell}\,\frac{\Gamma(\ell+1+\mathrm{i}\eta)}{\Gamma(n+\ell+2+\mathrm{i}\eta)} \times$$
$$\times\, \mathrm{e}^{\mathrm{i}\phi}{}_2F_1(n+1,\, -\ell+\mathrm{i}\eta;\, n+2+\ell+\mathrm{i}\eta;\, \mathrm{e}^{2\mathrm{i}\phi}), \tag{10.48}$$

for the irregular function, where the angle $\phi$ is defined by the formula: $\mathrm{e}^{\mathrm{i}\phi} = (b+\mathrm{i}k)/(b-\mathrm{i}k)$. It should be noted that the symmetric Green's matrix in a truncated basis has $(N+1)(N+2)/2$ elements but those will be fully determined by the elements of the two sequences $\mathcal{F}_\ell(k,n)$ and $\mathcal{H}^+_\ell(k,n)$, i.e. $2(N+1)$ numbers only need to be supplied. Because the first argument in the hypergeometric function occurring in (10.47) is a negative integer, formula (10.47) reduces to a polynomial (Pollaczek polynomial ) and the computation of $\mathcal{F}_\ell(k,n)$ is straightforward. The computation of $\mathcal{H}^+_\ell(k,n)$ has to be broken into two stages. Firstly, we note that $\mathcal{H}^+_\ell(k,0)$ can be computed directly since the first argument in the hypergeometric function in (10.48) is equal to one and in this particular case the hypergeometric function ${}_2F_1(1,b;c;z)$ could be evaluated by summation of a continued fraction to be specified in a moment. To obtain $\mathcal{H}^+_\ell(k,n)$ for $n>0$, we define the ratio of two neighboring functions

$$I_n = -\mathcal{H}^+_\ell(k,n+1)/\mathcal{H}^+_\ell(k,n) \tag{10.49}$$

which in view of (10.48) can be written as

$$I_n = \mathrm{e}^{\mathrm{i}\phi}\frac{[(n+1)(n+2\ell+2)]^{1/2}}{n+\ell+2+\mathrm{i}\eta} \times$$
$$\times\frac{{}_2F_1(-\ell+\mathrm{i}\eta,\, n+2;\, n+3+\ell+\mathrm{i}\eta;\, \mathrm{e}^{2\mathrm{i}\phi})}{{}_2F_1(-\ell+\mathrm{i}\eta,\, n+1;\, n+2+\ell+\mathrm{i}\eta;\, \mathrm{e}^{2\mathrm{i}\phi})}. \tag{10.50}$$

It is evident from (10.50) that the sequence $I_n$, $n=1,2,...,N$ is given in terms of the ratio of two hypergeometric functions of the type ${}_2F_1(a,b+1;c+1;z)/{}_2F_1(a,b;c;z)$ but this ratio can be very efficiently computed by summation of a continued fraction discovered by Gauss. Quite generally, we have (cf. [Jones *et al.* (1980)])

$$\frac{{}_2F_1(a,\, b;\, c;\, z)}{{}_2F_1(a,\, b+1;\, c+1;\, z)} = 1 + \cfrac{1}{a_1 + \cfrac{1}{a_2 + \cfrac{1}{a_3 + \cdots}}} \tag{10.51}$$

where the coefficients $a_j$ depend upon $z$ and are given by the following expressions

$$a_{2n} = z\,\frac{(b+n)(a-c-n)}{(c+2n-1)(c+2n)}; \qquad n = 1, 2, ...$$

$$a_{2n+1} = z\,\frac{(a+n)(b-c-n)}{(c+2n+1)(c+2n)}; \qquad n = 0, 1, 2, ...$$

Continued fraction summation method allows also to compute the hypergeometric function ${}_2F_1(a,\, 1;\, c;\, z)$ by setting in formula (10.51) $b = 0$ and replacing $c + 1$ by $c$. The summation procedure in (10.51) is known to be convergent in all of the complex z-plane away from the cut on the real axis running from one to infinity. Having computed $\mathcal{H}_\ell^+(k, 0)$ and the sequence in (10.50) we can generate $\mathcal{H}_\ell^+(k, n)$ for all values of $n$. The Green's function is well behaved even for $k = 0$ but the limiting procedure $k \to 0$ requires some care because at this particular value of $k$ the parameters of the hypergeometric function in (10.37) as well as in (10.50) become infinite while at the same time the argument value coincides with the branching point singularity. To calculate the limiting value of (10.40) for $k \to 0$ we use the familiar series representation of the hypergeometric function taking the asymptotic values of the gamma functions occurring in this series. The resulting series can be expressed in terms of the irregular confluent hypergeometric function $U(a, c; x)$ defined in [Abramowitz *et al.* (1965)], and, the final answer is

$$\big\langle s_{\ell 0}(b)|G_\ell^+(0)|s_{\ell 0}(b)\big\rangle = -(2/b)\,\xi^{2\ell+1}\,U(2\ell+2,\, 2\ell+2,\, \xi), \tag{10.52}$$

where $\xi = -4\mu\alpha Z/b$. Similar procedure applied to the ratio of two hypergeometric functions yields the $k \to 0$ limit of (10.50)

$$\lim_{k\to 0} I_n = \sqrt{(n+1)(n+2\ell+2)}\,\frac{U(2\ell+n+3,\, 2\ell+2;\, \xi)}{U(2\ell+n+2,\, 2\ell+2;\, \xi)}. \tag{10.53}$$

For integer values of the parameters the confluent function $U(a, c; x)$ can be expressed in terms of the incomplete gamma function $\Gamma(a, x)$ defined in

[Abramowitz *et al.* (1965)]. The required expressions are

$$U(2\ell+1,\, 2\ell+2;\, \xi) = \frac{1}{\xi^{2\ell+1}}; \tag{10.54}$$

$$U(2\ell+2,\, 2\ell+2;\, \xi) = \frac{1}{(2\ell+1)!}\left\{\sum_{m=0}^{2\ell} \frac{(-1)^m\, m!}{\xi^{m+1}} - \mathrm{e}^{\xi}\Gamma(0,\xi)\right\}. \tag{10.55}$$

The incomplete gamma function $\Gamma(0,\xi)$ has a logarithmic singularity at $\xi = 0$ and also a branch cut along the negative real axis of the complex $\xi$ plane. For $\xi > 0$ the function $\Gamma(0,\xi)$ is real while for $\xi < 0$ it will be complex but in either case is given in terms of the appropriate exponential integral function, i.e. $E_1(\xi)$ or $Ei(-\xi)$, both defined in [Abramowitz *et al.* (1965)]

$$\Gamma(0,\xi) = \begin{cases} E_1(\xi), & \text{for } \xi > 0, \\ -Ei(-\xi) \pm \mathrm{i}\pi, & \text{for } \xi < 0, \end{cases}$$

where the choice of sign of the imaginary part is determined by the outgoing (+), or ingoing (-) boundary condition imposed upon the Green's function ($G^+$ and $G^-$, respectively). The remaining confluent functions $U(a,c;x)$ are generated from a rapidly converging three-term recurrence relation

$$\begin{aligned} n(2\ell+1+n)\, U(2\ell+2+n,\, 2\ell+2;\, \xi) &= \\ = (2n+2\ell+\xi)\, U(2\ell+1+n,\, 2\ell+2;\, \xi) &- U(2\ell+n,\, 2\ell+2;\, \xi) \end{aligned}$$

where $n = 1, 2, 3, \ldots$ and (10.54)–(10.55) providing the starting values. This completes the $k \to 0$ limiting procedure and the Green's matrix has been determined in the whole complex k-plane.

In many applications the Lippmann–Schwinger equation (10.17) is replaced by the Schrödinger equation

$$G_0^{-1}\, |\chi\rangle = V\, |\chi\rangle\,, \tag{10.56}$$

whose discrete form

$$\det\left|\mathbf{G}_0^{-1} - \mathbf{V}\right| = 0, \tag{10.57}$$

instead of the resolvent involves the inverse Green's matrix in the truncated basis. The underlying variational principle in this case has got to be of the Kohn type, and differs from the Schwinger's principle in that the appropriate functional approximates $T - V$ instead of the $T$ operator. The advantage of using (10.56) is twofold: in contrast with (10.17), no matrix multiplication is required and the $G^{-1}$ matrix possessing a band structure is much simpler than the resolvent. Indeed, in the truncated $G^{-1}_{ij}$ matrix the non-vanishing elements, are

$$G^{-1}_{j-1\,j} = u_{j-1}; \qquad G^{-1}_{jj} = d_j; \qquad G^{-1}_{j\,j+1} = u_j, \tag{10.58}$$

with $j = 1, 2, ..., (N-1)$ so that in (10.58) the element $G^{-1}_{NN}$ has been left out as the latter requires a more careful treatment. Clearly, if we did not do that, trying the simplest approximation $G^{-1}_{NN} = d_N$, the truncated matrix necessarily would have been real which is inconsistent with the outgoing wave boundary condition. Actually, the only place where the appropriated boundary condition can be imposed is the element $G^{-1}_{NN}$ and we want now to determine that quantity (cf. [ Kónya *et al.* (1997)]). The simplest way to introduce the boundary condition is to stipulate consistency between the truncated resolvent equation

$$\sum_{k=0}^{N} G^{-1}_{ik}\, G_{kj} = \delta_{ij} \tag{10.59}$$

and the recurrence relation (10.42). Indeed, setting $i = j = N$ in the above resolvent equation and using (10.58), we have

$$u_{N-1}\, G_{N\,N-1} + G^{-1}_{NN}\, G_{NN} = 1, \tag{10.60}$$

whereas the recurrence (10.42), yields

$$u_{N-1}\, G_{N\,N-1} + d_N\, G_{NN} + u_N\, G_{N\,N+1} = 1, \tag{10.61}$$

so that by subtracting (10.61) from (10.60), we immediately arrive at the sought for formula expressing $G^{-1}_{NN}$ in terms of the Green's matrix

$$G^{-1}_{NN} = d_N + u_N\, G_{N\,N+1}/G_{NN} = d_N - u_N\, I_N, \tag{10.62}$$

which ensures that the boundary condition of $G_{ij}$ has been properly included into the truncated matrix $G^{-1}_{ij}$. As a cross-check, the resolvent equation (10.59) can be verified numerically by computing the determinant of the matrix on the left hand side in (10.59) whose exact value is one. Another very interesting cross-check is to set the potential $V$ equal to zero in (10.57). In this case the solutions of (10.57) should yield the unperturbed Coulomb spectrum already for $N = 1$. We wish to present here two different ways of choosing the Sturmian basis and either of them could be used for solving the two-body problem. The eigenenergy is obtained by

Table 10.1 Convergence of the variational method for the energy dependent basis. The potential (6.20) has been used with $\bar{a} = (0.44 + i0.83)\, fm$

$\ell = 1$

| $N$ | $\Gamma$ (MeV) | $\epsilon$ (MeV) |
|---|---|---|
| 24 | 0.102960 | -0.102644 |
| 48 | 0.101546 | -0.103374 |
| 72 | 0.101323 | -0.103387 |
| 96 | 0.101290 | -0.103399 |
| 120 | 0.101293 | -0.103403 |
| 144 | 0.101294 | -0.103402 |
| exact | 0.101294 | -0.103402 |

$\ell = 2$

| $N$ | $\Gamma$ (keV) | $\epsilon$ (keV) |
|---|---|---|
| 24 | 2.21629 | -0.461631 |
| 48 | 2.30971 | -0.506972 |
| 72 | 2.32850 | -0.508945 |
| 96 | 2.32673 | -0.506782 |
| 120 | 2.32599 | -0.507692 |
| 144 | 2.32639 | -0.507489 |
| exact | 2.32627 | -0.507563 |

$\ell = 3$

| $N$ | $\Gamma$ (eV) | $\epsilon$ (eV) |
|---|---|---|
| 24 | 5.16058 | 0.765385 |
| 48 | 4.79706 | 0.238994 |
| 72 | 4.84227 | 0.272577 |
| 96 | 4.82296 | 0.260934 |
| 120 | 4.82911 | 0.263764 |
| 144 | 4.82819 | 0.262051 |
| exact | 4.82841 | 0.262231 |

either solving the Lippmann–Schwinger equation (10.20), or the equivalent Schrödinger equation (10.56).

*(i) Fixed basis.* This is the simplest Sturmian basis. The parameter $b$

Table 10.2 Convergence of Sturmian expansion method for the potential (6.20) with $\bar{a} = (0.44 + i0.83)\ fm$ for the energy independent basis with a choice of three different values of the parameter $b$.

$\ell = 1$

| | $b = 0.1\ fm^{-1}$ | | $b = 0.2\ fm^{-1}$ | | $b = 0.3\ fm^{-1}$ | |
|---|---|---|---|---|---|---|
| $N$ | $\Gamma$ (MeV) | $\epsilon$ (MeV) | $\Gamma$ (MeV) | $\epsilon$ (MeV) | $\Gamma$ (MeV) | $\epsilon$ (MeV) |
| 24 | 0.099439 | -0.104817 | 0.100684 | -0.103467 | 0.101241 | -0.103492 |
| 48 | 0.100768 | -0.103656 | 0.101351 | -0.103384 | 0.101301 | -0.103404 |
| 72 | 0.101414 | -0.103505 | 0.101304 | -0.103403 | 0.101293 | -0.103402 |
| 96 | 0.101340 | -0.103375 | 0.101297 | -0.103402 | 0.101294 | -0.103402 |
| 120 | 0.101275 | -0.103406 | 0.101297 | -0.103402 | 0.101281 | -0.103408 |
| 144 | 0.101303 | -0.103402 | 0.101294 | -0.103402 | 0.101051 | -0.103564 |
| exact | | | 0.101294 | -0.103402 | | |

$\ell = 2$

| | $b = 0.1\ fm^{-1}$ | | $b = 0.2\ fm^{-1}$ | | $b = 0.3\ fm^{-1}$ | |
|---|---|---|---|---|---|---|
| $N$ | $\Gamma$ (keV) | $\epsilon$ (keV) | $\Gamma$ (keV) | $\epsilon$ (keV) | $\Gamma$ (keV) | $\epsilon$ (keV) |
| 24 | 2.24870 | -0.450656 | 2.31844 | -0.500041 | 2.32334 | -0.507363 |
| 48 | 2.31140 | -0.503290 | 2.32727 | -0.507437 | 2.32633 | -0.507603 |
| 72 | 2.32598 | -0.508957 | 2.32638 | -0.507583 | 2.32626 | -0.507565 |
| 96 | 2.32724 | -0.507151 | 2.32629 | -0.507565 | 2.32628 | -0.507568 |
| 120 | 2.32592 | -0.507552 | 2.32627 | -0.507564 | 2.32429 | -0.509106 |
| 144 | 2.32640 | -0.507558 | 2.32626 | -0.507582 | 2.26608 | -0.532752 |
| exact | | | 2.32627 | -0.507563 | | |

$\ell = 3$

| | $b = 0.1\ fm^{-1}$ | | $b = 0.2\ fm^{-1}$ | | $b = 0.3\ fm^{-1}$ | |
|---|---|---|---|---|---|---|
| $N$ | $\Gamma$ (eV) | $\epsilon$ (eV) | $\Gamma$ (eV) | $\epsilon$ (eV) | $\Gamma$ (eV) | $\epsilon$ (eV) |
| 24 | 5.14615 | 0.311039 | 4.84911 | 0.268331 | 5.23616 | 0.191862 |
| 48 | 4.83523 | 0.278112 | 4.82839 | 0.261995 | 4.82815 | 0.262397 |
| 72 | 4.82257 | 0.262846 | 4.82822 | 0.262382 | 4.82823 | 0.262442 |
| 96 | 4.82917 | 0.262174 | 4.82824 | 0.262431 | 4.82836 | 0.262379 |
| 120 | 4.82792 | 0.262641 | 4.82824 | 0.262439 | 4.81504 | 0.252704 |
| 144 | 4.82830 | 0.262386 | 4.82836 | 0.261866 | 4.62929 | 0.169112 |
| exact | | | 4.82841 | 0.262231 | | |

is arbitrarily chosen and we set

$$f_j = S_{\ell j}(b, r),$$

*(ii) Energy-dependent basis.* Here a specific choice of the parameter $b$ has been made, namely $b = -ik$ where $Im\, k > 0$ and we set

$$f_j = S_{\ell j}(-ik, r).$$

The necessity to dispose of the parameter $b$ in the fixed basis case is a disadvantageous feature. Of course, with an infinite basis since we have a complete set the value of $b$ is irrelevant and could be anything, however, with a truncated basis this is no longer true and the convergence of the method depends upon the choice of $b$. This problem does not exist in the energy-dependent basis, and, besides that, this choice leads to a considerable simplification of the whole scheme (cf. [Deloff *et al.* (1980)]) because the Green's matrix $\mathbf{G}$ and its inverse becomes diagonal

$$G_{nm} = \frac{1}{2ik}\frac{1}{m + \ell + 1 + i\eta}\delta_{nm},$$

and has a very simple form. All that needs to be calculated is just the potential matrix $\mathbf{V}$. On the other hand, the energy-dependent scheme works only in the upper half of the k-plane, i.e. it is not suitable for locating Gamow states.

The performance of these two calculational schemes has been tested numerically, similarly as before, for the Saxon–Woods potential (6.20) and the computations have been performed for kaonic Sulphur $^{32}$S for $\ell$ equal $1, 2$ and 3. The resulting energy shifts (with respect to pure Coulomb position) and widths have been displayed in Table 10.1 for the energy-dependent basis and in Table 10.2 for the energy-independent basis. The last entry called "exact" in Tables 10.1 and 10.2 refers to the results obtained by the matching method. In Table 10.2 three different values of the parameter $b$ have been tried. When the shifts and widths are small the value of $b$ close to the unperturbed Coulomb momentum $b \sim |k_0|$ appears to be a reasonable choice. Some authors [ Kónya *et al.* (1997); Kónya *et al.* (2000)] have been using extra weighting factors in the decomposition of the unit operator into a complete set of Sturmians which is supposed to accelerate the convergence rate. However, the price one has to pay for it is that the weighting factors involve new adjustable parameters introducing additional element of arbitrariness. Therefore, this has not been attempted here. In general, the calculated energies as functions of $N$ after a few damped oscillations converge to the exact value. It is apparent from Table 10.1 and Table 10.2 that for $N$ greater than about 50 the Sturmian method is accurate up to three leading digits in the calculated shifts and widths which ought to be quite adequate for all practical applications. It should be noted that the potential matrix could be calculated in either coordinate space, or in momentum space, whichever is more convenient and this is perhaps the main advantage of the Sturmian method that it is applicable in both the momentum space and the coordinate space. Nevertheless, the only paper we are aware of, in which the Sturmian method has been actually used for computing energy shifts and widths in hadronic atoms is [Yan *et al.* (1997)].

## 10.3 Fredholm Integral Equation Method

We have shown in Chapter 5 that the two-potential eigenvalue problem takes the form of a homogeneous Fredholm integral equation (5.10)

$$\phi_\ell(s,k,r) = -s \int_0^\infty \langle r|G_\ell^+(k)|r'\rangle\, v(r')\phi_\ell(s,k,r')\, \mathrm{d}r', \tag{10.63}$$

the kernel of which is given as the product of the Coulomb Green's function $G_\ell^+(k)$ and the nuclear potential. The parameter $s$ is regarded as given and we wish to find such values of $k$ for which the equation (10.63) has a non-trivial solution. The homogeneous Fredholm equation (10.63) is known to have solutions only for a particular set of $k$ values $k = \{k_n\}\,, (n = 1, 2, ...)$ providing the eigenenergies $E_n = k_n^2/2\mu$.

Equation (10.63) can be easily solved by standard methods, i.e. (i) the integral in (10.63) is approximated by a sum according to the rule

$$\int_0^\infty f(x)dx = \sum_{i=1}^{N} w_i f(x_i), \tag{10.64}$$

where $w_i, x_i$ are, respectively the integration weights and nodes; and (ii) the kernel of the integral equation is symmetrized by multiplying the equation (10.63) by $\sqrt{v(r)}$ and introducing $\sqrt{v(r)}\phi_\ell(s,k,r)$ as the new unknown function instead of $\phi_\ell(s,k,r)$. In result of the discretization envisaged in (10.64), the task of solving (10.63) is reduced to solving a system of homogeneous linear algebraic equations

$$\sum_{j=1}^{N} \left[\delta_{ij} + s\, v_i\, \langle x_i|G_\ell^+(k)|x_j\rangle\; v_j\right]\; C_j = 0, \tag{10.65}$$

where $x_< = \min(x_i, x_j), x_> = \max(x_i, x_j), v_i = \sqrt{w_i\, v(x_i)}$, and $C_j =$

$v_j \phi_\ell(s,k,x_j)$ . The symmetrization procedure can be carried out also when the potential $v(x_i)$ takes negative values and in such a case we use the decomposition $v(x_i) = -\sqrt{-v(x_i)}\,\sqrt{-v(x_i)}$. The necessary and sufficient condition for (10.65) to have a non-trivial solution is vanishing of the Fredholm determinant

$$F_l^+(s,k) = \det\left|\delta_{ij} + s\, v_i\, \langle x_i|G_\ell^+(k)|x_j\rangle\, v_j\right| = 0. \tag{10.66}$$

It should be reminded that the Fredholm determinant (10.66) is nothing else but the Jost function of the two-potential problem under consideration. For solving (10.66) only the Coulomb Green's function needs to be supplied. To proceed further the latter could be calculated directly from formula (5.11) and with the advent (cf. [Tomphson *et al.* (1985)]) of fast and accurate procedures calculating the Coulomb wave functions for complex arguments this is perfectly feasible. Since the Green's function (10.30) is defined in the whole k-plane away from the cut along negative imaginary semi-axis, the presented method could be used also for Gamow states. The numerical locating of zeros of the Fredholm determinant (10.66) is straightforward and could be done by using the Mueller algorithm mentioned above.

There is also a second alternative which uses Sturmian representation of the Green's function introduced in (10.30). With the aid of the Lippmann–Schwinger equation $G = G_0 + G_0 V_C G$ the Green's function could be written as a sum of two terms: the free Green's function plus the correction due to Coulomb interaction. The first term, $G_0$, is well known and can be cast in a closed form, while the second term can be immediately evaluated using (10.30) and the orthogonality relation (10.25). We obtain

$$\begin{aligned}\langle r|G_\ell^+(k)|r'\rangle &= -krr'\, h_\ell^+(kr_>)\, j_\ell(kr_<) - \\ &- \frac{\eta}{2k}\sum_{N=0}^{\infty}\frac{S_{\ell N}(-\mathrm{i}k,r)\; S_{\ell N}(-\mathrm{i}k,r')}{(N+\ell+1)(N+\ell+1+\mathrm{i}\eta)},\end{aligned} \tag{10.67}$$

and formula (10.67) is still exact. The second term representing Coulomb correction has a denominator which increases as $N^2$ for large $N$ and therefore formula (10.67) converges much more rapidly than the original expression (10.30). To avoid possible misunderstanding we wish to make it clear that using the word correction we did not mean to imply that this term is small as compared with the free resolvent. Actually, the sum contains all the poles of the Coulomb Green's function and close to the Coulomb bound states this term is the dominant contributor. The representation

(10.67) is very useful because the Green's function is expressed in terms of elementary functions and is readily amenable for computation.

Perhaps the greatest advantage of the Fredholm method lies with the simplicity of coding: all that is needed is the potential evaluated at the mesh points to feed the $v_i$ array and a routine calculating Coulomb wave functions at the integration mesh points. A practical estimate of the error can be obtained by performing the computations for $N_{max}$ mesh points and then repeating the computations with $N_{max}$ increased by say 30%. The difference between the two answers should provide a good estimate of the error and gives an indication what value of $N_{max}$ is required to attain the desired accuracy.

The performance of the presented calculational scheme has been tested numerically for the Saxon–Woods potential (6.20). We have used a Gaussian mesh in the interval $[0,1]$ with logarithmic mapping of the form $r = \log[(1+x)/(1-x)]$, transforming the integrals from $[0,\infty]$ range to $[0,1]$ range. The computations have been performed for kaonic Sulphur $^{32}$S for $\ell$ equal $1,2$ and 3. The resulting energy shifts (with respect to pure Coulomb position) and widths have been displayed in Table 10.3. The last entry called "exact", similarly as before, refers to results obtained by the matching method. As seen from Table 10.3 for $N_{max}$ greater than about 48 the Fredholm method is accurate up to three leading digits what is quite enough in all practical applications. We have used two methods to calculate the Green's function: one was based on formula (5.11) where the Coulomb wave functions were computed by using the code developed in [Tomphson *et al.* (1985)], whereas the other uses directly formula (10.67). Taking about fifty terms in the summation in formula (10.67) turns out to be sufficient for making the results presented in Table 10.3 identical for both methods to within four leading digits or better.

Table 10.3 Convergence of the Fredholm method for increasing mesh size. The potential (6.20) has been used with $\bar{a} = (0.44 + i0.83)\,fm$

$\ell = 1$

| Mesh | $\Gamma$ (MeV) | $\epsilon$ (MeV) |
|---|---|---|
| 24 | 0.101066 | -0.103303 |
| 48 | 0.101163 | -0.103362 |
| 72 | 0.101220 | -0.103384 |
| 96 | 0.101248 | -0.103392 |
| 120 | 0.101263 | -0.103396 |
| 144 | 0.101271 | -0.103398 |
| exact | 0.101294 | -0.103402 |

$\ell = 2$

| Mesh | $\Gamma$ (keV) | $\epsilon$ (keV) |
|---|---|---|
| 24 | 2.32275 | -0.512695 |
| 48 | 2.32511 | -0.508567 |
| 72 | 2.32549 | -0.508025 |
| 96 | 2.32571 | -0.507840 |
| 120 | 2.32586 | -0.507751 |
| 144 | 2.32595 | -0.507700 |
| exact | 2.32627 | -0.507563 |

$\ell = 3$

| Mesh | $\Gamma$ (eV) | $\epsilon$ (eV) |
|---|---|---|
| 24 | 4.79711 | 0.224782 |
| 48 | 4.83310 | 0.256618 |
| 72 | 4.83272 | 0.260499 |
| 96 | 4.83136 | 0.261517 |
| 120 | 4.83043 | 0.261903 |
| 144 | 4.82983 | 0.262086 |
| exact | 4.82841 | 0.262231 |

## 10.4 Momentum Space Methods

In a chargeless case it has been appreciated for a long time that momentum space methods offer a number of distinct advantages to performing bound state calculations:

*(i)* The wave equation in momentum space has a more general character in that it allows including arbitrary non-local interactions. In addition to that, relativistic effects can be treated quite simply by making appropriate modifications in the kinetic energy operator which is diagonal.

*(ii)* It turns out that, in contrast to what is the case in coordinate space, the momentum space wave functions corresponding to Gamow states are square integrable [Hernandez *et al.* (1984)].

*(iii)* The bound state problem can be reduced to an algebraic eigenvalue problem which can be solved by standard methods exhaustively described in many textbooks. Ready-to-use routines for solving such problems can be found in [Smith *et al.* (1976); Wilkinson *et al.* (1971)].

*(iv)* As a consequence of (iii) the solution of the eigenvalue problem provides us not only with the ground state energy but the whole bound state energy spectrum is supplied in a single calculation.

*(v)* In contrast to the coordinate space methods, there is no need for root-searching of the characteristic equation. In fact, this would be usually a very poor computational method for finding the eigenvalues.

*(vi)* A rather attractive feature is the simplicity of coding.

The inclusion of the Coulomb interaction introduces a far non-trivial complication connected with the presence of the on-shell singularity in the momentum space representation of the Coulomb potential. Therefore, a careful reconsideration of the bound state problem is in order.

The two-potential wave equation in the momentum space, takes the form

$$(p^2-k^2)\phi_\ell(s,k,p) = \int_0^\infty \left[v_\ell^C(p,p') + s v_\ell(p,p')\right] \phi_\ell(s,k,p')\, p'^2\, \mathrm{d}p', \quad (10.68)$$

where $\phi_\ell(s,k,p)$ denotes the bound state momentum space wave function. The potentials occurring in (10.68) are obtained by taking Fourier transforms of the Coulomb and the nuclear potential, respectively

$$v_\ell^C(p,p') = 2\mu\alpha Z \frac{2}{\pi}\int_0^\infty j_\ell(pr)\, j_\ell(p')\, r\mathrm{d}r = \frac{2\mu\alpha Z}{\pi p p'} Q_\ell(z_{pp'}); \quad (10.69)$$

$$v_\ell(p,p') = \frac{2}{\pi}\int_0^\infty j_\ell(pr)\, v(r)\, j_\ell(p')\, r^2\mathrm{d}r, \quad (10.70)$$

where $Q_\ell(z_{pp'})$ is Legendre polynomial of the second kind and $z_{pp'} = (p^2 + p'^2)/(2pp')$. For $p = p'$ we have $z_{pp'} = 1$ and the Legendre polynomial exhibits a logarithmic singularity. Since the Coulomb potential occurs only under the integral in (10.68) the singularity is integrable and, in fact, can be eliminated altogether by a subtraction invented by Lande and described by [Kwon *et al.* (1978)] To remove the singularity the integral containing the Coulomb potential is brought to the following form

$$\int_0^\infty v_\ell^C(p,p')\phi_\ell(s,k,p')\,p'^2\mathrm{d}p' = \phi_\ell(s,k,p)p^2 S_\ell(p) + \\ + \int_0^\infty v_\ell^C(p,p')\left[\phi_\ell(s,k,p')\,p'^2 - \phi_\ell(s,k,p)\,\frac{p^2}{P_\ell(z_{pp'})}\right]\mathrm{d}p', \tag{10.71}$$

where $P_\ell(z_{pp'})$ denotes Legendre polynomial. The integral on the right hand side of (10.71) is non-singular and the function $S_\ell(p)$ representing the counter term can be explicitly evaluated

$$S_\ell(p) = \int_0^\infty \frac{v_\ell^C(p,p')}{P_\ell(z_{pp'})}\,\mathrm{d}p' = \frac{2\mu\alpha Z}{p}\left(\frac{\pi}{2} - I_\ell\right), \tag{10.72}$$

where the numerical constants $I_\ell$ have been tabulated by [Kwon *et al.* (1978)]. Inserting (10.71) into (10.68), the integral equation (10.68) can be converted to a matrix equation by replacing the integral over $p'$ with a sum over $M$ mesh points $p_i$ with the corresponding weights $w_i$. Abbreviating $p_j\phi_\ell(s,k,p_j)\sqrt{w_j}$ as $X_j$, we end up with an algebraic eigenvalue problem

$$\sum_{j=0}^{M}(A_{ij} - k^2\,\delta_{ij})X_j = 0 \tag{10.73}$$

where the matrix $A_{ij}$, is

$$A_{ij} = \left[1 - S_\ell(p_i) + \sum_{\nu=1}^{M}(1-\delta_{i\nu})v_\ell^C(p_i,p_\nu)\,w_\nu/P_\ell(z_{i\nu})\right]p_i^2\,\delta_{ij} \\ - p_i\sqrt{w_i}\left[(1-\delta_{ij})v_\ell^C(p_i,p_j) + s v_\ell(p_i,p_j)\right]p_j\sqrt{w_j}. \tag{10.74}$$

As a result of the subtraction presented in (10.71), the diagonal element of $v_\ell^C(p_i,p_j)$ never enters $A_{ij}$ and thereby the singularity has been neutralized. It should be noted that the matrix $A_{ij}$ is symmetric and therefore if the the nuclear potential is real all the eigenvalues $k^2$ will necessarily also be real. Since for Gamow states the corresponding values of $k^2$ are in general complex, it is quite clear that the above scheme is applicable only for bound states ($Im\,k > 0$).

The advantageous feature of the momentum space approach is that the bound state problem reduces to a well known algebraic eigenvalue problem which can be solved by standard methods so that most of the computational problems which could possibly arise are likely to be well recognized and documented in the mathematical literature.

In order to be able to locate the Gamow states as well, the appropriate boundary condition has to be incorporated into the wave equation and the simplest way to accomplish that is to resort to the Lippmann–Schwinger equation placing the boundary condition into the Green's function. The homogeneous Lippmann–Schwinger equation equivalent to (10.68), takes the form

$$\phi_\ell(s,k,p) = \int_0^\infty \left[v_\ell^C(p,p') + s v_\ell(p,p')\right] \frac{\phi_\ell(s,k,p')p'^2\,\mathrm{d}p'}{p'^2 - k^2}. \tag{10.75}$$

Similarly as its ancestor (10.68), this equation is valid only in the upper part of the k-plane but the Cauchy integral in (10.75) could be analytically continued to the lower half of the k-plane with the aid of the Plemelj formula [Muskhelishvili (1953)]. Thus, for $Im\, k < 0$, (10.75) reads

$$\begin{aligned}\phi_\ell(s,k,p) = &\int_0^\infty \left[v_\ell^C(p,p') + s v_\ell(p,p')\right] \frac{\phi_\ell(s,k,p')p'^2\,\mathrm{d}p'}{p'^2 - k^2} \\ &+ \mathrm{i}\pi k \left[v_\ell^C(p,k) + s v_\ell(p,k)\right] \phi_\ell(s,k,k),\end{aligned} \tag{10.76}$$

and, owing to a different boundary condition, even for real potentials we have now a non-hermitian eigenvalue problem whose eigenvalues in general will be complex as appropriate for Gamow states. For a chargeless case this would be indeed the sought for solution of the problem for the equation (10.76) could be solved in a similar manner as (10.68) by reducing it to a matrix form. However when the Coulomb potential is present (10.76) *cannot be solved* because the Coulomb potential in (10.76) occurs also outside the integral and the corresponding singularity cannot be removed by subtraction. Actually, this result is not at all surprising because the on-shell singularity has been a major difficulty in momentum space scattering problems and Gamow states constitute just a small subclass of scattering states.

In contrast with the coordinate space where the introduction of Coulomb Green's function brings significant simplifications, there is not much point in using the momentum space Green's function since the latter has a logarithmic singularity on-shell (cf. [Schwinger (1964)]). Actually, this is evident from the perturbative expansion (10.31) remembering that $G_0$ is diagonal in momentum space. Since $V_C$ exhibits a logarithmic singularity this singularity will be present in every term containing $V_C$. It is still true that by introducing the Coulomb Green's function the Coulomb potential can be eliminated from the Lippmann–Schwinger equation but this does not help to remove the logarithmic singularity which would be reintroduced via the Green's function.

Although formally the Lippmann–Schwinger equation (10.75) and the Schrödinger equation (10.68) are equivalent and either of them could be solved numerically, but only (10.68) could be reduced to the algebraic eigenvalue problem. The procedure to solve (10.75) is different, one has to derive the Fredholm determinant and subsequently locate its roots which is numerically a more complicated task and gives only one root at a time. For the Schrödinger equation the rank of the secular determinant is equal to the size of the mesh (say,$N$) so that this determinant is a polynomial of the order $N$ in $k^2$. Therefore there are exactly $N$ roots corresponding to $N$ levels. By contrast, the Fredholm determinant corresponding to the Lippmann–Schwinger equation is not a polynomial in $k^2$ but has a more complicated energy dependence and in general there is no upper limit for the number of roots.

We have tested the performance of the momentum space method using the same Saxon–Woods potential (6.20) as before together with the Schrödinger equation (10.68). Of course, this is a local potential and must be first transformed to the momentum representation using formula (10.70). This is not quite trivial and some care has be exercised in calculating the integral in (10.70). Since the involved momenta could assume arbitrarily large values the corresponding integrand becomes then a rapidly oscillating function and in consequence direct integration might be quite difficult. To avoid unnecessary complications we followed here a procedure outlined in [Eisenstein *et al.* (1976)] where the integral (10.70) was computed in two stages. As a preliminary step the potential $v(r)$ was Fourier transformed by calculating the following integral

$$\tilde{v}(q) = \int_0^\infty j_0(qr)\, v(r)\, r^2\, \mathrm{d}r.$$

The point is that the Fourier transform $\tilde{v}(q)$ for a Saxon–Woods shape

Table 10.4 Convergence of the momentum space approach for increasing mesh size. The potential (6.20) has been used with $\bar{a} = (0.44 + i0.83)\,fm$.

$\ell = 1$

| Mesh | Γ (MeV) | ϵ (MeV) |
|---|---|---|
| 24 | 0.105646 | -0.092679 |
| 48 | 0.100011 | -0.103799 |
| 72 | 0.101181 | -0.103338 |
| 96 | 0.101279 | -0.103390 |
| 120 | 0.101290 | -0.103399 |
| 144 | 0.101293 | -0.103400 |
| exact | 0.101294 | -0.103402 |

$\ell = 2$

| Mesh | Γ (keV) | ϵ (keV) |
|---|---|---|
| 24 | 2.20149 | -0.79389 |
| 48 | 2.33744 | -0.51721 |
| 72 | 2.32701 | -0.51346 |
| 96 | 2.32601 | -0.50992 |
| 120 | 2.32606 | -0.50875 |
| 144 | 2.32613 | -0.50826 |
| exact | 2.32627 | -0.507563 |

$\ell = 3$

| Mesh | Γ (eV) | ϵ (eV) |
|---|---|---|
| 24 | 4.89257 | -465.296 |
| 48 | 4.78715 | -59.854 |
| 72 | 4.82038 | -17.733 |
| 96 | 4.82538 | -7.369 |
| 120 | 4.82687 | -3.657 |
| 144 | 4.82748 | -2.010 |
| exact | 4.82841 | 0.262231 |

could be obtained in an analytic form

$$\int_0^\infty \frac{\sin(qr)\, r\,\mathrm{d}r}{\exp[(r-c)/a]+1} = \pi a c \left\{ -\frac{\cos(qc)}{\sinh(\pi a q)} + \right.$$

$$\left. + \frac{\pi a}{c}\frac{\sin(qc)\coth(\pi a q)}{\sinh(\pi a q)} - \frac{2a^2 q}{\pi c}\sum_{n=1}^{\infty}(-\mathrm{e}^{-c/a})^n \frac{n}{[(qa)^2+n^2]^2} \right\}$$

where the series converges very fast owing to the exponential factor (cf. [Eisenstein *et al.* (1976)] for details). Knowing $\tilde{v}(q)$, the potential $v_\ell(p,p')$ immediately follows by projecting out the partial wave with orbital momentum $\ell$

$$v_\ell(p,p') = \frac{1}{\pi}\int_{-1}^{1} \tilde{v}(q)\, P_\ell(z)\, \mathrm{d}z,$$

where $q^2 = p^2 + p'^2 - 2zpp'$ and this integral can now be calculated numerically.

As indicated in [Kwon *et al.* (1978)] a delicate point is the correct choice of the mesh points distribution in (10.74) feeding sufficient number of points in the regions where the integrand varies rapidly. It is far from obvious how to do that in view of the strong imbalance in the ranges and strengths in the Coulomb and the nuclear potential and some experimentation is required. We have used several distributions trying to reproduce first the pure Coulomb spectrum which is well known and we could determine precisely the errors in each case. The nuclear interaction was less sensitive to the choice of the mesh. Ultimately, the mapping $p_i = \mu\alpha Z \tan(x_i\pi/2)$ where $x_i$ are Gauss–Legendre knots from the interval $[0,1]$ resulted in maximal reduction of error and was adopted for all computations presented hereafter. The results of our calculations are displayed in Table 10.4 for $\ell = 1, 2, 3$. It turns out that the rate of convergence for increasing mesh size is no worse than in the variational or Fredholm equation approach presented before. The only exception is the $\ell = 3$ case where even with the largest mesh the energy shift is incorrect. The absolute value of the energy shift is here merely an indication of the error which is much bigger than the real shift caused by the nuclear interaction. For $\ell = 2$ these relations are inverted, the error is of the same order but the shift is by a factor $10^3$ bigger and therefore the precision is adequate. It should be emphasized that the failure of the momentum space approach to reproduce correctly the $\ell = 3$ energy shift from the practical point of view is completely irrelevant as such small shifts cannot be measured experimentally. The only measurable quantities for the upper levels might be the widths and, as seen from Table 10.4, they can be computed rather satisfactorily despite the fact that they are quite small. It is in order to reiterate that even if the potential is non-local the variational method always provides a very reliable alternative to the momentum space approach. Therefore, if needs arises the upper level shifts could be always computed variationally.

# Chapter 11

# Examples

For a number of specific potential shapes the regular solution $\phi_\ell(s,k,r)$ can be obtained in an analytic form. These potentials, however, are somewhat artificial, and, indeed different from the potentials used in all practical applications, i.e. for the actual fitting of the hadronic atoms data. Nevertheless, there are at least two good reasons why such potentials might deserve attention. Firstly, they illustrate the general formalism developed in this work, and, secondly, the analytically solvable cases are always very useful for testing various computer codes. We are going to consider here four exactly solvable models: rank-one separable potential, delta-shell potential, square-well potential, and, a cut-off Coulomb potential. The advantageous feature of these models is that the Coulomb wave functions which will be present in the final expressions can be easily computed by elementary methods.

## 11.1 Rank-One Separable Potential

This is a particular case of the non-local potential in which $V_\ell(r,r')$ is assumed to be of a product form so that the two independent variables are separated, *viz.*

$$2\mu V_\ell(r,r') = -v_\ell(r)\,\lambda_\ell\,v_\ell(r'), \tag{11.1}$$

where the function $v_\ell(r)$, hereafter assumed to be real, is referred to as the form factor of the potential and the complex parameter $\lambda_\ell$ measures the strength of the potential. Similarly to (11.1) one defines a separable potential of rank $N$: just by taking a sum of $N$ terms each of them in a product form. However, going from rank-one to rank $N$ is a trivial matter and to simplify notation we shall stick to the form (11.1). Inserting the potential (11.1) into the wave equation (2.1) and multiplying this equation by $v_\ell(r)$, after integration over $r$, we obtain the well-known (cf. [Kok *et al.* (1981); Kok *et al.* (1982)] and references therein) bound state condition

$$1 + \lambda_\ell \int_0^\infty v_\ell(r) \langle r|G_\ell^+(k)|r'\rangle \, v_\ell(r')\mathrm{d}r\mathrm{d}r' = 0. \tag{11.2}$$

To bring this equation to a more manageable form we shall separate out the Coulomb correction in the Green's function with the aid of the representation (10.67). Using (10.67) in (11.2), for $Im\, k > 0$, one has

$$1 - \lambda_\ell \int_0^\infty \frac{\tilde{v}_\ell(p)^2 p^2 \, \mathrm{d}p}{p^2 - k^2 - \mathrm{i}\epsilon} + \frac{\eta}{2k} \sum_{N=0}^\infty \frac{\langle -\mathrm{i}kN|2\mu V_\ell|-\mathrm{i}kN\rangle}{(N+\ell+1)(N+\ell+1+\mathrm{i}\eta)} = 0, \tag{11.3}$$

where $\tilde{v}_\ell(p)$ is the form factor in momentum representation

$$\tilde{v}_\ell(p) = (2/\pi)^{\frac{1}{2}} \int_0^\infty r j_\ell(r) v_\ell(r) \, \mathrm{d}r,$$

and the Sturmian representation of the potential simplifies to the form

$$\begin{aligned} \langle -\mathrm{i}kN|2\mu V_\ell|-\mathrm{i}kN\rangle &= -\,\lambda_\ell \left[\int_0^\infty S_{\ell N}(-\mathrm{i}k, r) v_\ell(r) \, \mathrm{d}r\right]^2 = \\ &= -\,\lambda_\ell \left[\int_0^\infty \tilde{S}_{\ell N}(-\mathrm{i}k, p) \tilde{v}_\ell(p) \, \mathrm{d}p\right]^2 . \end{aligned} \tag{11.4}$$

In the Coulomb-free case the sum in (11.3) is absent and one recovers the well-known condition for the bound state supported by the separable potential (11.1). Coulomb level shifts for a given $\lambda_\ell$, can be determined by solving numerically eq. (11.3).

We wish now to evaluate the K-matrix and its low-energy expansion parameters. To this end we shall need the extended form of eq. (5.31), *viz.*

$$\begin{aligned} \psi_\ell^K(\lambda_\ell, k, r) &= \phi_\ell(k, r) - \\ &- \lambda_\ell \int_0^\infty \langle r|G_\ell(k)|r'\rangle \, v_\ell(r') \, v_\ell(r'') \, \psi_\ell^K(\lambda_\ell, k, r'') \, \mathrm{d}r' \, \mathrm{d}r'', \end{aligned} \tag{11.5}$$

and the corresponding formula (5.32) for the K-matrix

$$\mathcal{K}_\ell(\lambda_\ell, k) = -\lambda_\ell \int_0^\infty \phi_\ell(k, r) v_\ell(r) \, \mathrm{d}r \int_0^\infty v_\ell(r') \psi_\ell^K(\lambda_\ell, k, r') \, \mathrm{d}r', \tag{11.6}$$

where in both expressions (11.5) and (11.6) the separable form (11.1) has been used. From these two equations, we obtain the ultimate formula for the inverse of the K-matrix

$$\frac{1}{\mathcal{K}_\ell(\lambda_\ell, k)} = -\frac{\displaystyle\int_0^\infty v_\ell(r)\,\langle r|G_\ell(k)|r'\rangle\, v_\ell(r')\,\mathrm{d}r\,\mathrm{d}r' + \lambda^{-1}}{\left[\displaystyle\int_0^\infty v_\ell(r)\phi_\ell(k,r)\,\mathrm{d}r\right]^2}, \tag{11.7}$$

where the Green's function is given by (5.28). The right hand side of (11.7) can now be expanded in a power series in $k^2$, and the coefficients in the two lowest order terms provide us with the expressions, respectively, for the scattering length

$$\frac{1}{A_\ell(\lambda_\ell)} = -\frac{\displaystyle\int_0^\infty v_\ell(r)\,\langle r|G_\ell(0)|r'\rangle\, v_\ell(r')\,\mathrm{d}r\,\mathrm{d}r' + \lambda_\ell^{-1}}{\left[\displaystyle\int_0^\infty v_\ell(r)\phi_\ell(0,r)\,\mathrm{d}r\right]^2}, \tag{11.8}$$

and for the effective range

$$\begin{aligned}\tfrac{1}{2}R_\ell(\lambda_\ell) = &-\frac{2}{A_\ell(\lambda_\ell)}\frac{\int_0^\infty v_\ell(r)\dot{\phi}_\ell(0,r)\,\mathrm{d}r}{\int_0^\infty v_\ell(r)\phi_\ell(0,r)\,\mathrm{d}r} - \\ &- \frac{\int_0^\infty v_\ell(r)\left\langle r|\dot{G}_\ell(0)|r'\right\rangle v_\ell(r')\,\mathrm{d}r\,\mathrm{d}r'}{\left[\int_0^\infty v_\ell(r)\phi_\ell(0,r)\,\mathrm{d}r\right]^2},\end{aligned} \tag{11.9}$$

where

$$\begin{aligned}\langle r'|G_\ell^+(0)|r\rangle &= -\phi_\ell(0,r_<)\,\theta_\ell(0,r_>);\\ \left\langle r'|\dot{G}_\ell^+(0)|r\right\rangle &= -\phi_\ell(0,r_<)\,\dot{\theta}_\ell(0,r_>) - \dot{\phi}_\ell(0,r_<)\,\theta_\ell(0,r_<),\end{aligned}$$

and dot denotes derivative with respect to $k^2$. Comparing the above formula (11.9) for the effective range with our general expression (6.10), we notice that in (11.9) the term proportional to $1/A_\ell^2$ is absent.

The T-matrix may be obtained directly from the integral equation

$$\begin{aligned}\psi_\ell^T(\lambda_\ell, k, r) = &F_\ell(\eta, k\,r) - \\ &- \lambda_\ell \int_0^\infty \langle r\,|G_\ell^+(k)|\,r'\rangle\, v_\ell(r')\, v_\ell(r'')\, \psi_\ell^T(\lambda_\ell, k, r'')\,\mathrm{d}r'\,\mathrm{d}r'',\end{aligned} \tag{11.10}$$

where the Green's function is given in (4.23). The nuclear T-matrix follows immediately by going to the asymptotic limit in (11.10) and the final expression is

$$T_\ell^{cs}(k) = \mathrm{e}^{\mathrm{i}\delta_\ell^{cs}(k)} \sin\delta_\ell^{cs}(k) = \\ = \frac{(2\mu\lambda_\ell/k)\left[\int_0^\infty v_\ell(r)\, F_\ell(\eta, k\, r)\,\mathrm{d}r\right]^2}{1+\lambda_\ell \int_0^\infty v_\ell(r)\, \langle r|G_\ell^+(k)|r'\rangle\, v_\ell(r')\,\mathrm{d}r\,\mathrm{d}r'} \tag{11.11}$$

Separable plus Coulomb – the two-potential problem has been studied in many papers (cf. [Kok *et al.* (1981); Kok *et al.* (1982); de Maag *et al.* (1984)] and references therein) where analytic and elegant solutions have been obtained. The authors always used eq. (11.2) to calculate bound states but, in general, the calculation of the double integral presents a difficult task. In many cases even if an analytic answer could be produced, the result involves special functions whose computation might be very complicated. Therefore, our formula (11.3) is particularly convenient for computational purposes and does not require special functions. The other useful result are the general and explicit expressions (11.8) and (11.9) giving the low-energy expansion parameters for an arbitrary separable potential (11.1).

## 11.2 Delta-Shell Potential

This peculiar potential may be simultaneously regarded as local and non-local and is defined as

$$V(r) = V_0\, R\, \delta(r-R), \tag{11.12}$$

with $V_0$ and $R$ being, respectively, depth and range parameters, so that $s = -2\mu V_0 R^2$ and $v(r) = \delta(r-R)/R$. The local potential (11.12) may be also written as

$$2\mu V_\ell(r,r') = V(r)\,\delta(r-r') = \delta(r-R)\,(s/R)\,\delta(r'-R), \tag{11.13}$$

where the non-local character is apparent. It should be noted that the form factor $v_\ell(r) = \delta(r-R)$ and the strength $\lambda_\ell = s/R$ are both independent of $\ell$. The other pathological feature of the delta-shell potential (11.12) is that the derivative of the corresponding wave function suffers a discontinuity at $r = R$ what renders the matching method void. The bound state condition will be obtained by calculating this discontinuity. Inserting the delta-shell potential (11.12) into (2.3) and integrating the resulting equation in the interval $[R-\epsilon, R+\epsilon]$, we obtain

$$u'_\ell(s,k,R-\epsilon) - u'_\ell(s,k,R+\epsilon) = s\, u_\ell(s,k,R). \tag{11.14}$$

If we can find such $k$ value for which the condition (11.14) is satisfied, then this $k$ is the eigenvalue. Using (10.1) for calculating the derivatives and going to the limit $\epsilon \to 0$, the bound state condition takes the ultimate form

$$R\frac{\phi'_\ell(k,R)}{\phi_\ell(k,R)} - R\frac{f_\ell^{+\prime}(k,R)}{f_\ell^+(k,R)} = s, \tag{11.15}$$

involving Coulomb functions only. Locating the roots of the equation (11.15) is greatly facilitated by the fact that both logarithmic derivatives of the Coulomb wave functions occurring in (11.15) can be expediently computed by continued fraction summation.

The same bound state condition can be obtained from Fredholm equation approach. Indeed, using the delta-shell potential (11.12) in (10.63), we conclude that a non-trivial solution exists only if the following condition is satisfied

$$1 + (s/R)\left\langle R|G_\ell^+(k)|R\right\rangle = 0, \tag{11.16}$$

and using (5.11) in (11.16), we arrive at (11.15).

The K-matrix is obtained from (5.31) and (5.32), and, the final result is

$$\frac{1}{\mathcal{K}_\ell(s,k)} = -\frac{\theta_\ell(k,R)}{\phi_\ell(k,R)} + \frac{R}{s}\frac{1}{\phi_\ell(k,R)^2}, \tag{11.17}$$

and by expanding the right hand side of (11.17) in powers of $k^2$ we obtain the scattering length and the effective range. We have

$$\frac{1}{A_\ell(s)} = -\frac{\theta_\ell(0,R)}{\phi_\ell(0,R)} + \frac{R}{s}\frac{1}{\phi_\ell(0,R)^2}; \tag{11.18}$$

$$\tfrac{1}{2}R_\ell(s) = -\frac{\dot\theta_\ell(0,R)\,\phi_\ell(0,R) + \theta_\ell(0,R)\,\dot\phi_\ell(0,R)}{\phi_\ell(0,R)^2} + \frac{2}{A_\ell(s)}\frac{\dot\phi_\ell(0,R)}{\phi_\ell(0,R)}. \tag{11.19}$$

It is a trivial exercise to demonstrate that the bound state condition (11.15) and the expressions for the low-energy scattering parameters (11.18)–(11.19) immediately follow from the corresponding expressions derived for a separable potential in which we set $v_\ell(r) = \delta(r-R)$ and $\lambda_\ell = s/R$.

The scattering problem has a rather trivial solution. The Lippmann-Schwinger equation

$$\psi_\ell(k,r) = F_\ell(\eta, k\,r) - V_0\, R \int_0^\infty \langle r\,|G_\ell^+(k)|r'\rangle\, \delta(r'-R)\psi_\ell(k,r')\mathrm{d}r', \tag{11.20}$$

can be solved at once and by going to the asymptotic limit, we obtain the nuclear T-matrix

$$T_\ell^{cs}(k) = \mathrm{e}^{\mathrm{i}\delta_\ell^{cs}(k)}\, \sin\delta_\ell^{cs}(k) = \frac{s/\rho\, F_\ell(\eta,\rho)^2}{1 - s/\rho\, F_\ell(\eta,\rho)\, H_\ell^+(\eta,\rho)} \tag{11.21}$$

where $\rho = k\,R$. Owing to the Wronskian condition $\mathrm{W}[\mathrm{H}_\ell^+,\mathrm{F}_\ell]$=1, the denominator in (11.21) can be cast to the form

$$1 - \frac{s}{\rho}\, F_\ell(\eta,\rho)\, H_\ell^+(\eta,\rho) = F_\ell(\eta,\rho)\, H_\ell^+(\eta,\rho)\, \left[\frac{F_\ell'(\eta,\rho)}{F_\ell(\eta,\rho)} - \frac{H_\ell^{+\prime}(\eta,\rho)}{H_\ell^+(\eta,\rho)} - \frac{s}{\rho}\right],$$

from which the bound state condition (11.15) immediately follows.

## 11.3 Square-Well Potential

For the square well potential $V(r) = V_0\,\theta(R-r)$ with $V_0$ and $R$ being, respectively, the depth and range parameters, the regular solution for $r \le R$ is $\phi_\ell(s,k,r) = \phi_\ell(\varkappa, r)$ where $\varkappa = \sqrt{k^2 + 2s/R^2}$ and $s = -\mu V_0 R^2$. The matching method yields the following bound state condition

$$\frac{\phi_\ell'(\varkappa, R)}{\phi_\ell(\varkappa, R)} = \frac{f_\ell^{+\prime}(k,R)}{f_\ell^+(k,R)}; \tag{11.22}$$

where both sides of (11.22) can be rapidly computed using continued fraction representation of the Coulomb functions. The scattering length is obtained from (6.5) or (6.6) setting

$$\mathcal{L}_\ell(s) = R\frac{\phi_\ell'(\sqrt{2s}/R, R)}{\phi_\ell(\sqrt{2s}/R, R)};$$

which is readily calculated by continued fraction summation. The calculation of the effective range is more complicated and perhaps the quickest way is to expand the inverse of the K-matrix in powers of $k^2$, which yields

$$\tfrac{1}{2}R_\ell(s) = \frac{1}{A_\ell(s)}\left\{\frac{\dot\theta_\ell'(0,R) - \theta_\ell(0,R)\mathcal{D}_\ell(s)}{\theta_\ell'(0,R) - \theta_\ell(0,R)\mathcal{D}_\ell(s)} - \frac{\dot\phi_\ell'(0,R) - \phi_\ell(0,R)\mathcal{D}_\ell(s)}{\phi_\ell'(0,R) - \phi_\ell(0,R)\mathcal{D}_\ell(s)}\right\},$$

where $\mathcal{D}_\ell(s)$ is

$$\mathcal{D}_\ell(s) = \lim_{\epsilon\to 0}\left\{\frac{\phi'_\ell(\sqrt{2s}/R+\epsilon,R)}{\phi_\ell(\sqrt{2s}/R+\epsilon,R)} - \frac{\phi'_\ell(\sqrt{2s}/R-\epsilon,R)}{\phi_\ell(\sqrt{2s}/R-\epsilon,R)}\right\}\frac{1}{2\epsilon},$$

and can be easily computed.

To solve the scattering problem for any k, the wave function $\psi_\ell(k,r)$ can be written as

$$\psi_\ell(k,r) = \begin{cases} A\,F_\ell(\bar\eta,\,\varkappa r) & \text{for } r \le R \\ F_\ell(\eta,\,kr) + T^{cs}_\ell(k)\,H^+_\ell(\eta,\,kr) & \text{for } r \ge R, \end{cases} \tag{11.23}$$

where $\bar\eta = k\eta/\varkappa$. Matching at r=R the two pieces of the wave function (11.23) and doing the same with the derivative, we obtain two algebraic equations for the two unknowns: A and $T^{cs}_\ell$. This gives the ultimate formula for the nuclear T-matrix

$$T^{co}_\ell(k) = e^{i\delta^{cs}_\ell(k)}\,\sin\delta^{cs}_\ell(k) = -\frac{\varkappa F'_\ell(\bar\eta,\bar\rho)\,F_\ell(\eta,\rho) - kF_\ell(\bar\eta,\bar\rho)\,F'_\ell(\eta,\rho)}{\varkappa F'_\ell(\bar\eta,\bar\rho)\,H^+_\ell(\eta,\rho) - kF_\ell(\bar\eta,\bar\rho)\,H^{+\prime}_\ell(\eta,\rho)}.$$

where $\rho = k\,R$ and $\bar\rho = \varkappa\,R$.

## 11.4 Cut-off Coulomb Potential

This potential is defined as

$$V(r) = V_0\,\theta(R-r)\,R/r,$$

with $V_0$ and $R$ being, respectively, depth and range parameters, so that $s = -2\mu V_0 R^2$. The regular solution for $r \le R$ can be expressed in terms of Coulomb regular wave function in which the Coulomb parameter $\alpha Z$ has been replaced by $\alpha Z + s/(2\mu R)$. Therefore, the bound state condition takes a simple form

$$\frac{f^{+\prime}_\ell(k,R)}{f^+_\ell(k,R)} = \left\{\frac{\phi'_\ell(k,R)}{\phi_\ell(k,R)}\right\}_{\alpha Z\to\alpha Z+s/(2\mu R)},$$

and both sides of the above equation can be calculated by the continued fraction summation method. The scattering length is obtained from (6.5) or (6.6) using in the following expression for the logarithmic derivative $\mathcal{L}_\ell(s)$

$$\mathcal{L}_\ell(s) = \left\{\frac{\phi'_\ell(0,R)}{\phi_\ell(0,R)}\right\}_{\alpha Z\to\alpha Z+s/(2\mu R)},$$

where the regular Coulomb wave functions are given in (3.43)–(3.44). To calculate the effective range from (6.10) one needs the integral $I_\ell(s)$ defined in (6.11) which can be computed from the formula

$$I_\ell(s) = \left\{ \frac{w\left[\dot{\phi}_\ell(0,R), \phi_\ell(0,R)\right]}{\phi_\ell(0,R)^2} \right\}_{\alpha Z \to \alpha Z + s/(2\mu R)},$$

where dot denotes derivative with respect to $k^2$.

## 11.5 Bound States in Extended-Charge Coulomb Potential

In atomic physics the finite-size of the nuclear charge distribution induces a small correction to the binding energy which can be calculated perturbatively. Since the experimentally observed hadronic atom levels are peripheral, this effect is in most instances also small. However, for deeply bound states and especially for heavy hadrons, the finite size effect is large and requires a more refined treatment. The problem we wish to take on is to find the spectrum of the Hamiltonian involving the Coulomb potential that has been modified due to finite extend of the nuclear charge distribution, i.e.

$$V_C(r) = -\frac{\alpha Z}{r} \to -\alpha \int \frac{\rho(\mathbf{x})}{|\mathbf{r} - \mathbf{x}|} \mathrm{d}^3 x,$$

where $\rho$ denotes the charge density normalized to Z. In a particular case when the nuclear charge is assumed to be uniformly distributed, the appropriate bound state condition may be obtained in an analytic form. For uniform $\rho$ the above integral is easily done and the point-charge Coulomb potential will be replaced by

$$V_C(r) = \begin{cases} \dfrac{\alpha Z}{R}\left(\dfrac{r^2}{2R^2} - \dfrac{3}{2}\right) & \text{if } r \le R, \\ -\dfrac{\alpha Z}{r} & \text{if } r \ge R, \end{cases} \tag{11.24}$$

where R denotes the radius of the uniform charge distribution. As seen from (11.24), for $r \le R$ we have a harmonic oscillator potential while for $r \ge R$ we have the standard point-like Coulomb potential. Since the exact solutions for either of them are available, the bound state radial wave function $u_\ell$(k,r) for the potential (11.24) takes the form

$$u_\ell(k,\,r) = \begin{cases} f_\ell^+(k,R)\,\chi_\ell(k,r) & \text{for } r \le R, \\ f_\ell^+(k,r)\,\chi_\ell(k,R) & \text{for } r \ge R, \end{cases} \tag{11.25}$$

where $\chi_\ell(k,r)$ is the harmonic oscillator wave function

$$\chi_\ell(k,\,r) = r^{\ell+1}\,\mathrm{e}^{-\frac{1}{2}\xi\,\frac{r^2}{R^2}}\,M(a,\,b,\,\xi\,\tfrac{r^2}{R^2}), \tag{11.26}$$

with the parameters

$$\begin{aligned} \xi &= \sqrt{\mu\alpha\,Z\,R}, \\ a &= \frac{1}{4}[2\ell + 3(1-\xi) - k^2\,R^2/\xi], \\ b &= \ell + \frac{3}{2}. \end{aligned}$$

Clearly, the wave function (11.25) is continuous at r=R and the additional necessary requirement that its derivative be continuous at r=R gives the sought for bound state condition

$$R\,\frac{f_\ell^{+\prime}(k,R)}{f_\ell^+(k,R)} = R\,\frac{\chi_\ell'(k,R)}{\chi_\ell(k,R)}. \tag{11.27}$$

We recall that the left hand side of (11.27) can be readily calculated by continued fraction summation. The right hand side of (11.27) is

$$R\,\frac{\chi_\ell'(k,R)}{\chi_\ell(k,R)} = \ell + 1 - \xi + 2\xi\frac{a}{b}\,\frac{M(a+1,\,b+1,\,\xi)}{M(a,\,b,\,\xi)}, \tag{11.28}$$

and the particular ratio of the two confluent hypergeometric functions occurring in (11.28) may be very efficiently computed with the aid of the well known [Jones *et al.* (1980)] continued fraction representation

$$\frac{M(a+1,\,b+1,\,x)}{M(a,\,b,\,x)} = \cfrac{b}{b - \cfrac{(b-a)x}{b+1+\cfrac{(a+1)x}{b+2-\cfrac{(b-a+1)x}{b+3+\cfrac{(a+2)x}{b+4-\cfrac{(b-a+2)x}{b+5+\cdots}}}}}}. \tag{11.29}$$

The bound state condition (11.27) is now amenable for computations and involves no special functions.

# Chapter 12

# Chiral Theory Primer

So far hadronic atom theory has been formulated at a purely phenomenological level, but now we are going to present more ambitious developments starting from first principles. In this approach as fundamental objects, mutually interacting spin-half quarks and spin-one gluons are introduced and current understanding of strong interaction phenomena will be based on a theory called Quantun Chromodynamics (QCD). Indeed, QCD is conceptually very satisfactory, it is renormalizable and has had much success in the high energy regime where the asymptotic freedom allows the use of perturbative methods. By contrast, low-energies correspond to the confinement regime and practical solutions are difficult to come by. There are at least three reasons for that. Firstly, the strong coupling constant then becomes large which precludes perturbative methods. Secondly, the theory is non-linear due to gluon self interaction. Thirdly, the QCD Lagrangian describes the interaction between quarks and gluons whereas the relevant degrees of freedom are the hadrons and their bound states.

To overcome these difficulties, instead of dealing with QCD directly, various QCD inspired models have been proposed and the most promising of them are based on the Effective Field Theory (EFT) concept. The very idea of EFT is born from of the observation that it is a common feature of many physical systems that their behavior dramatically simplifies when examined at low energies (or temperatures) compared to system's own characteristic scales. Quite often the number of states that can then participate is very small and their interactions get weaker and weaker as the energies tend to zero. Effective Lagrangians supply practical means to exploit the above simplicity. The guiding idea belongs to Weinberg, who noted that there is no loss of generality in using field theory to account for the low-energy behavior of any system. The argument is that field theory as such ensures merely the basic properties like unitarity, invariance under specific symmetry, cluster structure, *etc.* but apart from that has very little content. When all these basic features are implanted by construction to the Lagrangian, it is not surprising that the ensuing theory must provide a good description of the system in the low-energy limit. Accordingly, EFT have long proven to be a powerful tool in particle and condensed matter physics. The same may well be true with nuclear physics where phenomenological approaches have had a long and venerated history and seems that it is only a matter of time when the EFT methods will prevail in nuclear physics.

The crucial step in constructing EFT is the recognition of the different energy scales involved in the problem at hand because this permits to study the low-energy dynamics independently of the details of the high-energy interactions. Setting the appropriate energy scale $\Lambda$, EFT is then tailored in such a way that only the relevant degrees of freedom are accounted for so that states with $m << \Lambda$ only enter, while the heavier excitations with $M >> \Lambda$ are regarded as being integrated out from the action and do not appear explicitly in the effective theory. For an assigned value of $\Lambda$, this procedure gives rise to a sequence of non-renormalizable interactions and it is convenient to organize them as an expansion in powers of the energy divided by $\Lambda$. The coefficients, which take the form of low-energy coupling constants, present in this expansion contain the information concerning the heavier excitations. Although EFT contains an infinite number of terms but this does not upset the renormalizability of the theory because at a given order in the energy expansion the low-energy Lagrangian involves a finite number of couplings which allows an order-by-order renormalization.

Some people say that EFT provide a practical alternative to the Theory of Everything. Indeed, even if the latter was available it would be highly impractical to consider all the dynamical degrees of freedom always on equal footing. Therefore, in the low-energy phenomenology it might be more convenient to device EFT that mimics the fundamental theory and integrate out the heavy degrees of freedom that cannot be accessed directly in the considered energy range. In some cases the information about the structure at small distances can be transmitted from the fundamental theory by the "matching" conditions to the effective theory. When this is not possible, usually due to a breakdown of the perturbative expansion, this information enters by fitting the experimental data. In the case of non-perturbative QCD, expressed in terms of interacting quarks and gluons,indexgluon effective theories involving only hadrons rely entirely on the internal symmetries of QCD. The most prominent of them is known as Chiral Perturbation Theory (ChPT) because the key ingredient used in that construction is the spontaneously (and explicitly) broken chiral symmetry of the underlying QCD. This theory has its roots in the current algebra developed in the 1960s, and was founded by the pioneering works of Weinberg whose results were subsequently systematized and extended in the classical papers by Gasser and Leutwyler.

In hadronic atom theory we are dealing with energies very close to threshold and these systems are non-relativistic. Therefore, they are ideally suited for applying EFT methods. To demonstrate how EFT works in practice, it is instructive to examine a simple exactly soluble example and compare it with the standard quantum mechanical approach (we follow here a very lucid presentation given by [Holstein (1999)]). As we shall see, these two methods are completely equivalent: given a set of parameters specifying the interaction, we can calculate the low energy scattering parameters so that the latter can be subsequently used in the master equation to compute hadronic atom level shifts and widths. While this does not appear yet to be very exciting that two phenomenological approaches turn out to be equivalent, it has to be noted that ChPT can offer much more in that it provides means to calculate the input parameters. The shifts and widths may be computed from first principles in an absolute calculations. We shall illustrate this point by the example of the pionium atom.

## 12.1 Quantum Mechanics: Zero-Range Potential

In order to facilitate the introduction to the EFT methods, let us begin first with a very simple model where we have two particles A and B interacting via a zero-range potential. The strong interaction potential is adopted in the form

$$2\mu V(r) = -(v/R^2)\delta(r - R), \tag{12.1}$$

with two parameters: the coupling constant $v$ and a range $R$. The latter is assumed to be very small and eventually we wish to examine the limit $R \to 0$. We are here interested in the scattering problem where in $R \to 0$ limit only s-waves will be important and in the following the orbital momentum suffix will be suppressed. A good starting point is the Coulomb free case when the wave function $u(r)$ for $A + B$ scattering satisfies the integral

Fig. 12.1 The vertex corresponding to contact interaction.

equation

$$u(r) = \sin pr + 2\mu \int_0^\infty \langle r\,|g_0^{(+)}(p)|\,r'\rangle,\ V(r')\,u(r')\mathrm{d}r' \tag{12.2}$$

where the free propagation Green's function is

$$\langle r\,|g_0^{(+)}(p)|\,r'\rangle = -\mathrm{e}^{\mathrm{i}pr_>}\,\frac{\sin pr_<}{p}. \tag{12.3}$$

It is a trivial matter to obtain the solution of (12.2) and for $r \geq R$, we have

$$u(r) = \sin pr - (v/R^2)\,\langle r\,|g_0^{(+)}(p)|\,R\rangle\,u(R). \tag{12.4}$$

With (12.4) in hand we can calculate the scattering amplitude

$$f_s(p) = \frac{\mathrm{e}^{\mathrm{i}\delta}\sin\delta}{p} = -\frac{2\mu}{p}\int_0^\infty \sin pr\,V(r)\,u(r)\mathrm{d}r = \frac{v\,\sin pR}{p^2R^2}\,u(R) \tag{12.5}$$

and inserting $u(R)$ from (12.4), the scattering amplitude is

$$f_s(p) = \left(\frac{\sin pR}{pR}\right)^2 \frac{v}{1 + (v/R^2)\,\langle R|g_0^{(+)}(p)|\,R\rangle}. \tag{12.6}$$

Now, going to the $R \to 0$ limit, we can cast the scattering amplitude to the form

$$f_s(p) = \frac{v}{1 - v\,\Pi(p)} = v + v\,\Pi(p)\,v + v\,\Pi(p)\,v\,\Pi(p)\,v + \cdots, \tag{12.7}$$

where $\Pi(p)$ is the propagator taking particles $A, B$ from a zero separation initial state, to a zero separation final state, whose explicit form is

$$\Pi(p) = -\lim_{R\to 0}\frac{\langle R|g_0^{(+)}(p)|\,R\rangle}{R^2} = \frac{1}{R} + \mathrm{i}p + O(R). \tag{12.8}$$

This propagator is divergent and must be defined via some sort of regularization. The scattering amplitude (12.7) can be represented as an infinite sum of Feynman diagrams where the lowest order diagram is just the vertex

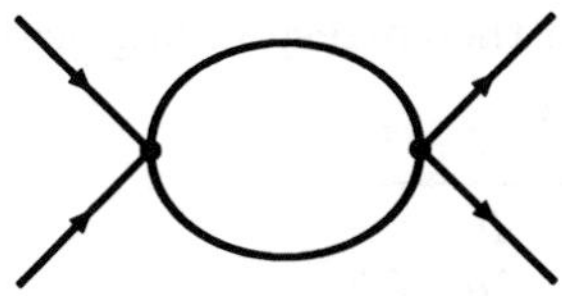

Fig. 12.2 One loop contribution to the scattering amplitude.

$v$ shown in Fig. 12.1. The propagator is visualized as a loop and the appropriate second order diagram is presented in Fig. 12.2. Finally, the complete scattering amplitude (12.7) given as a sum of a geometric series is displayed in Fig. 12.3. Using (12.8) in (12.7) we can calculate the scattering length

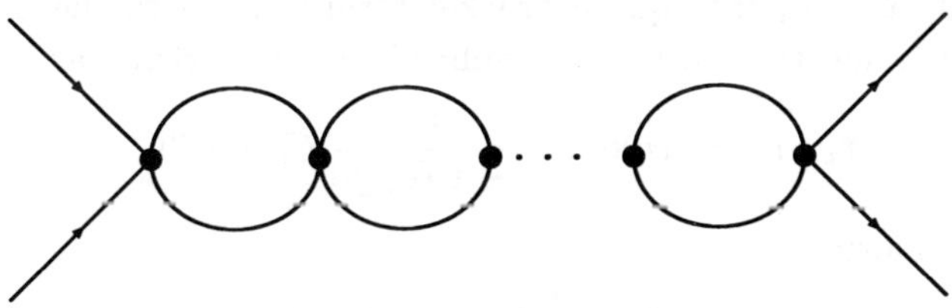

Fig. 12.3 Infinite sum of loop terms.

$$a = f_s(0) = \frac{v}{1 - v\,\Pi(0)}, \tag{12.9}$$

or

$$\frac{1}{a} = \frac{1}{v} - \Pi(0). \tag{12.10}$$

In the next step we include the Coulomb field. To this end, we introduce Coulomb distorted wave function $u_C(r)$ and instead of (12.2), we now have

$$u_C(r) = F_0(\eta, pr) + 2\mu \int_0^\infty \langle r\,|G_0^{(+)}(p)|\,r'\rangle\, V(r')\, u_C(r')\mathrm{d}r', \tag{12.11}$$

where the Coulomb Green's function is

$$\langle r\,|G_0^{(+)}(p)|\,r'\rangle = -H_0^+(\eta, pr_>)\,\frac{F_0(\eta, pr_<)}{p}. \tag{12.12}$$

The solution of eq. (12.11) for $r \geq R$, becomes

$$u_C(r) = F_0(\eta, pr) - (v/R^2)\,\langle r\,|G_0^{(+)}(p)|\,R\rangle\, u_C(R), \tag{12.13}$$

and can be used to obtain the corresponding scattering amplitude

$$\begin{aligned} f_{sc}(p) &= \mathrm{e}^{2\mathrm{i}\sigma_0} \frac{\mathrm{e}^{\mathrm{i}\delta^c} \sin \delta^c}{p} = \\ &= -\mathrm{e}^{2\mathrm{i}\sigma_0} \frac{2\mu}{p} \int_0^\infty F_0(\eta, pr)\, V(r)\, u_C(r) \mathrm{d}r, \end{aligned} \tag{12.14}$$

including all Coulomb effects. Given the explicit solution (12.13), formula (12.14) yields

$$f_{sc}(p) = \mathrm{e}^{2\mathrm{i}\sigma_0} \left[\frac{F_0(\eta, pR)}{pR}\right]^2 \frac{v}{1 + (v/R^2)\, \langle R | G_0^{(+)}(p) | R\rangle}. \tag{12.15}$$

In the limit $R \to 0$, the square bracket term reduces to the Sommerfeld factor $C_0(\eta)^2$, and the scattering amplitude can be written as

$$f_s(p) = C_0(\eta) \mathrm{e}^{\mathrm{i}\sigma_0} \frac{v}{1 - v\, \Pi_C(p)} C_0(\eta) \mathrm{e}^{\mathrm{i}\sigma_0}, \tag{12.16}$$

or in a series form

$$\begin{aligned} f_{sc}(p) = C_0(\eta) \mathrm{e}^{\mathrm{i}\sigma_0} &[v + v\, \Pi_C(p)\, v + \\ &+ v\, \Pi_C(p)\, v\, \Pi_C(p)\, v + \cdots ]\, C_0(\eta) \mathrm{e}^{\mathrm{i}\sigma_0}, \end{aligned} \tag{12.17}$$

where, similarly as before, we have introduced the Coulomb corrected propagator $\Pi_C(p)$

$$\begin{aligned} \Pi_C(p) &= - \lim_{R\to 0} \frac{\langle R | G_0^{(+)}(p) | R\rangle}{R^2} = \\ &= p \lim_{R \to 0} \frac{F_0(\eta, pR)}{pR} \frac{H_0^+(\eta, pR)}{pR}. \end{aligned} \tag{12.18}$$

It is apparent from (12.17) that there are Coulomb corrections to every

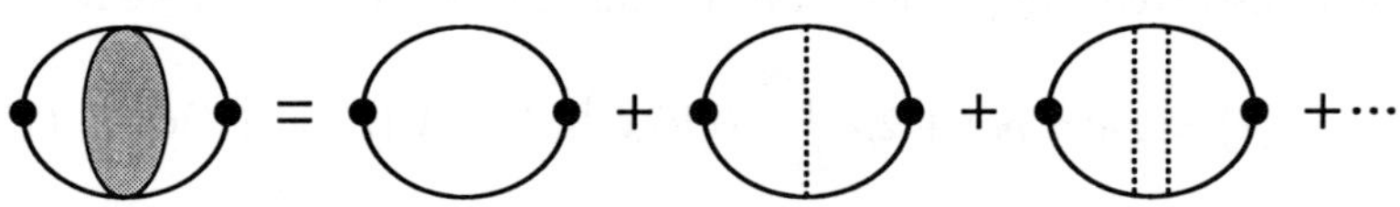

Fig. 12.4 Coulomb corrections to the propagator.

loop in the multiple scattering series and at the top of it there would be also initial and final state interaction caused by the Coulomb field. The loop correction has been shown in Fig. 12.4 where the propagator $\Pi_C(p)$ results from a summation of diagrams involving photon exchanges. Similar summation must be employed to account for the initial and the final state interaction, when the particles are not affected by the strong interaction, but are exposed to the Coulomb field. The diagrams contributing to the final state interaction are presented in Fig. 12.5 and an analogous expansion may be obtained for the initial state interaction. Similarly as in the case

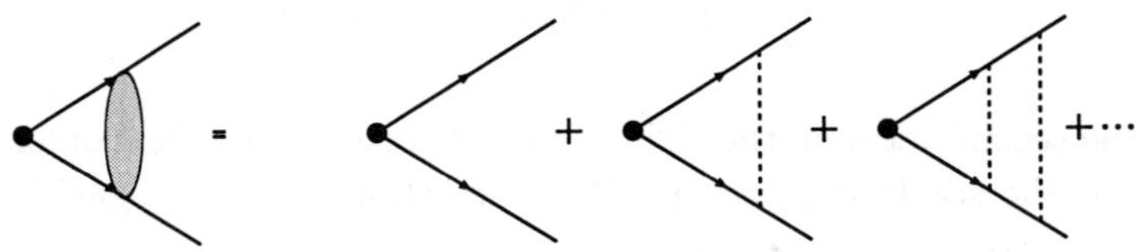

Fig. 12.5 Final state Coulomb interaction.

of neutral particle scattering, the Coulomb corrected propagator diverges for $R \to 0$. The nature of this divergence can be examined by using the appropriate expressions for the Coulomb wave functions valid for small $\rho$. Writing, (12.18) in the form

$$\Pi_C(p) = p \lim_{R\to 0} \frac{\Phi_0(p,R)}{pR} \frac{F_0^+(p,R)}{pRF_0^+(p)},$$

and using (3.5) and (3.13), in the limiting case of small $R$, we obtain

$$\Pi_C(p) = \frac{1}{R} + g_0^+(p) + 2p\eta\left(\ln\frac{2R}{a_B} + 2\gamma - 1\right) + O(R). \qquad (12.19)$$

This formula holds for both, attractive and repulsive Coulomb interaction. We wish to recall that the function $g_0^+(p)$ in absence of Coulomb field reduces to $ip$, in which case from (12.19) we immediately retrieve formula (12.8). However, now in addition to the previous $1/R$ singularity, formula (12.19) has also a logarithmic singularity of Coulomb origin. By contrast, Coulomb corrections associated with the initial and the final state interactions, are both finite. The complete multiple scattering series (12.16) can be now represented graphically in the form of an infinite sum of Coulomb corrected bubbles as displayed in Fig. 12.6. The initial and final state Coulomb interactions have been also shown there as the grey vertical ovals.

The bound states are identified with the poles of the scattering amplitude and to locate the poles it will be convenient to calculate the Coulomb corrected K-matrix. Using (12.16), we obtain

$$\frac{1}{\mathcal{K}(p)} = C_0(\eta)^2(p\cot\delta^c - \mathrm{i}p) + g_0^+(p) = \frac{1}{v} - \Pi_C(p) + g_0^{(+)}(p). \quad (12.20)$$

We note that by inserting (12.19) in (12.20), the function $g_0^{(+)}(p)$ is canceled. Hence, the resulting K-matrix does not depend upon $p$ anymore, and, in fact, what we eventually obtain is the Coulomb corrected scattering length. Accordingly, formula (12.20), gives

$$\frac{1}{a_c} = \frac{1}{a} + 2\mu\alpha Z\left(\ln\frac{2R}{a_B} + 2\gamma - 1\right), \tag{12.21}$$

where the second term on the right hand side constitutes the Coulomb correction to the scattering length. This correction diverges logarithmically but *all* the divergent terms have been accommodated in $a_c$ and therefore a possible way out is to regard $a_c$ instead of $v$ as a sole parameter characterizing the strength of the contact interaction. This model would be equivalent to adopting the view that for short ranged interactions the K-matrix near threshold is a slowly varying function of energy and can be replaced by a constant. In consequence, the bound state problem consists in solving the

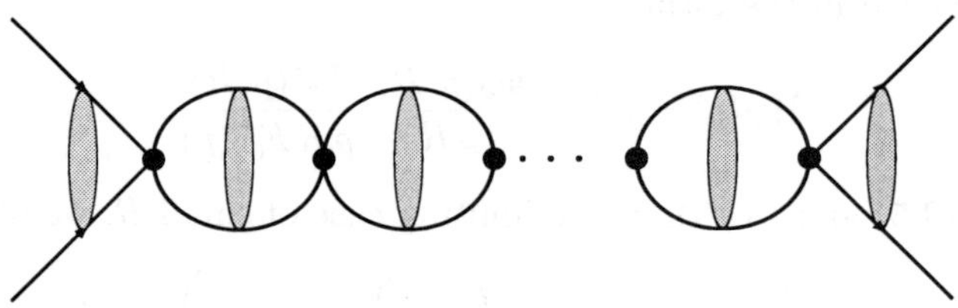

Fig. 12.6 Infinite sum of loops with Coulomb interaction.

familiar equation

$$g_0^+(k_n) = \frac{1}{a_c}, \qquad n = 1, 2, 3, ..., \tag{12.22}$$

which has already been discussed at length in Chapter 5. The solutions $k_n$ of eq. (12.22) yield the corresponding binding energies $E_n = k_n^2/2\mu$.

## 12.2 Effective Field Theory Approach

We turn now to the approach based on Effective Field Theory and our objective is to demonstrate its equivalence with quantum mechanics. Therefore, we are going to tackle the very same problem of $A + B$ scattering as before. It will be appropriate to consider first the situation when we have two chargeless particles $A, B$ and associated with them fields $\Psi_A, \Psi_B$, respectively. The particles are interacting only via a local strong interaction, so that the effective Lagrangian can be written as

$$\mathcal{L} = \sum_{i=A}^{B} \Psi_i^\dagger (i\frac{\partial}{\partial t} + \frac{\nabla^2}{2m_i})\Psi_i - c_0 \Psi_A^\dagger \Psi_A \Psi_B^\dagger \Psi_B + \dots \tag{12.23}$$

where $c_0$ is a coupling constant. The corresponding T-matrix $T_s(p)$ follows in result of a summation of an infinite sequence of Feynman diagrams shown in Fig. 12.3

$$T_s(p) = -\frac{4\pi}{2\mu} f_s(p) = c_0 + c_0^2 G_0(p) + c_0^3 G_0^2(p) + \dots = \frac{c_0}{1 - c_0\, G_0(p)} \tag{12.24}$$

where $G_0(p)$ is the amplitude for particles $A, B$ to travel from zero separation, to zero separation, whose explicit form is

$$G_0(p) = \lim_{\mathbf{r}',\mathbf{r} \to 0} 2\mu \int \frac{d^3q}{(2\pi)^3} \frac{e^{-i\mathbf{q}\cdot\mathbf{r}'} e^{i\mathbf{q}\cdot\mathbf{r}}}{\mathbf{p}^2 - \mathbf{q}^2 + i\epsilon} = -\frac{2\mu}{4\pi}\,\Pi(p), \tag{12.25}$$

obtained as the limiting value of the free Green's function considered in Chapter 4. We already know that the function $G_0(p)$ is divergent and to make it meaningful some sort of regularization is required. This can be achieved in a number of ways, but perhaps the simplest thing to do, is to introduce a cutoff regularization in the above integral with $q_{max} = \Lambda$. In this case, (12.25) gives

$$\Pi(p) = -\frac{2}{\pi} \int_0^\infty \frac{dq\, q^2}{p^2 - q^2 + i\epsilon} = \frac{2\Lambda}{\pi} + ip, \tag{12.26}$$

and when $2\Lambda/\pi$ is called $1/R$, we end up with the same formula as that obtained by partial wave expansion in (12.8). Actually, the linear divergence for $\Lambda \to \infty$ could be made harmless by introduction of an appropriate counter-term in the Lagrangian, which renormalizes $c_0$ to $c_0(\Lambda)$ in such a way that in the end all $\Lambda$ dependent terms are canceled. The scattering amplitude, is

$$f_s(p) = \frac{1}{[-4\pi/2\mu c_0(\Lambda)] - \Pi(p)}, \tag{12.27}$$

and going to $p \to 0$ limit, we identify the scattering length as

$$\frac{1}{a} = -\frac{4\pi}{2\mu\, c_0(\Lambda)} - \frac{2\Lambda}{\pi}, \tag{12.28}$$

and upon setting $v = -2\mu c_0(\Lambda)/(4\pi)$, this expression becomes identical with (12.10).

For charged particles $A, B$ (carrying charges, respectively, $e_A$ and $e_B$) we have essentially the same Lagrangian but in order to include the interaction with electromagnetic field the Lagrangian must be gauge invariant. For the time being, the Lagrangian (12.23) is only invariant under a global gauge transformation

$$\Psi_I' = \exp(-\mathrm{i}e_I\alpha)\Psi_I, \qquad I = A, B$$

where $\alpha$ is a real number. In order to have invariance under a local gauge transformation where $\alpha$ is a function of $(\boldsymbol{x}, t)$, one has to introduce the photon field $A_\mu$ in the standard manner, transforming as

$$A_\mu' = A_\mu + \partial_\mu \alpha.$$

To stipulate gauge invariance of the Lagrangian, the derivatives in (12.23) must be covariant, i.e. the following replacements need to be made

$$\begin{aligned} \nabla \Psi_I &\to (\nabla - \mathrm{i}e_I \boldsymbol{A})\Psi_I, &\qquad I = A, B \\ \mathrm{i}\frac{\partial}{\partial t}\Psi_I &\to (\mathrm{i}\frac{\partial}{\partial t} - \mathrm{i}e_I A_0)\Psi_I, &\qquad I = A, B. \end{aligned}$$

Since we are here interested in the low energy phenomena, the effects of transverse photons will be negligible, owing to the fact that their coupling is proportional to the velocities of particles $A, B$. Therefore, we shall concentrate our attention on the longitudinal photons. With the photon field present, there will be more diagrams for we have to include also diagrams in which static photons may be exchanged. The bubble sum must be extended to comprise diagrams with static photons exchanges between each of the lines, as has been already discussed in the preceding section. We also know that as a consequence of adding the diagrams with exchanged photons, the propagator will be modified and, at the top of that, both, the initial and final state Coulomb interaction will appear in the form of extra multiplicative factors. Including these corrections, the Coulomb modified scattering amplitude is

$$f_{cs}(p) = \mathrm{e}^{\mathrm{i}\sigma_0} C_0(\eta) \frac{1}{[-4\pi/2\mu c_0(\Lambda)] - \Pi_C(p)} \mathrm{e}^{\mathrm{i}\sigma_0} C_0(\eta), \qquad (12.29)$$

where $\Pi_C(p)$ is the Coulomb corrected propagator. The presence of Coulomb field complicates considerably the evaluation of the propagator, and for the repulsive Coulomb potential instead of (12.25), we have

$$\Pi_C(p) = -\lim_{\mathbf{r}',\mathbf{r}\to 0} 4\pi \int \frac{\mathrm{d}^3 q}{(2\pi)^3} \frac{\chi_{\mathbf{q}}^{(+)}(\mathbf{r}')\chi_{\mathbf{q}}^{(+)*}(\mathbf{r})}{\mathbf{p}^2 - \mathbf{q}^2 + \mathrm{i}\epsilon} = \\ = -\frac{2}{\pi}\int_0^\infty \frac{\mathrm{d}q\, q^2\, C_0^2(\eta_q)}{p^2 - q^2 + \mathrm{i}\epsilon}, \tag{12.30}$$

where $\eta_q = \mu e_A\, e_B/q > 0$. For attractive Coulomb interaction we have to add also the contribution from the discrete spectrum and for this case, the corresponding expression reads

$$\Pi_C(p) = -\frac{2}{\pi}\int_0^\infty \frac{\mathrm{d}q\, q^2\, C_0^2(-|\eta_q|)}{p^2 - q^2 + \mathrm{i}\epsilon} - \frac{4}{a_B}\sum_{j=1}^{\infty}\frac{\eta^2}{j\,(\eta^2 + j^2)}, \tag{12.31}$$

with $\eta = \mu e_A e_B/p$ and we wish to recall that now both $\eta$ and $\eta_q$ are negative. By using the identity

$$C_0^2(-|\eta_q|) = -C_0^2(|\eta_q|) + 2\pi\eta_q\, \coth \pi\eta_q$$

in the integrand of (12.31), the integral involving the hyperbolic cotangent function can be evaluated in the complex q-plane by contour method and results in a sum of the residues at the poles. This sum cancels exactly the contribution from the bound states, and the propagator is given as

$$\Pi_C(p) = \frac{2}{\pi}\int_0^\infty \frac{\mathrm{d}q\, q^2\, C_0^2(|\eta_q|)}{p^2 - q^2 + \mathrm{i}\epsilon} - 2\mathrm{i}p\, \pi\eta\, \coth \pi\eta, \tag{12.32}$$

so that now the same integral occurs in (12.32) and in (12.30). This integral diverges but by introducing a cut-off can be regularized and the final expression for the propagator, is

$$\Pi_C(p) = \frac{2\Lambda}{\pi} + g_0^+(p) - 2p\eta\,(\ln \Lambda a_B - \gamma), \tag{12.33}$$

valid for both, attractive and repulsive Coulomb interaction. The Coulomb propagator (12.33), in addition to linear divergence, has also a logarithmic singularity of Coulomb origin, in complete agreement with formula (12.19). Using (12.33) in (12.29), and introducing the Coulomb corrected K-matrix

$$f_{cs}(p) = \mathrm{e}^{\mathrm{i}\sigma_0}\, C_0(\eta)\, \frac{1}{1/\mathcal{K}(p) - g_0^+(p)}\, \mathrm{e}^{\mathrm{i}\sigma_0}\, C_0(\eta), \tag{12.34}$$

we find that the K-matrix is here just equal to the scattering length $\mathcal{K}(p) = a_c$ and the latter can be written as

$$\frac{1}{a_c} = \frac{1}{a} + 2p\eta\,(\ln \Lambda a_B - \gamma), \tag{12.35}$$

in good accord with the zero-range potential scattering result (12.21). The scattering amplitude (12.34) has poles, which may correspond to bound states, at $p = p_n$ such that $g_0^+(p_n) = 1/a_c$.

## 12.3 Chiral Perturbation Theory

Having assembled all the necessary tools we are now ready to present a relativistic application of the above techniques to the low-energy interactions of pions and nucleons. Actually, this example illustrates a very useful, and experimentally successful chapter of QCD in the low-energy limit. More importantly, this example is also meant to demonstrate that effective Lagrangian can remain predictive even if it is not possible to assign their effective couplings from an underlying theory.

Our starting point will be now at last QCD, i.e. a theory where spin-half quarks interact with spin-one gluons. The QCD Lagrangian density takes a very simple and elegant form

$$\mathcal{L}_{QCD} = -\frac{1}{4}\,G^a_{\mu\nu}G^{\mu\nu}_a - \sum_f \bar{q}_f(\not{D} + \mathcal{M})q_f, \tag{12.36}$$

where

$$G^a_{\mu\nu} = \partial_\mu G^a_\nu - \partial_\nu G^a_\mu + g f^a{}_{bc}\, G^b_\mu G^c_\nu$$

is the field strength tensor for the gluon fields, $G^a_\mu$. Here $a = 1, \ldots, 8$ labels the generators of $SU_c(3)$, and the subscript "c" stands for "color", which is the name given to the strong charge. This is the the gauge symmetry group of the theory, for which the $f^a{}_{bc}$ are the structure constants. The quarks are represented by Dirac spinors, $q_f$, where $f = 1, \cdots, 6$ enumerates the six currently known flavors of quarks. Their masses enter (12.36) through a six-by-six diagonal mass matrix $\mathcal{M}$ in flavor space. Arranged in order of increasing mass, the different quarks are: $u, d, s, c, b$ and $t$. To give some sort of feeling how much they differ, we list here their masses that fill the mass matrix $\mathcal{M}$, quoted after the current Particle Data Group review [Hagiwara *et al.* (2002)]:

$$
\begin{aligned}
m_u &= 1.5 - 4.5 \text{ MeV}, \\
m_d &= 5.0 - 8.5 \text{ MeV}, \\
m_s &= 80 - 155 \text{ MeV}, \\
m_c &= 1.0 - 1.4 \text{ GeV}, \\
m_b &= 4.0 - 4.5 \text{ GeV}, \\
m_t &= 174.3 \pm 5.1 \text{ GeV}.
\end{aligned}
$$

All of these quarks are assumed to transform in the three-dimensional representation of the color symmetry group, $SU_c(3)$, and so their covariant derivative, appearing in the combination $\not{D} = \gamma^\mu D_\mu$ in the Lagrangian, is: $D_\mu q_f = \partial_\mu q_f - \mathrm{i}\frac{1}{2}\, g\, G^a_\mu \lambda_a q_f$. The eight matrices, $\lambda_a$, denote the three-by-three Gell-Mann matrices, which act on the (suppressed) color index of each of the quarks. The explicit form for these matrices (not required in what follows) may be found in [Hagiwara *et al.* (2002)]. In all of these expressions $g$ represents the coupling constant whose value determines the magnitude of the quark-gluon and gluon-gluon couplings. The strong interactions as given by the above Lagrangian density are supposed to produce the binding of the quarks and gluons into hadronic bound states, that are observed experimentally, such as protons, neutrons, pions, kaons, *etc.* In Table 12.1 we have collected the masses and some of the quantum numbers for all of the lightest hadrons whose masses are below 1 GeV. They form a small fraction of states that have masses bigger than 1 GeV. The striking feature apparent in the presented above quark mass spectrum is that the two lightest quarks, $u$ and $d$, have significantly smaller mass than the remaining quarks. This observation suggests that it might be useful to study QCD dynamics in the limit of vanishing $(u, d)$ quark masses regarding their masses, divided by the scale $\Lambda$ typical for strong interactions, as small corrections to be treated perturbatively. The assumption that the $(u, d)$ quarks are massless brings considerable simplification because the QCD Lagrangian contains then no terms that would mix the right- and left-handed components of the $(u, d)$ quark fields

$$q_R = \tfrac{1}{2}(1 + \gamma_5)q; \qquad q_L = \tfrac{1}{2}(1 - \gamma_5)q.$$

The Dirac matrices $\frac{1}{2}(1 + \gamma_5)$ and $\frac{1}{2}(1 - \gamma_5)$ introduced above project, respectively, onto the left-handed and right-handed part of each of the quarks $(u, d)$. The QCD Lagrangian with massless $(u, d)$ quarks acquires a very

Table 12.1 Masses, Quantum Numbers and the quark content of the Lightest Hadrons (below 1 GeV).

| Particle | Quark Content | Mass (GeV) | Spin | Isospin |
|---|---|---|---|---|
| $\pi^-$, $\pi^+$, $\pi^0$ | $d\bar{u}$, $u\bar{d}$, $(u\bar{u}, d\bar{d})$ | 0.140, 0.135 | 0 | 1 |
| $K^+$, $K^0$ | $u\bar{s}$, $d\bar{s}$ | 0.494, 0.498 | 0 | $\frac{1}{2}$ |
| $K^-$, $\overline{K}^0$ | $s\bar{u}$, $s\bar{d}$ | 0.494, 0.498 | 0 | $\frac{1}{2}$ |
| $\eta$ | $u\bar{u}, d\bar{d}, s\bar{s}$ | 0.547 | 0 | 0 |
| $\rho^-$, $\rho^+$, $\rho^0$ | $d\bar{u}$, $u\bar{d}$, $(u\bar{u}, d\bar{d})$ | 0.770 | 1 | 1 |
| $\omega$ | $u\bar{u}, d\bar{d}, s\bar{s}$ | 0.783 | 1 | 0 |
| $K^{*+}$, $K^{*0}$ | $u\bar{s}$, $d\bar{s}$ | 0.892, 0.896 | 1 | $\frac{1}{2}$ |
| $K^{*-}$, $\overline{K}^{*0}$ | $s\bar{u}$, $s\bar{d}$ | 0.892, 0.896 | 1 | $\frac{1}{2}$ |
| $\eta'$ | $u\bar{u}, d\bar{d}, s\bar{s}$ | 0.958 | 0 | 0 |
| $f_0$ | $u\bar{u}, d\bar{d}, s\bar{s}$ | 0.980 | 0 | 0 |
| $a_0$ | $u\bar{u}, d\bar{d}, s\bar{s}$ | 0.985 | 0 | 1 |
| $p$, $n$ | $uud$, $ddu$ | 0.938, 0.940 | $\frac{1}{2}$ | $\frac{1}{2}$ |

useful symmetry: it becomes invariant under independent transformations of the right and left-handed quarks fields

$$q_R \to V_R\, q_R, \qquad q_L \to V_L\, q_L,$$

where $V_R, V_L$ are arbitrary two-by-two unitary matrices having unit determinant. The underlying group of symmetries thus obtained is $G = SU(2)_L \times SU(2)_R$, with the indices referring to the handedness of the quarks, on which the group transformations act. The symmetry which treats the left- and right-handed fermions differently is called a chiral symmetry. If the chiral symmetry was extended to an arbitrary number of flavors, the symmetry group would be $G = SU(N_f)_L \times SU(N_f)_R$ with $N_f$ denoting the number of flavors.

According to Noether theorem, the vector and axial vector currents associated with such symmetry of the Lagrangian, are

$$V_a^\mu = \bar{q}\gamma^\mu \tfrac{1}{2}\lambda_a q, \qquad A_a^\mu = \bar{q}\gamma^\mu\gamma_5 \tfrac{1}{2}\lambda_a q, \qquad a = 1, \dots, \mathrm{N}_f^2 - 1 \quad (12.37)$$

where the appropriate $SU(N_f)$ Gell-Mann matrices $\lambda_1$, $\lambda_2$, ... form a complete set of traceless, Hermitean $\mathrm{N}_f \times \mathrm{N}_f$ matrices. In addition to (12.36), there would be also their singlet counterparts

$$V_0^\mu = \bar{q}\gamma^\mu q, \qquad A_0^\mu = \bar{q}\gamma^\mu\gamma_5 q. \qquad (12.38)$$

The axial singlet current is anomalous and despite the symmetry of the Lagrangian $A_0^\mu$ is not conserved. Formally, the largest symmetry group one might get for massless quarks is therefore that generated by the conserved currents: $V_a^\mu, V_0^\mu$ and $A_a^\mu$ and comprises of pairs of $\mathrm{N}_f \times \mathrm{N}_f$ matrices, $(V_R, V_L)$, such that $\det(V_R V_L^{-1}) = 1$, i.e.

$$\mathrm{G}_{max} = SU(N_f)_R \times SU(N_f)_L \times U(1)_V.$$

As Nature would like to have it, the quark masses are non-zero and the presence of mass term in the QCD Lagrangian destroys chiral symmetry. Accordingly, the divergence of the currents will depend upon the quark masses

$$\partial_\mu V_a^\mu = \tfrac{1}{2}\mathrm{i}\,\bar{q}(\mathcal{M}\lambda_a - \lambda_a\mathcal{M})q. \tag{12.39}$$

We have already seen that no two quark masses are the same, which means that only the diagonal vector currents can be conserved: $\bar{u}\gamma^\mu u$, $\bar{d}\gamma^\mu d, \ldots$ and there will be $\mathrm{N}_f$ exactly conserved charges – one *per* each quark. flavor Taking appropriate linear combinations of those, allows to retrieve the exact conservation of baryon number, electric charge, strangeness, charm, *etc.* The ultimate exact symmetry group of real QCD is then the subgroup generated by the diagonal vector currents

$$\mathrm{G}_{QCD} = \mathrm{U}(1)^{N_f} \subset \mathrm{G}_{max}.$$

To proceed further in the quest of a manageable theory based on QCD, approximate symmetries have to be considered. Since the u quark is the lightest one, the simplest possibility is to set its mass equal to zero keeping the remaining masses at their physical values. The assumption $m_u \to 0$ makes the QCD Lagrangian invariant with respect to the chiral transformation $u \to \exp(\mathrm{i}\omega\gamma_5)$ where $\omega$ is a constant but the new symmetry of the Lagrangian does not enlarge the symmetry group. The axial anomaly is here the culprit and all the same the axial current $\bar{u}\gamma^\mu\gamma_5 u$ is not conserved.

Since setting a single quark mass equal to zero does not buy us anything, one may try making the two lightest quarks degenerate, assuming that their masses do not necessarily vanish but are taken to be equal. Although this assumption enlarges as expected the symmetry group which for $m_u = m_d$ becomes $SU(2)_R \times SU(2)_L \times U(1)^{N_f}$ but at the same time it leads a new difficulty. Indeed, if the latter symmetry was to be realized in the conventional way, as a consequence we would have to see nearly degenerate opposite parity doublets. They would be generated by the action of the conserved axial charge $Q_5$ which commutes with the Hamiltonian $H$. Every eigenstate of the Hamiltonian $|P\rangle$ with eigenvalue $E_P$ would have an opposite parity partner $Q_5|P\rangle$, but $H(Q_5|P\rangle) = Q_5(H|P\rangle) = E_P(Q_5|P\rangle$, so that these two states are necessarily degenerate. The existence of the predicted opposite parity degenerate doublets has not been confirmed by experiment even if allowance is given to the fact that the symmetry is only approximate so that degeneracy in practice means small level splitting between the opposite parity states. In the case of the nucleon with $J^P = \frac{1}{2}^+$ and mass about 0.94 GeV the closest state with $J^P = \frac{1}{2}^-$ lies almost 0.6 GeV higher in energy and they are far from forming approximately degenerate doublet. Similar pattern is observed for pions where the opposite parity partner is separated by an energy gap of the same size.

The resolution of the above difficulty has been achieved by postulating that the axial symmetry is spontaneously broken. In general terms, we speak about spontaneously broken symmetry whenever in a field theory the Hamiltonian H is invariant under some Lie group G but the the ground state is not. Denoting the generators of the group G as $Q_i$, the invariance under G means that

$$[Q_i, H] = 0$$

for all i. When the symmetry is spontaneously broken, the action of some of the generators on the ground state $|0\rangle$ gives a non-zero result, i.e.

$$Q_i|0\rangle \neq 0,$$

for at least one of the generators $Q_i$. Since the Hamiltonian commutes with all generators of the group G, we have

$$H\, Q_i\,|0\rangle = Q_i\, H|0\rangle = 0.$$

irrespectively whether the ground state respects the symmetry or not. If it does, the above equality follows trivially from the fact that $Q_i\,|0\rangle = 0$. However, if the symmetry is spontaneously broken and for some i $Q_i|0\rangle$ does not vanish, then this equation implies that all the states $(Q_i\,|0\rangle$ have zero energy. We conclude that in such a case we have states of zero energy, one state for each broken generators. In relativistic systems zero-energy state corresponds to a massless particle so as a consequence of spontaneously broken symmetry zero mass particles have to be generated. This is the essence of the Goldstone theorem and these massless particles are called Goldstone bosons.

The generators which leave the ground state invariant form a subalgebra: if both $Q_i$ and $Q_j$ annihilate the vacuum it is also the case for their commutator. This yields a subgroup $H \subset G$. If the symmetry is spontaneously broken we need to distinguish the two groups, one is the symmetry group of the Hamiltonian, G, and the second is the symmetry group of the vacuum H. When the groups G and H involve, respectively, $n_G$ and $n_H$ generators then for $n_G > n_H$ there are $n_G - n_H$ generators which belong to the quotient G/H of the two groups and which do not annihilate the ground state. The spectrum of the system contains $n_G - n_H$ independent states of zero-energy.

The above considerations remain useful also if the symmetry G is only approximate. In such a case the Goldstone particles become pseudo-Goldstone bosons. They are systematically light and weakly coupled at low-energies.

As mentioned above, the hypothesis that the chiral symmetry $G = SU(2)_L \times SU(2)_R \times U(1)^{N_f-1}$ is spontaneously broken alleviates the parity-doubling problem. In the limit where the two lightest quarks are made massless only spontaneous symmetry breaking is at work and the ground state of QCD is invariant under the subgroup $H = SU(2)_I \times U(1)^{N_f-1}$, such that $G/H = SU(2)_I$. Isospin symmetry can be understood at the quark level to consist of the diagonal subgroup of G, for which the left-handed and right-handed components of the (u,d) quarks are transformed the same way. Given this symmetry-breaking pattern, the spectrum must include three Goldstone bosons, with the quantum numbers of $(\pi^+, \pi^0, \pi^-)$. Indeed, the experimental fact that the pions are abnormally light when compared with the remaining particles makes them ideal candidates to be interpreted as the pseudo-Goldstone bosons.

The above scheme may be easily extended to incorporate into QCD the $SU(3)_V$ symmetry known as the eightfold way. The latter symmetry helped to group the observed mesonic and baryonic states into multiplets of $SU(3)_V$, such as singlets, octets, decouplets, *etc.* In QCD the eightfold way symmetry is obtained under the assumption that the masses of the (u,d,s) quarks vanish but the resulting chiral symmetry $G = SU(3)_L \times SU(3)_R \times U(1)^{N_f-2}$ is spontaneously broken. This mechanism reduces the symmetry group down to $H = SU(3)_V \times U(1)^{N_f-2}$ and generates an octet of massless pseudoscalar Goldstone bosons.

In fact, the chiral symmetry is broken not only spontaneously but also explicitly when the quark masses are assuming their physical values. Therefore, the pseudoscalar Goldstone bosons are no longer required to be massless but since their masses arise only on account of explicit symmetry breaking the pseudo-Goldstone meson masses will be proportional to the three light quark masses. The latter masses may be regarded as small symmetry breaking parameters.

The existence of a set of the pseudo-Goldstone bosons, much lighter than the other hadrons indicates that in the low-energy regime it might be convenient to replace the complicated structure of quark and gluon fields present in the QCD Lagrangian by an effective theory describing merely the degrees of freedom connected with the Goldstone bosons and incorporating all symmetries of the underlying QCD.

It is convenient to introduce the Goldstone boson fields as "angular" variables making them dimensionless. In the three-flavor sector $N_f = 3$ the pseudoscalar fields may be collected in a three-by-three unitary matrix U

$$U = \exp(i\sqrt{2}\Phi/F), \quad \Phi = \begin{pmatrix} \frac{\pi^0}{\sqrt{2}} + \frac{\eta_8}{\sqrt{6}} & \pi^+ & K^+ \\ \pi^- & -\frac{\pi^0}{\sqrt{2}} + \frac{\eta_8}{\sqrt{6}} & K^0 \\ K^- & \bar{K}^0 & -\frac{2\eta_8}{\sqrt{6}} \end{pmatrix} \tag{12.40}$$

but to see how it works in practice it will be sufficient to confine our attention to the two-flavor case $N_f = 2$ where U becomes a two-by-two unitary matrix

$$U = \exp\left\{\mathrm{i}\frac{\boldsymbol{\tau}\cdot\boldsymbol{\pi}}{F}\right\}, \quad \boldsymbol{\tau}\cdot\boldsymbol{\pi} = \begin{pmatrix} \pi^0 & \sqrt{2}\,\pi^+ \\ \sqrt{2}\,\pi^- & -\pi^0 \end{pmatrix}. \tag{12.41}$$

The boldfaced symbols represent vectors in isospin space: $\boldsymbol{\pi}$ denotes the isovector pion field and $\boldsymbol{\tau}$ is Pauli isospin vector. The constant F has been introduced for later convenience. Introducing chiral transformations for the left-handed and right-handed quark fields in the form

$$\psi_L \to L\,\psi_L; \qquad \psi_R \to R\,\psi_R,$$

the corresponding transformation of the field U, is

$$U \to L\,U\,R^\dagger.$$

We wish to construct the most general effective Lagrangian which is a function of U and $\partial_\mu$U that would be real, a Lorenz scalar and also invariant under the above transformation. Leaving out the derivatives for a moment, such Lagrangian would be of the form $Tr(\mathrm{UU}^\dagger \ldots \mathrm{UU}^\dagger)$, i.e. the trace of a product of interlacing U and $\mathrm{U}^\dagger$. However, since U is unitary, the condition $\mathrm{UU}^\dagger$=1 reduces all these terms to a constant and therefore a non-trivial Lagrangian does not exist without derivatives. This fact has profound consequences because in the limit of vanishing pion momenta the Goldstone bosons become decoupled and therefore their interactions are bound to be weak at low-energies. Terms involving higher order derivatives are therefore expected to be suppressed by powers od momenta and for low energy less important. Since the Goldstone bosons are pseudoscalar particles, parity conservation requires that only even number of derivatives are allowed in the Lagrangian. We infer then that the lowest order term in the partial derivatives can be written in the form manifestly invariant under chiral rotations

$$Tr(\partial^\mu U\, \partial_\mu U^\dagger) \rightarrow Tr(L\partial^\mu U\, R^\dagger\, R\, \partial_\mu U^\dagger\, L^\dagger) = Tr(\partial^\mu U\, \partial_\mu U^\dagger).$$

This expression can be used in the effective chiral Lagrangian and at two-derivative order this Lagrangian is also unique. We infer that the lowest order in field derivatives chiral Lagrangian can be written as

$$\mathcal{L}_2 = \frac{F^2}{4}\, Tr(\partial^\mu U\, \partial_\mu U^\dagger), \tag{12.42}$$

where the subscript 2 indicates two-derivative order. The field theory characterized by this Lagrangian is sometimes refeed to as the non-linear $\sigma$ model. The coefficient in front of the two-derivative term (12.42) and the scaling of the pion field (12.41) have been fixed by requiring that the kinetic term for the pions in the effective Lagrangian has the standard normalization for scalar field. Expanding the matrix U in powers of the pion field, we obtain

$$U = 1 + \mathrm{i}\frac{(\boldsymbol{\tau}\cdot\boldsymbol{\pi})}{F} - \tfrac{1}{2}\frac{(\boldsymbol{\tau}\cdot\boldsymbol{\pi})(\boldsymbol{\tau}\cdot\boldsymbol{\pi})}{F^2} + \cdots, \tag{12.43}$$

and the spatial derivative of the above expression is given as

$$\partial^\mu U = \mathrm{i}\frac{(\boldsymbol{\tau}\cdot\partial^\mu\boldsymbol{\pi})}{F} - \frac{(\boldsymbol{\tau}\cdot\partial^\mu\boldsymbol{\pi})(\boldsymbol{\tau}\cdot\boldsymbol{\pi})}{2\,F^2} - \frac{(\boldsymbol{\tau}\cdot\boldsymbol{\pi})(\boldsymbol{\tau}\cdot\partial^\mu\boldsymbol{\pi})}{2\,F^2} + \cdots. \tag{12.44}$$

Writing similar expansion for $\partial_\mu U^\dagger$, we can express the chiral Lagrangian (12.42) in terms of pion field

$$\mathcal{L}_2 = \tfrac{1}{2}\,\partial_\mu\boldsymbol{\pi}\cdot\partial^\mu\boldsymbol{\pi}(1 - \tfrac{1}{2}\,\frac{\boldsymbol{\pi}\cdot\boldsymbol{\pi}}{F^2} + \cdots), \tag{12.45}$$

where the lowest order term takes the familiar form of the free pion Lagrangian.

The Noether currents associated with the chiral SU(2)×SU(2) symmetry of the Lagrangian (12.42) can be also written in terms of the pion field. The vector ($\boldsymbol{V}_\mu$) and axial vector ($\boldsymbol{A}_\mu$) currents, are

$$\boldsymbol{V}_\mu = \mathrm{i}\,\frac{F^2}{4}\,Tr(\boldsymbol{\tau}[\partial_\mu U,\, U^\dagger]); \tag{12.46a}$$

$$\boldsymbol{A}_\mu = \mathrm{i}\,\frac{F^2}{4}\,Tr(\boldsymbol{\tau}\{\partial_\mu U,\, U^\dagger\}), \tag{12.46b}$$

where square bracket denotes a commutator and, respectively, a curly bracket an anticommutator. Using (12.43) and (12.44), the currents (12.46) can be expressed in terms of the pion field

$$\boldsymbol{V}_\mu = \boldsymbol{\pi} \times \partial_\mu \boldsymbol{\pi} + \cdots\,, \tag{12.47a}$$

$$\boldsymbol{A}_\mu = -F\,\partial_\mu \boldsymbol{\pi} + \cdots\,. \tag{12.47b}$$

The axial current depends upon the coupling constant F which gives the possibility to pin down its value. Indeed, from (12.47b) we have

$$A^1_\mu - \mathrm{i}\,A^2_\mu = -F\,\partial_\mu(\pi^1 - \mathrm{i}\,\pi^2) = -\sqrt{2}\,F\,\partial_\mu \pi^-$$

which gives

$$\langle 0|A^1_\mu - \mathrm{i}\,A^2_\mu|\pi^+(p)\rangle = \mathrm{i}\,\sqrt{2}\,F\,p_\mu. \tag{12.48}$$

The expression on the left hand side of (12.48) can be recognized as the familiar strong interaction factor present in the $\pi^+$ decay matrix element, and therefore the coupling constant F is identified as the pion decay constant $F_\pi$. The latter can be obtain from the experimentally known pion decay rate and we have $F=F_\pi= 92.4$ MeV.

To make the effective Lagrangian realistic a term reflecting explicit chiral symmetry breaking of QCD must be included. Such term is indispensable because without it the Goldstone bosons would be massless. Staying in the two-flavor sector, the mass term in QCD $\bar{q}\mathcal{M}q$ where the matrix $\mathcal{M}$ can be written as

$$\mathcal{M} = \begin{pmatrix} m_u & 0 \\ 0 & m_d \end{pmatrix} = \tfrac{1}{2}(m_u + m_d) + \tfrac{1}{2}(m_u - m_d)\,\tau_3 \tag{12.49}$$

mixes the right- and the left-handed components of the quark field and the symmetry $\mathrm{SU}_L(2)\times\mathrm{SU}(2)_R$ is violated. The first diagonal term in (12.49) is still isospin invariant whereas the term proportional to the difference of the quark masses contains the isospin breaking operator $\tau_3$. The structure of the mass term in QCD is such that the chiral symmetry would have been restored only if $\mathcal{M}$ had transformed as

$$\mathcal{M} \to L\mathcal{M}R^\dagger$$

and this property must be mimicked by the effective theory. The building blocks in constructing the effective Lagrangian are now $U, \partial U, \partial^2 U, \cdots, \mathcal{M}$ and this Lagrangian must be invariant under the transformation $U \to LUR^\dagger$ and $\mathcal{M} \to L\mathcal{M}R^\dagger$ — exactly as its QCD prototype. Since the pseudoscalar mesons are very light, the mass correction is expected to be small and the Lagrangian can be expanded in powers of $\mathcal{M}$. The first term of this expansion would be the chiral Lagrangian of the massless theory invariant under $\mathrm{SU}_L(2)\times\mathrm{SU}(2)_R$. The term linear in $\mathcal{M}$ may be a function of U and the derivatives of U but the latter are suppressed by powers of the momenta. If the Lagrangian is viewed as an expansion in both, the powers of $\mathcal{M}$ and powers of derivatives, then the leading term linear in $\mathcal{M}$ contains no derivatives of U. The requirement that the Lagrangian must be invariant under simultaneous chiral transformation of both, U and $\mathcal{M}$, reduces the infinite number of possible forms to just two: $\mathrm{Tr}(\mathcal{M}U^\dagger)$ and $\mathrm{Tr}(\mathcal{M}^\dagger U)$. The leading symmetry breaking term of the effective Lagrangian may contain an a linear combination of them but parity conservation introduces additional restriction, namely that a change of $\pi$ into $-\pi$ should leave the Lagrangian intact. This property translates into $\mathrm{U}\leftrightarrow\mathrm{U}^\dagger$ symmetry of the effective theory and therefore the leading chiral symmetry breaking term of the Lagrangian is

$$\mathcal{L}_{sb} = \tfrac{1}{2}F^2\, B\, Tr[\mathcal{M}(U + U^\dagger)], \tag{12.50}$$

where B is a new low-energy constant. Since the sum $\mathrm{U}+\mathrm{U}^\dagger$ contains only even powers of $\boldsymbol{\tau}$, after taking the trace in (12.50) the term proportional to $\tau_3$ gives no contribution and in the lowest order of the chiral symmetry breaking the effective Lagrangian respects isospin symmetry. To establish the value of the low-energy constant B, the symmetry breaking Lagrangian is expanded in powers of the pion field

$$\mathcal{L}_{sb} = (m_u + m_d)\, B \left[F^2 - \tfrac{1}{2}(\boldsymbol{\pi}\cdot\boldsymbol{\pi}) + \frac{(\boldsymbol{\pi}\cdot\boldsymbol{\pi})^2}{24F^2} + \cdots\right]. \tag{12.51}$$

The first term contributes only to the vacuum energy generated by the symmetry breaking, the second term is quadratic in the pion field and therefore can be identified with the mass term of the conventional pion

Lagrangian. The last term quadratic in $\boldsymbol{\pi}\cdot\boldsymbol{\pi}$ induces corrections to the pion-pion interaction associated with chiral symmetry breaking. The low-energy constant B may be eliminated in favor of the pion mass

$$m_\pi^2 = (m_u + m_d)B[1 + O(\mathcal{M})], \tag{12.52}$$

whose value cannot be calculated from ChPT and this number must be taken from experiment. This is the price to be paid for replacing QCD with its full complexity by a perfectly tractable effective theory that involves only hadronic degrees of freedom.

Differentiating the QCD Hamiltonian with respect to the quark mass $m_u$ gives $\bar{u}u$ and the vacuum expectation value of the latter operator gives the quark condensate $\langle 0|\bar{u}u|0\rangle$. These steps may be repeated also in the effective theory by taking a derivative of the vacuum energy. Calculating this derivative from (12.51) and subsequently taking a vacuum expectation value, by invoking the equivalence of the two theories one obtains

$$\langle 0|\bar{u}u|0\rangle = \langle 0|\bar{d}d|0\rangle = -F^2 B[1 + O(\mathcal{M})]. \tag{12.53}$$

When the constant B is eliminated between (12.52) and (12.53), we arrive at the familiar Gell-Mann, Oakes and Renner formula

$$F^2 m_\pi^2 = (m_u + m_d)\langle 0|\bar{u}u|0\rangle, \tag{12.54}$$

which shows that the origin of the pion mass is twofold: it results not only from the explicit chiral symmetry breaking via the non-zero quark mass but is also acquired from the spontaneous symmetry breaking owing to the fact that the quark condensate does not vanish.

The lowest order Lagrangian is now completely determined and may be used for calculating concrete physical processes. Expanding the Lagrangian $\mathcal{L}_2$ to fourth order in the pion field, we obtain

$$\begin{aligned}\mathcal{L}_2 = \tfrac{1}{2}[(\partial^\mu\boldsymbol{\pi}\cdot\partial_\mu\boldsymbol{\pi}) - m_\pi^2(\boldsymbol{\pi}\cdot\boldsymbol{\pi})] + \frac{m_\pi^2}{24F^2}(\boldsymbol{\pi}\cdot\boldsymbol{\pi})(\boldsymbol{\pi}\cdot\boldsymbol{\pi}) +\\ + \frac{1}{6F^2}[(\boldsymbol{\pi}\cdot\partial^\mu\boldsymbol{\pi})(\boldsymbol{\pi}\cdot\partial_\mu\boldsymbol{\pi}) - (\boldsymbol{\pi}\cdot\boldsymbol{\pi})(\partial^\mu\boldsymbol{\pi}\cdot\partial_\mu\boldsymbol{\pi})] + \cdots\end{aligned} \tag{12.55}$$

One of the most successful application of ChPT is the calculation of the $\pi\pi$ low-energy scattering parameters. To see how it works in practice we shall evaluate the $\pi\pi$ scattering amplitude at the tree level. Using the Lagrangian (12.55), it is a straightforward matter to calculate the three tree diagrams and the result is

$$\begin{aligned}&\langle \pi^\gamma(q_c)\pi^\delta(q_d)|T|\pi^\alpha(q_a)\pi^\beta(q_b)\rangle = \\ &= \frac{1}{F^2}[\delta^{\alpha\beta}\delta^{\gamma\delta}(s-m_\pi^2) + \delta^{\alpha\gamma}\delta^{\beta\delta}(t-m_\pi^2) + \delta^{\alpha\delta}\delta^{\gamma\beta}(u-m_\pi^2)],\end{aligned} \tag{12.56}$$

where s, t, u are the standard Mandelstam invariants. When isospin is conserved, a single function A(s, t, u) determines completely the $\pi\pi$ scattering amplitude and by permuting its arguments it is possible to account for the crossing symmetry. The relevant formula is

$$\begin{aligned}&\langle \pi^\gamma(q_c)\pi^\delta(q_d)|T|\pi^\alpha(q_a)\pi^\beta(q_b)\rangle = \\ &\quad = \delta^{\alpha\beta}\delta^{\gamma\delta}A(s,t,u) + \delta^{\alpha\gamma}\delta^{\beta\delta}A(t,s,u) + \delta^{\alpha\delta}\delta^{\gamma\beta}A(u,t,s),\end{aligned} \tag{12.57}$$

and by comparing (12.57) and (12.56) it is evident that the tree level result reads

$$A(s,t,u) = (s-m_\pi^2)/F^2. \tag{12.58}$$

It is more convenient to introduce the scattering amplitudes with assigned total isospin value I (where I = 0, 1, 2) and these are obtained as appropriate linear combinations

$$T^0(s,t,u) = 3A(s,t,u) + A(t,s,u) + A(u,t,s), \tag{12.59a}$$
$$T^1(s,t,u) = A(t,s,u) - A(u,t,s), \tag{12.59b}$$
$$T^2(s,t,u) = A(t,s,u) + A(u,t,s). \tag{12.59c}$$

In the c.m. system $s = 4(k^2+m_\pi^2), t = -2k^2(1-\cos\theta)$ and $u = -2k^2(1+\cos\theta)$, where k is the three-momentum and $\theta$ is the scattering angle and the partial wave amplitudes $t_\ell^I(s)$ together with the associated phase shifts $\delta_\ell^I$ are obtained from (12.59) by effecting partial wave projection

$$\begin{aligned}t_\ell^I(s) &= \sqrt{\frac{s}{s-4m_\pi^2}}\, \mathrm{e}^{\mathrm{i}\delta_\ell^I}\sin\delta_\ell^I = \\ &= \frac{1}{64\pi\sqrt{s}}\int_{-1}^{1}\mathrm{d}\cos\theta\, P_\ell(\cos\theta)\, T^I(s,t,u),\end{aligned} \tag{12.60}$$

and in the elastic region $4m_\pi^2 \le s \le 16m_\pi^2$ these phase shifts are real. Using (12.59) and (12.58) in (12.60), one obtains the phase shift at the tree level and going to the zero-momentum limit, the above formula yields the values of the $\pi\pi$ scattering length. Denoting the scattering length as $a_\ell^I$, one arrives at the remarkable result

$$a_0^0 = \frac{7m_\pi^2}{32\pi F^2}; \quad a_0^2 = -\frac{m_pi^2}{16\pi F^2}; \quad a_1^1 = -\frac{m_\pi^2}{24\pi F^2}, \tag{12.61}$$

first derived by Weinberg by current algebra methods.

On the other hand, this result certainly is not the whole story and in fact the calculated scattering amplitude is not satisfactory because it violates unitarity. As can be seen from (12.60), the partial wave amplitudes are bound to satisfy the unitarity constraint

$$\mathrm{Im}\, t_\ell^I(s) \geq \sqrt{\frac{s}{s-4m_\pi^2}}|t_\ell^I(s)|^2, \tag{12.62}$$

but at the tree-level the imaginary part of the $\pi\pi$ scattering amplitude derived from a Hermitean Lagrangian must necessarily vanish.

The recipe to alleviate this difficulty is well known: the imaginary part required to restore unitarity will appear as soon as the calculation is extended beyond the tree level by including loop corrections. However, loop amplitudes are in general divergent and therefore regularization and renormalization procedures turn out to be indispensable ingredients of the theory. The former allows to extract a finite part from a divergent integral while the latter ensures that the final result is independent upon the particular regularization procedure. Divergent integrals are quite common in field theories and in particular they occur in quantum electrodynamics where the diverging pieces can be all gathered and accommodated in two physical quantities namely, in the charge and in the mass. By contrast, theories like ChPT are classified as non-renormalizable because they generate interactions that are highly singular at short distances. As a result, the loop integrals are divergent in such a way that the calculation of multi-loop integrals leads to divergences with more and more complicated structures. In order to neutralize the infinities of the theory new sets of data are needed at each order to absorb the infinite bits to render the result finite. Since the number of input constants increases faster than the number of predictions, i.e. higher-order corrections to the scattering amplitude, non-renormalizable theories have been avoided, as much as possible.

The $\pi\pi$ scattering amplitude at order $\mathrm{p}^4$ receives contribution from the loop diagram resulting from the two-derivative Lagrangian but there will be also the tree diagram induced by the four-derivative term of the $\mathcal{L}_4$ Lagrangian, as shown in Fig. 12.7. In qualitative terms, the contribution

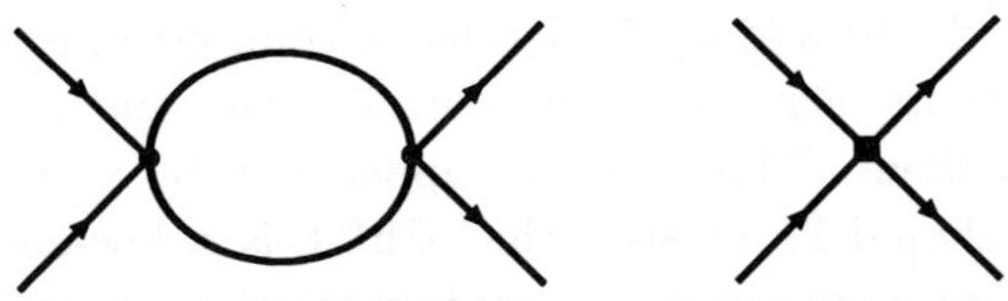

Fig. 12.7 The leading contributions of the order $p^4$ to the scattering amplitude: one-loop diagram from $\mathcal{L}_2$ and tree diagram from $\mathcal{L}_4$.

of these two diagrams to the scattering amplitude can be written as

$$p^4[A \log \frac{p^2}{\mu^2} + L(\mu)] \tag{12.63}$$

where the log term comes from the loop diagram while the constant term $L(\mu)$ results from the tree-diagram (cf. Fig. 12.7). The constant $\mu$ represents a cut-off and in the limit $\mu \to \infty$ the loop integral diverges. As there are no unknown parameters in $\mathcal{L}_2$ Lagrangian, it is possible to work out the coefficient A but this is not needed here. Although the regularized result is finite but in general it depends upon the unknown cut-off $\mu$. This dependence might disappear, however, if we insist that the derivative of the expression (12.63) with respect to $\mu$ vanishes. The constant L is made thereby $\mu$ dependent, or becomes renormalized, so that by construction the $\mu$ dependent terms are cancelled when the contributions of the two diagrams presented in Fig. 12.7 are added. This is the essence of renormalization where it is understood that the renormalized constant $L(\mu)$ is to be supplied from outside, e.g. from experiment.

In the actual practice the situation is more complicated but the idea is the same: in order to absorb the divergences from the loop integral at two-derivative level we need counterterms and those must be four-derivative in character. Gaser and Leutwyler have examined this problem devising the most general form for such an order four counterterm which in the two-flavor case contained eight low-energy constants [Gasser *et al.* (1984); Gasser *et al.* (1984)]. When Gasser and Leutwyler adjusted the empirical values of these constants from various experiments, ChPT became predictive at four-derivative level.

On the other hand, when one wishes to calculate two-loop diagrams from $\mathcal{L}_2$, or one-loop diagram from $\mathcal{L}_4$, then the above scheme is insufficient because new divergences will arise which are of six-derivative character. Although the renormalization scheme in principle could be extended beyond four-derivative order by adding six-derivative counterterms but the number of coupling constants increases at an alarming rate and $\mathcal{L}_6$ contains more than hundred of them. This certainly appears to be a large number but at this point it should be recalled that ChPT is a low-energy procedure involving an expansion in energy-momentum which means that graphs with more loops and/or more complicated vertices carry higher powers of the small expansion parameter $(p, m_\pi)$ and so are suppressed relative to tree-level and low-loop order diagrams. This is true, however, only if the energies considered are small, as compared to the chiral scale, and in this case the truncation of the chiral expansion at four-derivative level is justified. It is conceivable that with increasing energy higher order terms in the chiral expansion eventually become of importance which in consequence leads to a proliferation of unknown parameters at higher-derivative orders. This would be then a signal that the theory has been stretched beyond its range of applicability.

# PART II

# Comparison with Experiment

# Chapter 13

# Two-Meson Atomic Bound States

Two-meson systems present a considerable theoretical interest because they are the simplest to study the hadron-hadron interaction. On a quark level they involve $(q\bar{q}q\bar{q})$ and constitute the next level of complication after that of a single meson $(q\bar{q})$ and a single baryon $(qqq)$. Below the breaking scale of chiral symmetry the interactions between quarks and gluons ruled by QCD are highly non-perturbative but the spectrum of the theory is rather simple, containing only the octet of pseudoscalar mesons: $\pi$, K and $\eta$. At low-energy these pseudoscalar mesons are known to be weakly interacting, both among themselves and with the nucleons which suggests that QCD can be treated perturbatively when a suitable transformation of the degrees of freedom is effected. Such program has been realized by ChPT where the interaction of pseudoscalar mesons is described in terms of Goldstone boson fields interacting only through derivative couplings. This hypothesis justifies the soft behavior of pseudoscalar mesons interaction at low-energy and the realm of hadronic atoms is ideally suited for studying these interactions right at threshold. Experimentally the investigation of two-meson systems is very difficult but attempts to observe the signals of the production from two-meson exotic atoms have been undertaken or contemplated in several laboratories. The three pairs of pseudoscalar mesons $\pi^+\pi^-, \pi^\pm K^\mp$ and $K^+K^-$ seem to be particularly promising and they are discussed below.

## 13.1 Pionium

When two oppositely charged $\pi$ mesons are bound by electromagnetic forces we have to do with the lightest hadronic atom, $\pi^+\pi^-$, which has been called pionium. Although its basic properties were predicted already in 1960s [Uretsky *et al.* (1961); Bilenky *et al.* (1969)] but for many following years pionium was of purely academic interest. In recent years, however, the situation changed dramatically and this atomic system has become the focus of much theoretical and experimental activity. Since in the low-energy regime QCD has been successfully approached by introducing effective Lagrangians such as used in ChPT, the $\pi\pi$ scattering observables become of primary importance providing invaluable information concerning the explicit breaking of chiral symmetry in strong interactions. Particularly important is the region near threshold because in the chiral limit when the masses of the (u,d) quarks go to zero, the $\pi\pi$ scattering lengths are bound to vanish. The chiral symmetry is explicitly broken and the measurement of these threshold parameters gives the opportunity to verify directly the current concepts of chiral symmetry breaking. Experimentally, accurate $\pi\pi$ scattering cross sections are not available at threshold and therefore the investigation of pionium has been considered as a practicable alternative to infer the values of the $\pi\pi$ scattering lengths.

The fact that we can produce only pion beams but not targets is a major difficulty because $\pi\pi$ system can be investigated only by indirect methods. There are several possibilities for studying experimentally the low-energy $\pi\pi$ scattering (for a review of this subject cf. [Martin *et al.* (1976)]).

At present, the most reliable experimental measure of the low-energy $\pi\pi$ scattering parameters comes from the study of $K_{e4}$ decays ($K^+ \to \pi^+\pi^-e^+\nu_e$ and $K^- \to \pi^+\pi^-e^-\bar{\nu}_e$). In this method both final state pions are on the mass shell and are the only strongly interacting particles in the final state. Unfortunately, the range of the available c.m energies is not large and while theoretically it spans the interval from 280 MeV to 490 MeV but in practice the observed $\pi\pi$ effective mass spectra fall-off rapidly above 350 MeV so that this range is even smaller. The $\pi\pi$ scattering matrix element is included via Watson final state interaction theorem as a multiplicative factor in the decay matrix elements. The main advantage of this method is that it is model independent. The disadvantageous feature is the small branching ratio of the $K_{e4}$ decay channel which is only $\approx 4\times10^{-5}$. Until recently the only high-statistics experiment on $K_{e4}$ decay was that from Geneva-Saclay collaboration [Rosselet *et al.* (1977)] but now the new measurement at Brookhaven [Pislak *et al.* (2001)] has collected a $K_{e4}$ event sample more than ten times larger than all previous experiments combined.

Another possibility to extract the $\pi\pi$ cross section from experiment involves the measurement of pion induced pion production cross section and indeed this has been the most prolific source of information on $\pi\pi$ scattering. The power of this method owes to the fact that over a broad range of incident momenta the amplitudes of these reactions are dominated by a one-pion-exchange diagram so that the associated matrix element is proportional to the $\pi\pi$ scattering amplitude. The experimental double differential cross section provides then the input to the Goebel-Chew-Low extrapolation procedure [Goebel (1958)] used to determine the $\pi\pi$ cross section. The purpose here is to suppress the contribution from the background originating from the unwanted non-OPE processes and also to remove the off-shell effects. The extrapolation procedure takes advantage of the presence of the pion pole in the OPE contribution to the production amplitude. By admitting only such kinematic configurations where the absolute value the four-momentum transfer to the nucleon is close to the pion mass the OPE term receives much larger weighting than the background term and the latter could be thereby minimized. Therefore, staying as close to the pion pole as possible, we can isolate the OPE contribution from the background. Peripheral pion induced pion production has many attractive features and offers ample scope for experimental investigation: there are five different charge configurations and all of them make experimentally feasible channels giving access to an unlimited range of effective $\pi\pi$ masses. On the other hand, it has to be kept in mind that the extraction of the $\pi\pi$ scattering observables from $\pi N \rightarrow \pi\pi N$ data is not free from theoretical uncertainties connected with the lack of a reliable procedure controlling fully the size of the background term. Therefore, the $\pi\pi$ observables obtained this way cannot be regarded as model independent.

The determination of the low-energy $\pi\pi$ phase shifts and eventually the associated scattering lengths from the experimental $\pi\pi$ cross sections involves a sophisticated procedure based on the fixed momentum transfer dispersion relations. Pion-pion scattering amplitudes constrained by analyticity together with crossing and Bose symmetry were studied in the 1970s which resulted in a set of dispersion relations for the partial waves, known as Roy equations [Roy (1971)]. By imposing the unitarity constraint, the latter are cast into a set of coupled non-linear integral equations for the corresponding phase shifts. In the low-energy regime only the two s-waves (I=0,2) and the P wave (I=1) need to be considered and one is faced with the problem of solving a system of three coupled non-linear integral equations. The two s-wave $\pi\pi$ scattering lengths enter these equations as the subtraction constants and are of paramount importance in the analysis of the low-energy scattering data. Indeed, if they could be provided from outside to be used as input, the solutions of the Roy equations would generate the S and P wave phase shifts below 800 MeV to an amazing degree of accuracy (modern applications of Roy equation are reviewed by [Ananthanarayan *et al.* (2001)]). Ideally suited for this purpose would be the measurement of the energy shift and width of the 1s level in pionium which pins down the values of s-wave $\pi\pi$ scattering lengths.

Pionium atoms have been first observed at Dubna [Afanasyev *et al.* (1993)] and also at CERN [Santamarina (2001)] where presently a large collaboration is engaged in the DIRAC (DImeson Relativistic Atomic Complex) experiment to measure pionium life time in the ground state (cf. [Adeva *et al.* (1995)]). On the agenda of this experiment is also a measurement of the difference of the binding energies in the 2s and 2p levels.

The pionium atom presents us with a two-channel situation ($\pi^+\pi^-$ and $\pi^0\pi^0$) with the bound state pole located between the $\pi^0\pi^0$ and $\pi^+\pi^-$ thresholds. Therefore, the dominant contributor to the decay rate of pionium is the charge-exchange process $\pi^+\pi^- \to \pi^0\pi^0$. The decay rate $W_{n\ell}(\pi^0\pi^0)$ from the initial state specified by the principal and orbital quantum numbers $n$ and $\ell$, respectively, follows directly from the Fermi's Golden Rule

$$W_{n\ell}(\pi^0\pi^0) = 2\pi \int \frac{d^3k}{(2\pi)^3} \,|\, \langle\phi_f|V|\psi_{n\ell}\rangle \,|^2\, \delta(2\Delta m - \frac{k^2}{m_0}),$$

where $V$ is the strong interaction responsible for the decay, $\Delta m = m_c - m_0$ with $m_0$ and $m_c$ denoting, respectively, the masses of $\pi^0$ and $\pi^{\pm}$. At threshold the transition matrix element from the $\ell = 0$ states may be related to the s-wave scattering amplitude

$$\langle\phi_f|V|\psi_{n0}\rangle \approx \langle\phi_f|V|\phi_i\rangle\; \psi_{n0}(0) \approx -\frac{4\pi}{m_0}\frac{\sqrt{2}}{3}(a_2 - a_0)\psi_{n0}(0)$$

where $\psi_{n0}(0)$ is the coordinate space pionium wave function evaluated at the origin. It has been assumed that the transition potential is isospin invariant and has a much shorter range than the spread of the atomic wave function. In the above formula the strong s-wave $\pi\pi$ interaction enters through the s-wave scattering lengths $a_I$ where I=0,2 is the total isospin. The transition rate from the $\ell = 0$ states, takes the form

$$W_{n0}(\pi^0\pi^0) = \frac{16\pi}{9}\frac{\sqrt{2m_0\Delta m}}{m_0}(a_0 - a_2)^2|\psi_{n0}(0)|^2 \qquad (13.1)$$

and this expression was first given by [Uretsky *et al.* (1961)]. Its modern version contains a number of corrections, not only of electromagnetic origin, but also those induced by relativity and isospin symmetry breaking resulting from mass difference. Actually, the relativistic correction may be obtained directly from Fermi's formula using relativistic expression for the energies of the outgoing pions. In this case the phase space integration gives

$$\int_0^\infty \mathrm{d}k k^2\, \delta(2m_c - 2\sqrt{k^2+m_0^2}) =$$
$$= \tfrac{1}{2} m_c \sqrt{m_c^2 - m_0^2} \approx \tfrac{1}{2} m_0 \sqrt{2m_0 \Delta m}(1 + \frac{5\Delta m}{4m_0}),$$

so that the relativistic correction to the decay rate is of about 4%.

Another possible decay mode is into two photons. The corresponding rate can be derived from QED and is given as

$$W_{n0}(\gamma\gamma) = \frac{2\pi\alpha^2}{m_c^2} |\psi_{n0}(0)|^2. \tag{13.2}$$

This rate is about 250 times smaller than the strong interaction decay rate (13.1). Since the rate $W_{n0}(\gamma\gamma)$ is known, it has been suggested that the measurement of the ratio of the rates for the decay of $\pi^+\pi^-$, either in n$s$ bound states or low-lying continuum states, into $\pi^0\pi^0$ and $\gamma\gamma$, might used to determine the strong decay rate (13.1). Unfortunately, the search for a suitable reaction with $\pi^+\pi^-$ pair in the final state has not been successful so far.

Using in (13.1) for the scattering lengths $a_0$ and $a_2$ the theoretical values calculated from ChPT, the pionium life time in the ground state turns out to be of the order $3\times10^{-15}$ sec (cf. [Gasser *et al.* (2001); Gashi *et al.* (1998); Gashi *et al.* (2002)]). Since pionium life time is so short, the observation of pionium in the DIRAC experiment involves detection the $\pi^+\pi^-$ pairs resulting from the ionization of pionium taking place in the production target. The pions are first copiously produced in proton-nucleus (e.g. Ti or Ni) interactions using 24 GeV/c proton beam from CERN Proton Synchrotron. Some of the $\pi^+\pi^-$ pairs with sufficiently low relative momenta will form pionium atoms. After production in the target, these relativistic atoms may either decay into $\pi^0\pi^0$, or owing to the electromagnetic interaction with the target atoms, they may be excited or ionized. The $\pi^+\pi^-$ pairs resulting from ionized pionium atoms ("atomic" pairs) have low relative momenta (below 3 MeV/c) in their c.m. frame and in the lab. frame they travel at small opening angles with nearly equal laboratory velocities. The experimental setup takes advantage of the above characteristics and consists of a magnetic double arm spectrometer with a resolution on relative pair momenta of 1 MeV/c. The quantity of interest is the number of "atomic" pairs above the background of "free" pion pairs. The detection of the "atomic" pairs allows to determine pionium breakup probability which for a given target material and thickness is a known function of the pionium life time. The goal of the DIRAC experiment is to measure pionium life time with a precision at the level of 10 % which, as seen from (13.1), allows to determine $|a_0 - a_2|$ to within 5 %.

In order to infer the s-wave scattering lengths from the measurement of pionium life time, the theoretical expression for the life time must be worked out with an accuracy commensurable with the precision of DIRAC experiment. In particular, the electromagnetic corrections disregarded in (13.1) have to be accounted for. The calculations of these corrections have been performed in the quantum mechanical framework by introducing $\pi\pi$ potentials (cf. [Gashi *et al.* (1998); Gashi *et al.* (2002)]), and also by using field theory techniques based on the the ChPT Lagrangian. Since the latter leads to a manifestly Lorentz invariant theory where the treatment of electromagnetic bound state problems is difficult, a non-relativistic approach in the spirit of Caswell and Lepage [Caswell *et al.* (1986)], has been generally regarded as a more efficient alternative. The most complete treatment has been presented in [Gasser *et al.* (2001)] where further references may be found. In this framework the pionium decay width into a pair of neutral pions is calculated in two steps. In the first step the ChPT Lagrangian is modeled by a non-relativistic effective Lagrangian that describes strong and electromagnetic interactions of pions in the very low-energy regime. The coupling constants occurring in the effective Lagrangian are fixed by the requirement that the non-relativistic theory must be equivalent to ChPT and thereby also to QCD. Since the bound states and their decays are most easily handled within a Hamiltonian framework, in step two, the Hamiltonian approach based on the non-relativistic effective Lagrangian is developed. The pionium bound states are identified with the poles of the $\pi\pi$ scattering amplitude, which with the Hamiltonian in hand, can be obtained by standard methods of the non-relativistic two-channel scattering theory. The Hamiltonian can be written in the usual form as a sum of three terms: the kinetic energy $\hat{H}_0$, the Coulomb $\hat{V}_C$ interaction and the strong interaction $\hat{V}_S$, with $\hat{H} = \hat{H}_0 + \hat{V}_C + \hat{V}_S$. The hat notation indicates that these operators are 2×2 matrices in the two-channel space: the neutral channel $\pi^0\pi^0$ and the charged channel $\pi^+\pi^-$ denoted as "0" and "c", respectively. The Coulomb interaction is present only in the charged channel which can be enforced by using channel projection operators $\hat{P}_0$ and $\hat{P}_c$ and putting $\hat{V}_C = \hat{P}_c V_C \hat{P}_c$ where $V_C$ is Coulomb interaction. Introducing the free and the Coulomb Green operators defined as $\hat{g}(E) = (E - \hat{H}_0)^{-1}$ and $\hat{G}(E) = (E - \hat{H}_0 - \hat{V}_C)^{-1}$, respectively, the scattering operator $\hat{t}(E)$ obeys the Lippmann-Schwinger equation

$$\hat{t}(E) = \hat{V}_S + \hat{V}_S\, \hat{g}(E)\, \hat{t}(E). \tag{13.3}$$

With both the strong and the Coulomb interaction accounted for, the equation for the scattering operator $\hat{T}(E)$ involves Coulomb Green operator and reads

$$\hat{T}(E) = \hat{V}_S + \hat{V}_S\, \hat{G}(E)\, \hat{T}(E), \tag{13.4}$$

and the pole structure of the T-matrix is predominantly determined by the Coulomb interaction whereas $\hat{V}_S$ is merely responsible for a small shift of the pole in the complex E plane. The Coulomb pole corresponding to the ground state at $E = E_0$ can be exposed by writing the Coulomb Green operator as

$$\hat{G}(E) = \hat{\mathcal{G}}(E) + \hat{P}_c |\, \phi_0 \rangle \frac{1}{E - E_0} \langle \phi_0 \,| \hat{P}_c, \tag{13.5}$$

where $\hat{\mathcal{G}}(E)$ is the non-singular at $E_0$ part of the Green operator and $\phi_0$ denotes the Coulomb ground state. The presence of the projection operators in the pole term ensures that this singular term will not appear in the neutral channel. Inserting (13.5) in (13.4), after re-shuffling, the latter equation can be cast to the form

$$\hat{T}(E) = [1 - \hat{V}_S \hat{\mathcal{G}}(E)]^{-1} \hat{V}_S \left\{ 1 + \hat{P}_c |\, \phi_0 \rangle \frac{1}{E - E_0} \langle \phi_0 \,| \hat{P}_c\, \hat{T}(E) \right\} \tag{13.6}$$

and introducing a new operator $\hat{\tau}(E)$

$$\hat{\tau}(E) = [1 - \hat{V}_S \hat{\mathcal{G}}(E)]^{-1} \hat{V}_S \tag{13.7}$$

we have

$$\hat{T}(E) = \hat{\tau}(E) \left\{ 1 + \hat{P}_c |\, \phi_0 \rangle \frac{1}{E - E_0} \langle \phi_0 \,| \hat{P}_c\, \hat{T}(E) \right\}. \tag{13.8}$$

This formula sandwiched between the Coulomb ground states, yields

$$\langle \phi_0 | \hat{P}_c \hat{T}(E) \hat{P}_c | \phi_0 \rangle = \\ = \langle \phi_0 | \hat{P}_c \hat{\tau}(E) \hat{P}_c | \phi_0 \rangle \left\{ 1 + \frac{\langle \phi_0 | \hat{P}_c \hat{T}(E) \hat{P}_c | \phi_0 \rangle}{E - E_0} \right\}, \tag{13.9}$$

and the scattering matrix is obtained from the explicit solution of the above equation

$$\langle \phi_0 | \hat{P}_c \hat{T}(E) \hat{P}_c | \phi_0 \rangle = (E - E_0) \frac{\langle \phi_0 | \hat{P}_c \hat{\tau}(E) \hat{P}_c | \phi_0 \rangle}{E - E_0 - \langle \phi_0 | \hat{P}_c \hat{\tau}(E) \hat{P}_c | \phi_0 \rangle}. \tag{13.10}$$

It is evident from (13.10) that the operator $\hat{T}(E)$ is non-singular at $E = E_0$ but develops a pole at a shifted position and one is led to the ultimate eigenvalue equation

$$E - E_0 = \int \frac{d^3p}{(2\pi)^3} \frac{d^3q}{(2\pi)^3} \, \phi_0(\boldsymbol{p}) \, \langle c\boldsymbol{p} | \hat{\tau}(E) | c\boldsymbol{q} \rangle \, \phi_0(\boldsymbol{q}), \tag{13.11}$$

where $\phi_0(\boldsymbol{p})$ denotes the ground state Coulomb wave function in momentum space and the expectation value of the operator $\hat{\tau}$ must be evaluated in the charged channel. Since the level shift is small in comparison with the separation between the ground and the first excited Coulombic level, the bound state equation (13.11) may be solved for E iteratively. In fact, the first iteration where E is set equal to $E_0$ on the right hand side of (13.11) already ensures sufficient accuracy. The operator $\hat{\tau}$ is obtained from the equation

$$\hat{\tau}(E) = \hat{V}_S + \hat{V}_S \, \hat{\mathcal{G}}(E) \, \hat{\tau}(E), \tag{13.12}$$

whose formal solution is given by formula (13.7). It is worth noting that while the dependence upon the strong interaction is explicit in eq. (13.12), the Coulomb interaction enters via the ground state expectation value in (13.11) and is also present in the Coulomb Green operator $\hat{\mathcal{G}}$. In eq. (13.12) the most important is the strong interaction and the Coulomb interaction may be treated perturbatively. Accordingly, in the leading approximation the Coulomb interaction is neglected and the Coulomb Green operator is replaced by the free propagator $\hat{g}$. As a result of this simplification, the operator $\hat{\tau}$ satisfies eq. (13.3) and becomes identical with the Coulomb-free scattering operator $\hat{t}$. Invoking the small shift approximation, in (13.11) one may set $\hat{\tau}(E) \approx \hat{t}(E_0)$ where $\hat{t}$, calculated to all orders in the strong interaction, can be replaced by the scattering length. Since the effective

Lagrangian is equivalent to ChPT this scattering length must be the same in both approaches. Recalling that for contact interaction there is no dependence upon the off-shell momenta, in the leading approximation, one has

$$\langle \boldsymbol{p}|\hat{t}(E_0)|\boldsymbol{q}\rangle \approx -\frac{4\pi}{m_c}\,\frac{1}{\hat{K}^{-1}-\mathrm{i}\hat{Q}} \tag{13.13}$$

where $\hat{K}$ is the isospin invariant K-matrix evaluated at the $\pi^+\pi^-$ threshold and $\hat{Q}$ contains the channel momenta

$$\hat{K} = \begin{pmatrix} \frac{1}{3}a_0+\frac{2}{3}a_2 & \frac{\sqrt{2}}{3}(a_2-a_0) \\ \frac{\sqrt{2}}{3}(a_2-a_0) & \frac{2}{3}a_0+\frac{1}{3}a_2 \end{pmatrix}; \qquad \hat{Q} = \begin{pmatrix} q & 0 \\ 0 & 0 \end{pmatrix}, \tag{13.14}$$

with $q=\sqrt{m_c^2-m_0^2-\frac{1}{4}m_c^2\alpha^2}$. Also, owing to isospin conservation, the K-matrix has been expressed in terms of the scattering lengths $a_0$ and $a_2$. The T matrix in (13.11) can be now pulled outside the integral whose value is $|\psi_{00}(0)|^2 = (\frac{1}{2}m_c)^3\,\alpha^3/\pi$ where $\psi_{00}(0)$ denotes as before the configuration space Coulomb wave function at zero separation. Collecting all factors together in (13.11), we have

$$E - E_0 = -\frac{4\pi}{m_c}|\psi(0)|^2\left[\frac{2}{3}a_0+\frac{1}{3}a_2+\mathrm{i}\,q\,\frac{2}{9}(a_2-a_0)^2\right], \tag{13.15}$$

and the above expression provides a less important small shift of the ground state energy, as well as the decay width

$$\Gamma = -2\,\mathrm{Im}\,E = \frac{2}{9}\alpha^3\,p_t\,m_c^2(a_0-a_2)^2, \tag{13.16}$$

which is in agreement with (13.1). [1]

As mentioned above, in order to determine the scattering lengths via the measurement of pionium lifetime the theoretical expression for the decay width must be known with a precision that matches the accuracy of the lifetime measurement of DIRAC. Therefore, several authors have calculated perturbatively corrections to the leading order formula (13.15). These corrections come from the electromagnetic effects and from quark mass differences and the associated small expansion parameters are $\alpha$ and $(m_u-m_d)^2$, respectively. When each of them is counted as a small parameter of order

[1] In this book the scattering length has *always* the dimension of length. However, in the literature devoted to pions, the scattering length is traditionally given in inverse pion mass units and the symbols $a_0$ and $a_2$ are used there for *dimensionless* quantities corresponding to our $a_0\,m_c$ and $a_2\,m_c$, respectively.

$\delta$, the leading term in the decay width is of order $\delta^{7/2}$ but it is necessary to get the width up to and including terms of order $\delta^{9/2}$. It is believed that the contributions form the next-to-next-to-leading order are negligible even for future improved data and there is not much point going past $\delta^{9/2}$ order.

Field theoretical calculations of these corrections, however, are beyond the scope of this book, so we shall confine our attention to a potential model approach. The challenge we are addressing is to provide a simple procedure which for a given set of scattering lengths generated by ChPT allows to calculate pionium life time including electromagnetic corrections.

The pionium life time $\tau$ is the inverse of the decay rate $\Gamma$ corresponding to the dominant decay mode $\pi^+\pi^- \to \pi^0\pi^0$. The 1s level shift $\epsilon$ and the decay width $\Gamma$ for the atomic state of pionium are related to the complex $\pi^+\pi^-$ scattering length $A$ by the Deser et al. and Trueman [Deser *et al.* (1954)] formula [Deser *et al.* (1954); Trueman (1961)]

$$\epsilon + \mathrm{i}\,\tfrac{1}{2}\Gamma = \tfrac{1}{2}m_c^2\alpha^3\,A, \tag{13.17}$$

where the scattering length $A$ includes the Coulomb correction. The pionium life time $\tau$ obtained from (13.17) is

$$\tau = \frac{1}{m_c^2\alpha^3 \mathrm{Im}\,A} = \frac{0.01716\ \mathrm{fm}}{\mathrm{Im}\,A}\ \mathrm{fs}. \tag{13.18}$$

The Coulomb corrected complex scattering length $A$, in turn, can be expressed in terms of the appropriate two-channel K-matrix

$$A = K_{cc} + \mathrm{i}\,q\,\frac{K_{0c}\,K_{c0}}{1 - \mathrm{i}\,q\,K_{00}}, \tag{13.19}$$

where $q$ is the c.m. momentum of the $\pi^0\pi^0$ pair and it is understood that $q$ and also all the K-matrix elements are to be evaluated at the $\pi^+\,\pi^-$ threshold. To calculate the K-matrix we represent the strong interaction as a two-channel potential matrix

$$\begin{pmatrix} V_{00}(r) & V_{0c}(r) \\ V_{c0}(r) & V_{cc}(r) \end{pmatrix}, \tag{13.20}$$

where the different radial potentials operate in the s-waves. In addition to the strong interaction (13.20) there is also a Coulomb potential present in the charged channel responsible for the atomic binding. Instead of a two-channel Schrödinger equation, it will be more convenient for us to use as our dynamical framework a completely equivalent two-channel version of coordinate space Lippmann-Schwinger equation which takes the form of a set of two coupled integral equations. Taking first the incident wave in the neutral channel, we have

$$u_0(r) = \frac{\sin(qr)}{q} +$$
$$+m_0\, f \int_0^\infty \langle r\,|g_0^{(K)}(q)|\,r'\rangle\, [V_{00}(r')\, u_0(r') + V_{0c}(r')\, w_0(r')]\ \mathrm{d}r'; \quad (13.21a)$$

$$w_0(r) = m_c \int_0^\infty \langle r\,|G_0^{(K)}(0)|\,r'\rangle\, [V_{c0}(r')\, u_0(r') + V_{cc}(r')\, w_0(r')]\ \mathrm{d}r', \quad (13.21b)$$

where $(u_0, w_0)$ are the wave functions in the neutral and the charged channel, respectively, and the suffix 0 indicates to the type of the imposed boundary condition. Following [Gashi *et al.* (2002)] we have included in (13.21a) a relativistic correction which changes the mass $m_0$ into $m_0\, f = W - 2m_c^2/W$ where $W$ denotes the total energy in the c.m. frame. The factor $f$ is equal to one at threshold and therefore such factor is not needed in (13.21b). The kernel in eq. (13.21a) contains the s-wave part of the Green's function $\langle r\,|g_0^{(K)}(q)|\,r'\rangle$ corresponding to free propagation whose explicit form is

$$\langle r\,|g_0^{(K)}(q)|\,r'\rangle = -\frac{\sin(qr_<)}{q}\cos(qr_>). \quad (13.22)$$

In the charged channel we have to include the Coulomb interaction and this has been accomplished implicitly in (13.21) by introducing the s-wave projection of the Coulomb Green's function, $\langle r\,|G_0^{(K)}(0)|\,r'\rangle$. The Coulomb analog of (13.22) can be constructed by introducing the Coulomb wave functions $\phi_\ell(k,r)$ and $\theta_\ell(k,r)$ which are even and analytic functions of the momentum defined in [Lambert (1969)]. Similarly as the free Green's function (13.22), the Coulomb Green's function can be written in a product form

$$\langle r\,|G_0^{(K)}(0)|\,r'\rangle = -\phi_0(0,r_<)\,\theta_0(0,r_>), \quad (13.23)$$

and in the zero-momentum limit required in the problem at hand, the regular and irregular Coulomb wave functions introduced above become just proportional to the Bessel functions $J_1(x)$ and $Y_1(x)$, respectively

$$\phi_0(0,r) = \tfrac{1}{2}a_B\, x\, J_1(2x); \qquad \theta_0(0,r) = -\pi\, x\, Y_1(2x), \quad (13.24)$$

where $x = \sqrt{2r/a_B}$ with $a_B$ denoting the Bohr radius.

Given the solution of (13.21), the elastic and inelastic K-matrix elements, are obtained by going to the asymptotic limit $r \to \infty$ in (13.21)

$$K_{00} = -\frac{m_0 f}{q} \int_0^\infty \sin(q\,r)[V_{00}(r)u_0(r) + V_{0c}(r)w_0(r)]\,\mathrm{d}r; \quad (13.25a)$$

$$K_{0c} = -m_c \int_0^\infty \phi_0(0,r)[V_{c0}(r)u_0(r) + V_{cc}(r)w_0(r)]\,\mathrm{d}r. \quad (13.25b)$$

To obtain the two remaining elements of the K-matrix, the above procedure must be repeated taking this time the incident wave in the charged channel. With this boundary condition imposed, the resulting integral equations are

$$u_c(r) = m_0 f \int_0^\infty \langle r\,|g_0^{(K)}(q)|\,r'\rangle[V_{00}(r')u_c(r') + V_{0c}(r')w_c(r')]\,\mathrm{d}r'; (13.26a)$$

$$w_c(r) = \phi_0(0,r) +$$

$$+\,m_c \int_0^\infty \langle r\,|G_0^{(K)}(0)|\,r'\rangle[V_{c0}(r')u_c(r') + V_{cc}(r')w_c(r')]\,\mathrm{d}r', \quad (13.26b)$$

where the new wave functions are distinguished from the previous ones, by appending a different suffix, $c$. By going to the asymptotic limit in (13.26), similarly as before, we obtain the corresponding expressions for the elastic and the inelastic components of the K-matrix

$$K_{cc} = -m_c \int_0^\infty \phi_0(0,r)[V_{c0}(r)u_c(r) + V_{cc}(r)w_c(r)]\,\mathrm{d}r; \quad (13.27a)$$

$$K_{c0} = -\frac{m_0 f}{q} \int_0^\infty \sin(qr)[V_{00}(r)u_c(r) + V_{0c}(r)w_c(r)]\,\mathrm{d}r. \quad (13.27b)$$

It is worth noting that owing to the different channel masses and the inclusion of the relativistic correction, the K-matrix thus obtained is not symmetric but its off-diagonal elements instead of being just equal satisfy a slightly more complicated constraint $m_0\, f\, K_{0c} = m_c\, K_{c0}$.

It has been argued by many authors that the strong interaction potential (13.20) to a good approximation can be assumed to be isospin invariant. This assumption brings essential simplification because in such a case, the three potentials in (13.20) can be expressed as linear combinations of the potentials with isospin I=0 and isospin I=2, denoted in the following, respectively, as $V_0(r)$ and $V_2(r)$. The required even symmetry for two bosons excludes isospin I=1 state for s-waves. Using the Clebsch-Gordan coefficients appropriate for combining two i=1 states into eigenstate with total isospin I=0 and I=2, we obtain

$$V_{00}(r) = \frac{1}{3}V_0(r) + \frac{2}{3}V_2(r); \tag{13.28a}$$

$$V_{cc}(r) = \frac{2}{3}V_0(r) + \frac{1}{3}V_2(r); \tag{13.28b}$$

$$V_{0c}(r) = \frac{\sqrt{2}}{3}[V_2(r) - V_0(r)] = V_{c0}(r). \tag{13.28c}$$

In order to proceed further we have now to establish the I=0 and I=2 potentials by adjusting them to the corresponding scattering lengths deduced from ChPT calculations. It is far from obvious how to do that because even the simplest potential needs at least two parameters, e.g. a depth and a range, and it seems that there is not enough information to determine a potential having at the disposal just the scattering length. One possibility is to adjust the potential to the phase shifts over certain energy range taking the input phase shifts either from experiment or from a ChPT calculation. Another alternative is to take advantage of the fact that in the ChPT calculations the hadronic mass of the pion $m$ (assumed to be the same for $\pi^{\pm}$ and $\pi^0$) is a parameter whose value must be supplied and the inferred low-energy scattering parameters depend upon the adopted pion mass value. Particularly useful will be the values of the scattering length $a_0$ and $a_2$ resulting from a recent two-loop ChPT calculation (G. Colangelo quoted by [Gashi *et al.* (2002)]) for two values of the pion mass, i.e. for $m = m_0$ and $m = m_c$

$$a_0(m_0) = 0.2930 \text{ fm}; \qquad a_2(m_0) = -0.0614 \text{ fm}; \tag{13.29a}$$

$$a_0(m_c) = 0.3055 \text{ fm}; \qquad a_2(m_c) = -0.0632 \text{ fm}. \tag{13.29b}$$

We wish to adjust the potentials $V_0$ and $V_2$ in such a way that they reproduce the ChPT scattering parameters at threshold. With two free parameters this may be accomplished for each of the two values of the pion mass. Therefore, using a potential that contains only two adjustable parameters it is perfectly possible to reproduce the variation of the scattering length presented in (13.29). For each isospin value, we have now to satisfy two constraints and the two parameters specifying the potential can be determined. Our goal here is to calculate Coulomb corrections to pionium decay rate and in the description of close to threshold phenomena, the shape of the potentials $V_0(r)$ and $V_2(r)$ is not important. This can be seen by representing the scattering amplitude at low energies in the familiar form $\mathrm{Re}\, f_I(q) = a_I + b_I\,(q/m_c)^2 + \cdots$, where $\mathrm{b}_I$ is the slope parameter which may be sensitive to the shape of the potential. However, for energies comparable with pionium binding energy the factor $(q/m_c)^2 \sim 10^{-5}$ is so small that the dependence upon the slope parameter, and therefore also upon the shape of the potential, is strongly suppressed. In this situation the most convenient for illustrative purposes will be a delta-shell potential

$$V_I(r) = s_I\,\delta(r - R_I), \qquad I = 0, 2; \tag{13.30}$$

characterized by a dimensionless strength $s_I$ and a range $R_I$. When the interaction potential between two pions of mass $m$ is taken in the form (13.30), the scattering length $a_I$ takes a very simple form

$$a_I(m) = -R_I\,\frac{m\,s_I\,R_I}{1 + m\,s_I\,R_I}, \qquad I = 0, 2. \tag{13.31}$$

The requirement that this formula exactly reproduces the scattering lengths given in (13.29), pins down uniquely the values of the strength and the range parameters of the potentials (13.30), viz.

$$s_0 = -0.2373737; \quad R_0 = 1.204791 \text{ fm} \tag{13.32a}$$

$$s_2 = 0.4987821; \quad R_2 = 0.456031 \text{ fm}. \tag{13.32b}$$

Having determined the delta shell potentials (13.30), we may return to the problem of calculating the K-matrix. It is instructive to evaluate first the K-matrix without Coulomb corrections assuming isospin invariance. The K-matrix elements at the single unsplit threshold are then given as superpositions of $a_0$ and $a_2$. Using (13.14), we have

$$K = \begin{pmatrix} 0.0597 & -0.1738 \\ -0.1738 & 0.1826 \end{pmatrix}, \tag{13.33}$$

where all entries are in fm. The $\pi^+\pi^-$ complex scattering length resulting from the strong interaction only obtained from (13.33) is

$$A = (0.1825 + \mathrm{i}\,5.436 \times 10^{-3})\,\text{fm}. \tag{13.34}$$

Now, we are going to evaluate Coulomb corrections to the K-matrix (13.33) and to this end we insert the potentials (13.30) in the equations (13.21) and, subsequently in the equations (13.26). Owing to the delta function entering the potentials, the task of solving the system of integral equations (13.21), as well as (13.26), becomes quite trivial and reduces to that of solving a set of four inhomogeneous algebraic equations for four unknown quantities, the latter being the corresponding wave functions evaluated at $r = R_0$

and $r = R_2$. Similar simplifications take place in the expressions (13.25) and (13.27) defining the Coulomb corrected K-matrix whose elements are determined by the wave functions values at $r = R_0$ and $r = R_2$. Although the final expressions for the K-matrix might be obtained in analytic form but they would be rather messy and it is more convenient to solve the algebraic equations numerically. After simple computations, we obtain the the Coulomb corrected K-matrix at threshold ($W = 2m_c$)

$$K = \begin{pmatrix} 0.06296357 & -0.1842118 \\ -0.1730108 & 0.1830678 \end{pmatrix}, \tag{13.35}$$

where again all numbers are in fm. Using these values in (13.19), we are in the position to compute the Coulomb corrected complex scattering length in the charged channel, and the result is

$$A = (0.1830 + \mathrm{i}\, 5.7350 \times 10^{-3})\,\mathrm{fm}. \tag{13.36}$$

which gives thc pionium life time $\tau$

$$\tau = 2.992 \text{ fs}.$$

In addition to the Coulomb correction obtained here for point-like charges, there are further corrections which have not been included so far and might influence the value of the complex scattering length. The most important effect seems to be the reduction of the Coulomb attraction caused by the finite size of the charge distribution. This leads to a modification of the point-like Coulomb potential at small separations $V_C \to V_{fs}$ where the potential $V_{fs}(r)$ for uniform charge distribution has been given by [Oades (1970)]. This effect may be included by supplementing the potential $V_{cc}(r)$ by an extra piece $\Delta V_C(r)$ equal to the difference $V_{fs}(r) - V_C(r)$. The latter correction can be treated perturbatively which means that the Lippmann-Schwinger equations are left intact and the potential $\Delta V_C(r)$ is included only in (13.25b) and (13.27a). This procedure give rise to the following corrections

$$K_{0c} \to K_{0c} - m_c \int_0^\infty \phi_0(0, r)\, \Delta V_C(r)\, w_0(r) \mathrm{d}r, \tag{13.37a}$$

$$K_{cc} \to K_{cc} - m_c \int_0^\infty \phi_0(0, r)\, \Delta V_C(r)\, w_c(r) \mathrm{d}r, \tag{13.37b}$$

where the integration over r in (13.37) has to be performed numerically. Taking the charge distribution radius to be 0.856 fm, we get $K_{0c}$=-0.17282 fm and $K_{cc}$= 0.18186 fm which gives the ultimate value of the scattering length that includes the electromagnetic corrections

$$A = (0.1818 + \mathrm{i}\, 5.7287 \times 10^{-3})\ \mathrm{fm}. \tag{13.38}$$

The pionium life time turns out to be little affected and we get

$$\tau = 2.995\ \mathrm{fs},$$

in good accord with the value $\tau$=2.9±0.1 fs and $\tau$=2.92 fs obtained, respectively, in [Gasser *et al.* (2001)] and in [Gashi *et al.* (2002)].

In conclusion, it has to be stated clearly that the simple potentials (13.31) have been introduced to simulate $\pi\pi$ interaction at the energy scale of the order of a few keV but they certainly are not applicable outside this range, and, e.g. they predict wrong signs of the slope parameters.

## 13.2 K$\pi$ Atom

Recently, there has been also a considerable revival of interest in the $K^{\pm}\pi^{\mp}$ atomic bound states, although as yet they have not been observed experimentally. While the investigation of pionium allows to verify the correctness of the current views concerning the $SU(2)_L \times SU(2)_R$ chiral symmetry-breaking pattern, the study of K$\pi$ atoms might test our understanding of $SU(3)_L \times SU(3)_R$ symmetry breaking. There are indications that the size of the quark condensate may be flavor dependent and since K$\pi$ scattering is the simplest process involving strangeness it is well suited to explore the three-flavor sector. It is believed that the measurement of the lifetime of the K$\pi$ atom which is planed in the DIRAC experiment may shed some light on the flavor dependence. Apart from that, on a purely phenomenological level, such measurement supplying the value of the isovector scattering length may be of great value in the phase shift analysis because this threshold parameter can be used as a subtraction constant in K$\pi$ dispersion relations.

Despite a close resemblance to the $\pi\pi$ system, these two systems show also substantial differences: in the K$\pi$ case, in general, there are more states because K and $\pi$ are different particles and therefore all limitations imposed by Bose statistics must be lifted. The full crossing symmetry of the $\pi\pi$ scattering amplitude does not extend to the K$\pi$ case as the s$\leftrightarrow$t crossing operation generates a different process: $K\bar{K} \rightarrow \pi\pi$. Finally, the inelastic channels play a more important role since there is no restriction analogous to G parity and the K$\pi\pi$ states can couple to all K$\pi$ states with $\ell > 0$.

The available experimental information on K$\pi$ phase shifts accumulated in 1960s and 1970s is less detailed than that in the $\pi\pi$ case. The K$\pi$ differential cross-sections have been extracted by looking for a OPE signal in the pion production processes on nucleons using both, K$^+$, as well as K$^-$ beams with energies ranging from 2 to 13 GeV. Some of the studied processes are

$$\begin{aligned} K^+p \to\ & K^+\pi^-\Delta^{++} \\ & K^0\pi^0\Delta^{++} \\ & K^+\pi^+\Delta^0 \\ & K^0\pi^+p \\ & K^+\pi^+n \\ K^-p \to\ & K^-\pi^-\Delta^{++} \\ & K^-\pi^+n \\ & K^-\pi^0p \\ & \bar{K}^0\pi^-p \\ K^+n \to\ & K^+\pi^-p \\ & K^0\pi^+n \\ K^-n \to\ & K^-\pi^+p \end{aligned}$$

and the K$\pi$ scattering amplitudes in various charge combinations enter the corresponding production cross-sections. They are extracted by extrapolation to the pion pole. Isospin conservation allows to express the different K$\pi$ amplitudes as linear combinations of I=1/2 and I=3/2 amplitudes

$$T(K^\pm\pi^\pm \to K^\pm\pi^\pm) = T^{(3/2)};$$
$$T(K^\pm\pi^\mp \to K^\pm\pi^\mp) = \frac{1}{3}T^{(3/2)} + \frac{2}{3}T^{(1/2)};$$
$$T(K^+\pi^- \to K^0\pi^0) = T(K^-\pi^+ \to \bar{K}^0\pi^0) = \frac{\sqrt{2}}{3}(T^{(3/2)} - T^{(1/2)}).$$

The corresponding I=1/2 and I=3/2 phase shifts can be deduced from the scattering data by using dispersion relations techniques. Owing to the shortage of low-energy scattering data, the resulting threshold parameters have large uncertainties and the I=1/2 and I=3/2 scattering lengths spread over a wide range of possible values: $0.13 < a_{1/2} < 0.24$ and $-0.13 < a_{3/2} <$

-0.05 in units of inverse pion mass.

As mentioned above, the measurement of the lifetime of the $K^{\pm}\pi^{\mp}$ atom is of considerable importance because it provides a model independent way to infer the value of the difference of the two scattering lengths $|a_{1/2}-a_{3/2}|$. The dominant decay channel of $K^-\pi^+$ atom in the ground state is $\bar{K}^0\pi^0$ and the the decay probability can be obtained as before from the Fermi's Golden Rule. Assuming that the transition takes place from the $\ell = 0$ state, a formula analogous to (13.1) is obtained

$$W_{n0}(\pi^0\bar{K}^0) = \frac{1}{\tau_n} = \frac{8\pi}{9}\sqrt{\frac{2\Delta m}{\mu}}(a_{1/2}-a_{3/2})^2|\psi_{n0}(0)|^2, \qquad (13.39)$$

where $\tau_n$ is the life time of the state with the principal quantum number n, $\mu$ is the reduced mass of the atomic system and $\Delta m$ denotes the difference between the two thresholds.

The measurement of K$\pi$ threshold parameters in the DIRAC experiment is based on the idea put forward to investigate the pionium atom and uses the same experimental set up. The K$\pi$ atoms may be detected because they have a sizable probability to break-up when they travel in the target material where they have been produced. The ratio of the number of experimentally detected K$^{\pm}\pi^{\mp}$ pairs to the full number of produced atoms yields an empirical value of the break-up probability associated with a given target and the latter quantity is related in a known way to the lifetime of the K$\pi$ atom. This lifetime used in formula (13.39) gives the value of $|a_{1/2}-a_{3/2}|$. Using the theoretical values of the scattering lengths obtained from recent ChPT calculations this lifetime should be about 4.9 fs. In three years of running time of DIRAC experiment the authors of the proposal expect to collect a few thousands of K$\pi$ atomic pairs which should lead to a determination of $|a_{1/2}-a_{3/2}|$ with 10-15% accuracy.

## 13.3 Kaonium

Kaonium, or the $K^+K^-$ atomic bound state has not been observed experimentally. This system is similar to pionium but there are also differences. In particular, the threshold of the neutral channel $K^0\bar{K}^0$ lies about 8 MeV above the $K^+K^-$ threshold but the atomic state would be unstable owing to the open $\pi\pi$ and $\pi\eta$ channels. Therefore, the available phase space is much bigger than in the case of pionium and the width of the ground state may be expected to be substantially larger. To make a quantitative prediction one needs the value of the scattering length and the latter is an arithmetic average of the scattering lengths $A_I$ with the total isospin I=0 and I=1. These have not been measured and their theoretical estimates are not very accurate. For both isospin values there is a resonance located close to the $K^+K^-$ threshold: for I=0 this is the $f_0(980)$ whereas for the I=1 case this is the $a_0(984)$ resonance. These resonances are strongly coupled to the $K\bar{K}$ channel and dominate the threshold behavior of the $K\bar{K}$ scattering amplitude. The poles of the corresponding T-matrix associated with these resonances provide an estimate of the scattering length. Close to threshold, the T-matrix is

$$T_I = \frac{A_I}{1 - \mathrm{i}\, k\, A_I}; \qquad I = 0, 1$$

where k is the c.m. momentum. At the resonant energy the denominator vanishes which yields $A_I = -\mathrm{i}/k_r$ where $k_r = \sqrt{m_K(E_r - 2m_K - \mathrm{i}\,\frac{1}{2}\Gamma)}$ and $(E_r, \Gamma)$ denote the position and the width of the resonance, respectively. The total widths of $f_0(980)$ is between 40 and 100 MeV and that of $a_0(984)$ between 50 and 100 MeV and using these values, we get

$$-1.6 \leq \mathrm{Re}\, A_0 \leq -1.0; \qquad 0.8 \leq \mathrm{Im}\, A_0 \leq 0.9$$

for isospin I=0 and

$$-1.4 \leq \mathrm{Re}\, A_1 \leq -0.9; \qquad 0.8 \leq \mathrm{Im}\, A_1 \leq 1.1$$

for isospin I=1 where all entries are given in fermi. The 1s level shift $\epsilon$ and the width $\Gamma$ can be now evaluated from the Deser–Trueman formula

$$\epsilon + \mathrm{i}\tfrac{1}{2}\Gamma = m_k^2\, \alpha^3\, \tfrac{1}{2}(A_0 + A_1)$$

which gives

$$0.4 \ keV \leq \Gamma \leq 0.5 \ keV.$$

The calculated lifetime of kaonium is therefore three orders of magnitude shorter than that of pionium and is of the order $10^{-3}$ fs.

# Chapter 14

# Hadronic Hydrogen

By hadronic hydrogen one means an atomic system in which a negatively charged hadron ($\pi^-$, $K^-$, $\Sigma^-$, $\bar{p}$,...) forms a bound state with a proton. The binding occurs on account of the Coulomb attraction and the strong hadron-proton interaction leads to observable effects only in the states with the lowest (n,$\ell$) quantum numbers. Similarly as the two-meson case considered before hadronic-hydrogen constitutes a two-body problem and therefore is relatively simple.

Hadronic hydrogen atoms exhibit entirely different cascade features than other exotic atoms. In contrast with the remaining hadronic atoms with Z>1 hadronic hydrogen is deprived from electrons and as a result internal Auger effect is excluded as a possible deexcitation process. Instead, in the initial phase of the cascade the main deexcitation mechanism is connected with the external Auger effect. Denoting a generic hadron as $h^-$, we have a ionization process

$$(h^-p)_i + H_2 \rightarrow (h^-p)_f + e^- + H_2^+$$

in which the hadronic hydrogen undergoes a i$\rightarrow$f transition. The inelastic collisions with the target atoms lead to various Coulomb deexcitation processes

$$(h^-p)_i + p \rightarrow (h^-p)_f + p$$

and when the final principal quantum number decreases nearly half of the transition energy is transmitted into the kinetic energy of the hadronic hydrogen. The reverse process leads to deceleration and we have certain distribution of the kinetic energy.

At present, only Pionic Hydrogen, Kaonic Hydrogen and Antiprotonic Hydrogen have been observed experimentally and we are going to discuss these systems in sequence.

## 14.1 Pionic Hydrogen

The pion-nucleon interaction has been the subject of intensive studies since the discovery of the pion. Nevertheless, it is fair to say, that owing to two modern developments only quite recently these investigations have reached a high precision phase. Firstly, with the acquisition of the so called meson factories (LAMPF, PSI and TRIUMF) and the corresponding influx of the new high-accuracy $\pi$N scattering data a considerable progress has been achieved in the $\pi$N phase shifts analyses providing means to examine even such subtleties as isospin symmetry breaking effects. Secondly, these low-energy scattering experiments have been complemented by high-quality pionic X-ray measurements and the shift and the width of the ground state in pionic hydrogen and also in pionic deuterium have been determined in a series of experiments conducted at Paul Scherrer Institute by the ETH Zurich-Neuchatel-PSI collaboration. In this pionic hydrogen experiment the 3p-1s X-ray transition was measured, by combining a highly efficient stopping arrangement with a newly developed system of detectors. The pions from low-energy pion beam at PSI were degraded inside the cyclotron trap which is a magnetic field device designed to increase the stop density in the target cell at its center. The X-ray energies emitted from the deexcitation of the pionic hydrogen were determined using two bent crystal spectrometers together with a sophisticated detector system based on charge coupled devices. Comparing with earlier experiments considerable improvement has been achieved and the strong interaction shift of the 1s level has been determined with an accuracy of 0.5% while the error for the width (8%) was an order of magnitude larger. There is a new experiment under way with the ambitious goal of measuring both, the shift and the width with the precision at 1% level.

To proceed further it will be in order to fix some necessary notation and conventions used in the literature for pion-nucleon scattering, at the same time reminding the reader of some well-known features of $\pi$N interaction. The $\pi^{\pm}$p elastic and charge-exchange scattering takes place at all energies and inelastic channels become significant at incident pion energies above 300 MeV. The $\pi$N interaction is known to be nearly isospin invariant, and to the extend that charge invariance holds, the three amplitudes $F^{+}_{\pi}p$, $F^{-}_{\pi}p$ and $F_{SCX}$ (single-charge-exchange $\pi^{-}p \rightarrow \pi^{0}n$) may be expressed as superpositions of two amplitudes with total isospin 1/2 and 3/2. Therefore, the measurement of all three processes provides means of checking experimentally the isospin conservation. Besides total isospin amplitudes denoted as $F^{1/2}$ and $F^{3/2}$ it is convenient to introduce the isoscalar and isovector amplitudes $F^+$ and $F^-$, respectively, defined as

$$F^+ = \frac{1}{3}F^{1/2} + \frac{2}{3}F^{3/2}; \qquad F^- = \frac{1}{3}F^{1/2} - \frac{1}{3}F^{3/2};$$
$$F^{1/2} = F^+ + 2\,F^-; \qquad F^{3/2} = F^+ - F^-.$$

Using the above amplitudes, the different hadronic amplitudes are

$$\begin{aligned} F_{\pi^+ p} &= F^{3/2}; \\ F_{\pi^- p} &= \frac{1}{3}F^{3/2} + \frac{2}{3}F^{1/2}; \\ F_{SCX} &= \frac{\sqrt{2}}{3}[F^{3/2} - F^{1/2}]; \\ F_{\pi^0 n} &= \frac{2}{3}F^{3/2} + \frac{1}{3}F^{1/2}. \end{aligned}$$

In the zero-momentum limit ($q \to 0$) only the s-wave scattering length term survives in these amplitudes. This limiting procedure defines the I=1/2 and I=3/2 s-wave scattering lengths

$$\lim_{q \to 0} F^I(q) = a_{2I},$$

and similarly the isovector (isoscalar) scattering lengths

$$\lim_{q \to 0} F^{\pm}(q) = a^{\pm},$$

or using an alternative notation

$$\begin{aligned} \lim_{q \to 0} F^-(q) &= b_0 = \frac{1}{3}(a_1 + 2a_3); \\ \lim_{q \to 0} F^+(q) &= b_1 = \frac{1}{3}(a_3 - a_1). \end{aligned}$$

The $\pi^- p$ bound state is unstable and its dominant decay channel is $\pi^0 n$. Consequently, the pionic hydrogen presents us with a two-channel problem and in fact there is a good deal of similarity between the pionic hydrogen and the $\pi^+\pi^-$ system considered in the previous chapter. Therefore the same model as that developed for pionium could be in principle used here as well. On the other hand, having in mind the prospect of solving the pionic deuterium problem within the Faddeev approach, it will be more convenient for us to work with separable potentials of Yamaguchi type. The latter have been also used by a number of workers cf. [Peresypkin *et al.* (1974); Afnan *et al.* (1974); Mizutani *et al.* (1977)].

To get insight into the pionic hydrogen problem, let us consider then a two-channel situation, where the upper channel labeled as 1 corresponds to the neutral $\pi^0 n$ system and the lower channel labeled as 2, respectively, to the $\pi^- p$ system. We assume that the two-channel interaction respects isospin invariance and the isospin symmetry is broken only by the Coulomb potential operative in channel 2 and by the mass splitting within isospin multiplets. Since we wish to consider an atomic system it is essential to treat the Coulomb interaction exactly. To meet this requirement, we choose the two-channel Lippmann-Schwinger equation as our dynamical framework that in coordinate representation, takes the form

$$u_1(r) = \\ = \int_0^\infty \langle r\,|G_1^+(W)|\,r'\rangle\; [V_{11}(r',r'')\,u_1(r'') + V_{12}(r',r'')\,u_2(r'')]\;\mathrm{d}r'\;\mathrm{d}r''; \tag{14.1a}$$

$$u_2(r) = \\ = \int_0^\infty \langle r\,|G_2^+(W)|\,r'\rangle\; [V_{21}(r',r'')\,u_1(r'') + V_{22}(r',r'')\,u_2(r'')]\;\mathrm{d}r'\;\mathrm{d}r'', \tag{14.1b}$$

where we have assumed spherical symmetry of the problem and $u_j(r)$ denotes zero orbital momentum radial wave function in channel $j$ . The strong $\pi N$ interaction is adopted here in the form of a non-local potential matrix $V_{ij}$. In (14.1) we have introduced the Green matrix whose only non-vanishing diagonal elements are

$$\langle r\,|G_1^+(W)|\,r'\rangle = -\frac{2\mu_1}{p_1}\exp\left(\mathrm{i}p_1 r_>\right)\,\sin\left(p_1 r_<\right), \tag{14.2}$$

for the neutral channel, while in the charged channel we have to take into account the Coulomb interaction and the exact Green's function , in this case reads (cf. (4.23))

$$\langle r\,|G_2^+(W)|\,r'\rangle = -\frac{2\mu_2}{p_2}H_0^+(\eta, p_2\,r_>)\;F_0(\eta, p_2\,r_<), \tag{14.3}$$

where $r_< = \min(r, r')$, $r_> = \max(r, r')$. In (14.1)–(14.3) $W$ denotes the total c.m. energy (including the rest mass), $\mu_j$ are the reduced masses in the two channels and $p_j$ are the channel momenta: $p_j = \pm\sqrt{2\mu_j(W - E_j)}$ with

$E_j$ being the threshold energies and the sign ambiguity will be resolved in a moment. All masses here are assumed to take their physical values. In (14.3) $\eta = -\alpha\mu_2/p_2$ and $G_0, F_0$ denote the standard Coulomb wave functions for orbital momentum $\ell = 0$, defined in [Abramowitz *et al.* (1965)]. Finally, it should be noted that there is no ingoing wave in (14.1), as appropriate for a bound state problem. For energies above the neutral channel threshold but below the charged channel threshold the flux in channel 2 vanishes for large r. By contrast, the flux in channel 1 does not vanish at infinity for there will be an outgoing wave in channel 1 associated with the "leakage" of the unstable bound state system.

As mentioned above, to simplify matters, we assume that the interaction is separable, i.e. that the potential matrix is

$$V_{ij}(r, r') = -v(r)\, s_{ij}\, v(r'), \tag{14.4}$$

where the function $v(r)$ represents the shape of the potential and the dimensionless parameters $s_{ij}$ are the measure of the strength of the potential. Time reversal implies $s_{ij} = s_{ji}$. With separable potentials, the system of integral equations (14.1) can be solved analytically. To this end it is sufficient to multiply each of the equations by $v(r)$ and integrate over $r$. This gives a system of two homogeneous algebraic equations for the two unknown quantities

$$X_j = \int_0^\infty v(r)\, u_j(r)\, \mathrm{d}r, \quad j = 1, 2$$

and the latter will have a non-trivial solution if, and only if, the determinant of the system $D(W)$ vanishes. Expanding the determinant, we are led to the explicit bound state condition

$$\begin{aligned} D(W) = &\left(1 + s_{11} \langle v|G_1^+(W)|v\rangle\right)\left(1 + s_{22} \langle v|G_2^+(W)|v\rangle\right) \\ &- s_{12}^2 \langle v|G_1^+(W)|v\rangle \langle v|G_2^+(W)|v\rangle = 0, \end{aligned} \tag{14.5}$$

where we have introduced the abbreviation

$$\langle v|G_j^+(W)|v\rangle = \int_0^\infty v(r) \left\langle r \left| G_j^+(W) \right| r' \right\rangle v(r')\, \mathrm{d}r\, \mathrm{d}r'.$$

The determinant can vanish only at some particular value of the energy $W = E_B$ that will be interpreted as the bound state energy. Normally, knowing the underlying interaction, by solving (14.5) one obtains the binding energy. However, in the problem at issue we have a reversed situation: we know the binding energy from experiment and it is the interaction that we are after. In the case of the pionic hydrogen atom we have an unstable bound state in the charged channel and the binding energy will be a complex number. We set

$$E_B = E_2 + E_{1s} - (\epsilon + \mathrm{i}\tfrac{1}{2}\gamma) \tag{14.6}$$

where $E_{1s} = -\mu_2\alpha^2/2$ is the purely Coulombic 1s state binding energy. Since in our formalism there is no room for the radiative decay of the pionic hydrogen the partial width $\gamma$ is a fraction of the total width $\Gamma$ given by the formula $\gamma = \Gamma/(1+P^{-1})$ where $P$ is the Panofsky ratio. It has been shown by [Siegel *et al.* (1986)] that the effect of the $(\pi^-,\gamma)$ reaction on the accounted for hadronic channels is negligible. The experimental values for $\epsilon, \Gamma$ (cf. [Sigg *et al.* (1996)]) and P (cf. [Spuller *et al.* (1977)]) adopted in this book, are

$$\begin{aligned}\epsilon &= 7.108 \pm 0.013(\mathrm{stat}) \pm 0.034(\mathrm{syst})\, eV,\\ \Gamma &= 0.868 \pm 0.040(\mathrm{stat}) \pm 0.038(\mathrm{syst})\, eV,\\ P &= 1.546 \pm 0.009,\end{aligned}$$

and in the following we shall take $\epsilon = 7.108 \pm 0.047\, eV$ and $\gamma = 0.527 \pm 0.047\, eV$ as the input values. It has to be recalled that in this book we have defined $\epsilon$ in accordance with a different convention, so that our $\epsilon$ has opposite sign than that used by [Sigg *et al.* (1996)]. In our approach we have tacitly assumed that under perturbative treatment all electromagnetic corrections contribute the same amount to the purely Coulombic level and to the level shifted by strong interaction. More precisely, we are going to ignore the small effects caused by the distortion of the wave function. Accordingly, the electromagnetic corrections need not concern us here and they have been left out altogether but, of course, they would be indispensable for calculating the total displacement of the level from its Coulombic position.

The pole of the T-matrix that corresponds to the solution of (14.5) can be located on one of the four Riemann sheets as appropriate for a two-channel problem. This is also apparent from the mentioned above sign ambiguity in the definition of the channel momenta in (14.3). The right choice of the Riemann sheet is essential and this can be accomplished by proper adjustment of the signs of the imaginary parts of the channel momenta $p_j$. We are using here the standard enumeration of the Riemann sheets introduced in [Frazer *et al.* (1964)] , i.e

$$\begin{aligned} \text{sheet I:} &\quad Im\, p_1 > 0;\ Im\, p_2 > 0 \\ \text{sheet II:} &\quad Im\, p_1 < 0;\ Im\, p_2 > 0 \\ \text{sheet III:} &\quad Im\, p_1 < 0;\ Im\, p_2 < 0 \\ \text{sheet IV:} &\quad Im\, p_1 > 0;\ Im\, p_2 < 0. \end{aligned}$$

In the pionic hydrogen case, with an unstable bound state in channel 2, we have to enforce the pole to be located on the second sheet.

To proceed further we need some concrete shape factor $v(r)$ and our choice here is the exponential shape, i.e. we set

$$v(r) = \sqrt{\beta^3/\mu}\ \exp(-\beta r) \tag{14.7}$$

where $\mu$ is the reduced pion-nucleon mass in the case of exact isospin symmetry (we take average mass for each isospin multiplet) and $\beta$ is the inverse range parameter. With the exponential form (14.7), the potential (14.4) is identical with the familiar Yamaguchi potential and the Green's function matrix elements can be obtained in an analytic form. The final result is

$$\langle v|G_1^+(W)|v\rangle = -\frac{\mu_1}{\mu}\,\frac{1}{(1-\mathrm{i}p_1/\beta)^2} \tag{14.8}$$

for the neutral channel, while the corresponding formula for the charged channel reads

$$\langle v|G_2^+(W)|v\rangle = -\frac{\mu_2}{\mu}\,\frac{1}{(1-\mathrm{i}p_2/\beta)^2}\,\frac{{}_2F_1(1,\mathrm{i}\eta;\mathrm{i}\eta+2;z^2)}{\mathrm{i}\eta+1} \tag{14.9}$$

with $z = (\beta+\mathrm{i}p_2)/(\beta-\mathrm{i}p_2)$. The last fraction in (14.9) accounts for the Coulomb interaction and the symbol ${}_2F_1(a,b;c;z)$ denotes the hypergeometric function defined in [Abramowitz *et al.* (1965)]. The computation of the hypergeometric function entering (14.9) is greatly simplified owing to the fact that the first parameter is equal to unity in which case the continued fraction representation of ${}_2F_1(1,b;c;z)$ discovered by Gauss proves to be useful (a good account can be found in [Jones *et al.* (1980)]). The continued fraction summation converges in the whole of the complex $z$ plane away form the branch cut on the real axis running from one to infinity.

With exact isospin symmetry the three strength parameters $s_{11}, s_{12}, s_{22}$ are not independent and can be expressed in terms of isospin 1/2 and isospin 3/2 strengths denoted hereafter as $s_1$ and $s_3$, respectively. In the bound state condition (14.5) both the real and the imaginary part of $D(E_B)$ have to vanish simultaneously and that gives us two real equations. Since the bound state energy is known (cf. (14.6)), we put

$$s_{11} = (s_1 + 2s_3)/3; \quad s_{22} = (2s_1 + s_3)/3; \quad s_{12} = \sqrt{2}(s_3 - s_1)/3$$

in (14.5), and regarding $s_1$ and $s_3$ as our two unknowns, we arrive at two algebraic equations that can be solved analytically

$$s_1^2 \operatorname{Im} ac^* + s_1 \operatorname{Im}(ab^* - c) - \operatorname{Im} b = 0; \tag{14.10a}$$

$$s_3 = -\frac{1 + s_1 \operatorname{Re} a}{\operatorname{Re} b + s_1 \operatorname{Re} c}, \tag{14.10b}$$

where

$$a = (\langle v|G_1^+(E_B)|v\rangle + 2\,\langle v|G_2^+(E_B)|v\rangle)/3;$$
$$b = (2\,\langle v|G_1^+(E_B)|v\rangle + \langle v|G_2^+(E_B)|v\rangle)/3$$
$$c = \langle v|G_1^+(E_B)|v\rangle\,\langle v|G_2^+(E_B)|v\rangle\,.$$

With $s_1$ and $s_3$ in hand, the corresponding scattering lengths ($a_{2I}$ with I=1/2 and 3/2) are obtained from

$$a_{2I} = \frac{2}{\beta}\frac{s_{2I}}{1 - s_{2I}}. \tag{14.11}$$

For local potentials the method outlined above could be also applied but in such case it would be more convenient to use instead of (14.1) an equivalent set of two coupled Schrödinger equations. For fixed energy and the proper choice of the Riemann sheet, these differential equations can be integrated numerically and the bound state equation is obtained from the requirement of vanishing of the Wronskian determinant. The latter is again a function of the isospin 1/2 and isospin 3/2 strength parameters, or if one prefers, the corresponding potential depths. Although the bound state condition is defined then only numerically but from it one can get two real equations that can be solved numerically using standard procedures. With a local potential, however, the solution of the three-body problem becomes much more complicated and this is the main reason why we preferred to work with a separable potential.

Our calculational scheme is now complete and we shall present the results. Using as our input the experimental values of the pionic hydrogen level shift and width, the bound state equation has been solved analytically by adopting a number of "reasonable" values for $\beta$ and in our computations we have used the values from $2\,fm^{-1}$ to $10\,fm^{-1}$. Although we do not

not know the precise value of the range but there is no physical mechanism known that might generate long range forces in the $\pi N$ system, the longest range is unlikely to be bigger than 0.5 $fm$ and this sets the lower limit of acceptable $\beta$ values. In principle, there is no upper limit for $\beta$ but for $\beta > 10\, fm^{-1}$ we have in practice reached the limit of the zero-range forces and things change very little above that limit. The exact solutions of the

Table 14.1 The values of $\pi$N scattering lengths inferred from pionic hydrogen data. The experimental data in the last line are from [Gashi *et al.* (2000)].

| $\beta$ [fm$^{-1}$] | $a_1$ [$m_\pi^{-1}$] | $a_3$ [$10^{-1}\, m_\pi^{-1}$] |
|---|---|---|
| 2.0 | 0.1767± 0.0046 | -0.9377± 0.0852 |
| 3.0 | 0.1760± 0.0046 | -0.9306± 0.0846 |
| 4.0 | 0.1757± 0.0046 | -0.9263± 0.0841 |
| 5.0 | 0.1756± 0.0046 | -0.9228± 0.0837 |
| 6.0 | 0.1756± 0.0046 | -0.9197± 0.0834 |
| 7.0 | 0.1756± 0.0046 | -0.9167± 0.0830 |
| 8.0 | 0.1756± 0.0046 | -0.9138± 0.0827 |
| 9.0 | 0.1756± 0.0047 | -0.9110± 0.0823 |
| 10.0 | 0.1757± 0.0047 | -0.9082± 0.0820 |
| Deser | 0.1760 ± 0.0046 | -0.9258 ±0.0857 |
| expt | 0.1679 ± 0.0059 | -0.785 ± 0.034 |

bound state equation are presented in Table 14.1 where the errors reflect only the experimental uncertainty of our input, i.e. $\epsilon, \Gamma$ and the Panofsky ratio. Our isoscalar and isovector scattering lengths are in good agreement with the values extracted in [Sigg *et al.* (1996)]. This has been illustrated in Fig. 14.2 where we have compareda representative sample of our computations with the values obtained by Sigg *et al.* [Sigg *et al.* (1996)]. The solutions corresponding to $\beta$ spanning the range $2 - 10$ fm$^{-1}$ are located very close to each other in the ($b_1$, $b_0$) plane and putting more than three points on the plot might have obscured the picture. The error bars reflect only the experimental uncertainty of our input. As mentioned above, the bound state equation (14.5) yields a second order equation for $s_1$ and $s_3$ and therefore we have always two solutions (cf. (14.10)). Only one of them is presented in Table 14.1 whereas the second solution leads to both $b_0$ and $b_1$ positive and has had to be rejected.

As noted by [Sigg *et al.* (1996); Siegel *et al.* (1986)], at the energy value close to the unstable bound state in channel 2, the scattering amplitude in the open channel 1 shows a strong resonant behavior. For a separable potential, the s-wave scattering amplitude $f(W)$ in channel 1 may be easily calculated analytically. In order to tackle the scattering problem the system (14.1) must be modified and to satisfy the different boundary condition the right hand side in (14.1a) must be supplemented with the ingoing wave $\sin p_1 r$. As a result, we end up with an inhomogeneous system of equations for the quantities $X_j$ but the poles of the scattering amplitude appear at energies at which the determinant of the system (14.5) vanishes. Not surprisingly this condition is identical with the bound state equation. The scattering amplitude takes a simple form

$$f(W) = \frac{e^{i\delta}\sin\delta}{p_1} = \frac{\mu_1}{\mu}\frac{2}{\beta}\,\frac{s_{11} + (s_{11}s_{22} - s_{12}^2)\,\langle v|G_2^+(W)|v\rangle}{(1 + p_1^2/\beta^2)^2\, D(W)}, \tag{14.12}$$

where $\delta$ is the corresponding phase shift that for real $W$ below the $\pi^-$p threshold is a real number. The resonance is not of a Breit-Wigner shape but its position $E_r$ may be easily established from (14.12) as the energy at which the phase shift is equal to $\frac{1}{2}\pi$. Close to the resonant energy, i.e. for $W \approx E_r$ we have $\cot\delta \approx (W - E_r)/(\frac{1}{2}\Gamma_r)$ and this allows us to infer the value of the width $\Gamma_r$ of the resonance. The values of $(\epsilon, \gamma)$ been calculated in [Sigg *et al.* (1996)] by identifying them with $(E_2 + E_{1s} - E_r, \Gamma_r)$. In principle, the values of $(\epsilon, \gamma)$ obtained that way do not have to be identical with those determined from the position of the bound state pole. To check that point, we have repeated the procedure of [Sigg *et al.* (1996)] but using our separable potentials whose depths have been adjusted to reproduce the values of $(\epsilon, \gamma)$ obtained in [Sigg *et al.* (1996)]. We found that the two methods give nearly identical results and the differences in $(\epsilon, \gamma)$ did not exceed 1 meV. For illustration, in Fig.14.1 we show the behavior of $\sin\delta$ close to the resonance for the case of $\beta = 3\,fm^{-1}$ where the strengths parameters inferred from the pole location were found to be $s_1 = 0.271820$ and $s_3 = -0.245868$.

Before concluding our discussion of the pionic hydrogen we wish to mention one last thing, namely we are going to show how from (14.5) one can retrieve the Deser–Trueman formula (cf. [Deser *et al.* (1954)]). This task will be accomplished by obtaining an approximate solution of (14.5) and to this end (14.5) is cast to the form

$$1 + s_{\text{eff}}(W)\,\langle v|G_2^+(W)|v\rangle = 0, \tag{14.13}$$

where we have introduced an effective energy dependent complex strength

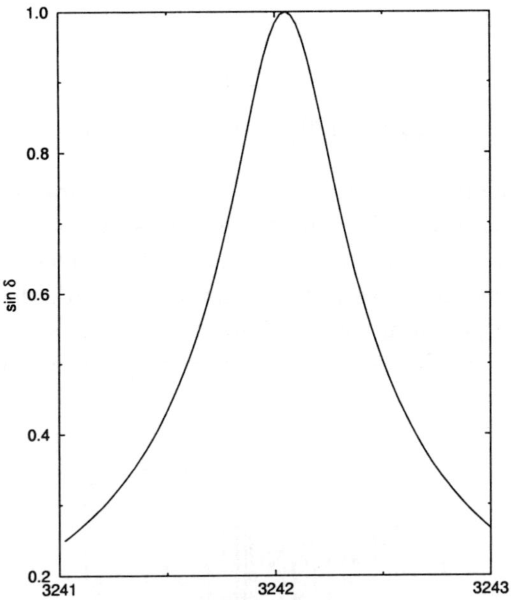


Fig. 14.1 $\sin\delta$ vs energy close to the resonance calculated from (14.12) for $\beta = 3\,fm^{-1}$.

parameter $s_{\text{eff}}$, defined as

$$s_{\text{eff}}(W) = s_{22} - \frac{s_{12}^2 \langle v|G_1^+(W)|v\rangle}{1 + s_{11}\langle v|G_1^+(W)|v\rangle}. \tag{14.14}$$

The complex $\pi^- p$ scattering length $a_{\pi p}$ can be expressed in terms of $s_{\text{eff}}(W)$ evaluated at threshold

$$a_{\pi p} = \frac{\mu_2}{\mu}\frac{2}{\beta}\frac{s_{\text{eff}}(E_2)}{1 - s_{\text{eff}}(E_2)}, \tag{14.15}$$

and the Coulomb corrected $\pi^-$p scattering length, denoted as $a_{\pi p}^c$, can be obtained from the exact formula derived in [van Haeringen (1975)]

$$\frac{1}{a_{\pi p}^c} = \frac{e^{\xi}}{a_{\pi p}} + 2\mu_2\alpha\,\text{Ei}(\xi), \tag{14.16}$$

where $\xi = 4\alpha\mu_2/\beta$ and $\text{Ei}(\xi)$ is the exponential integral function defined in [Abramowitz *et al.* (1965)]. It should be noted here that the zero-range limit ($\beta \to \infty$) does not exist in (14.16) because the function $\text{Ei}(\xi)$ for $\xi$=0 has a logarithmic singularity. For the case of $\beta = 3\,fm^{-1}$ just considered, we obtain

$$a_{\pi p} = 0.12081 + \mathrm{i}\,0.004441\, fm;$$
$$a^{c}_{\pi p} = 0.12068 + \mathrm{i}\,0.004458\, fm;$$

so that the Coulomb corrections do not exceed a fraction of a percent. However, in general, the Coulomb correction is model dependent, and, in particular, it is rather sensitive to the range of the nuclear potential what can be seen when the above result is juxtaposed with the $\pi$d case where the range of the potential was comparable with the size of the deuteron and, accordingly, the Coulomb correction to $\pi$d scattering length was much bigger (1.5%).

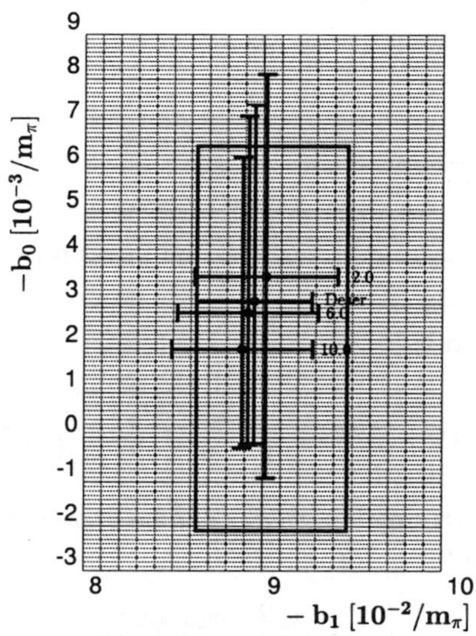


Fig. 14.2 The values of isoscalar and isovector scattering lengths obtained by solving the bound state equation (14.5) for $\beta$ equal, respectively, 2.0 fm$^{-1}$, 6.0 fm$^{-1}$, and 10.0 fm$^{-1}$ (indicated on the plot). The point marked as Deser has been obtained from (14.20). The rectangle corresponds to the values obtained in [Sigg *et al.* (1996)].

Since we wish to obtain an approximate solution of (14.13) that is located not far from the Coulomb bound state, we set $W = E_2 + E_{1s} + \delta E$ where $\delta E$ is a small displacement. To calculate $\delta E$ and derive the Deser–Trueman formula from (14.13), we have to assume that (*i*) the complex energy shift $\delta E = -\epsilon - \mathrm{i}\frac{1}{2}\gamma$ is small in comparison with Coulomb energy ($|\delta E/E_{1s}| << 1$), and, (*ii*) that the range of the strong interaction is small as compared with the Bohr radius ($\beta >> \mu_2\alpha$). Introducing a complex momentum $p_c = \sqrt{2\mu_2 E_{1s}} = \mathrm{i}\mu_2\alpha$ corresponding to the Coulomb bound state, we can see that when $p_2 \to p_c$ then $\mathrm{i}\eta \to -1$ and the Green's function (14.9) occurring in (14.13) becomes singular. This singularity is of paramount importance since it induces a zero in the nuclear S-matrix that is necessary to cancel the bound pole in the Coulomb S-matrix. As a result of this cancellation, the full S-matrix in the charged channel, which is a product of the Coulomb S-matrix and the nuclear S-matrix, remains finite at $p_2 = p_c$. In compliance with the small shift assumption, we set $p_2 = p_c + \delta p$ where $\delta p$ is supposed to be a small correction and since the most rapid variation in (14.9) arises on account of the pole term, we approximate $1 + \mathrm{i}\eta$ by $\delta p/p_c$. Apart from that, elsewhere we replace $p_2$ by $p_c$. The hypergeometric function for $\mathrm{i}\eta = -1$ reduces to a polynomial $1 - z^2$ and neglecting small terms of the order of $p_c/\beta$, from (14.13) we obtain

$$\delta p \approx -4\mathrm{i}\,\frac{p_c^2}{\beta}\,\frac{\mu_2}{\mu}\,s_{\mathrm{eff}}(E_2) \approx -2\mathrm{i}\,p_c^2\,a_{\pi p} \qquad (14.17)$$

where we have used (14.15) retaining only linear term in $a_{\pi p}$. The above result gives the Deser–Trueman formula [Deser *et al.* (1954)] in its standard form

$$\delta E \approx p_c\,\frac{\delta p}{\mu_2} \approx -2\mu_2^2\,\alpha^3\,a_{\pi p}, \qquad (14.18)$$

where, in view of the above discussion, it does not really matter whether we take $a_{\pi p}$ or $a^c_{\pi p}$. It is perhaps in order to recall that although the Deser–Trueman formula (14.18) has been derived here for a specific choice of the underlying interaction, but its validity is quite general. To examine the accuracy of Deser–Trueman formula we turn again to our previous example when $\beta = 3\,fm^{-1}$ and by computing $a_{\pi p}$ from (14.15) and inserting in (14.18), we obtain $(\epsilon, \gamma) = (7.024, 0.516)$ eV to be compared with our input values equal $(\epsilon, \gamma) = (7.108, 0.527)$ eV that ought to have been reproduced if formula (14.18) had been exact. It is a remarkable property of the Deser–Trueman formula that it is independent of the range of the underlying interaction and therefore the error in this formula must be of the same size as the uncertainty in the exact result caused by varying $\beta$. If one is prepared to tolerate such uncertainty formula (14.18) could be used to infer $a_1$ and $a_3$. Introducing a two-channel K-matrix, isospin invariance can be invoked to pin down its elements at the single unsplit threshold

$$K = \begin{pmatrix} \frac{1}{3}a_1 + \frac{2}{3}a_3 & \frac{\sqrt{2}}{3}(a_3 - a_1) \\ \frac{\sqrt{2}}{3}(a_3 - a_1) & \frac{2}{3}a_1 + \frac{1}{3}a_3 \end{pmatrix}$$

and the complex $\pi^- p$ scattering length takes the form

$$a_{\pi p} = K_{22} + \mathrm{i}\, p_t \frac{K_{12}^2}{1 - \mathrm{i}\, p_t K_{11}}, \tag{14.19}$$

where $p_t$ is the momentum in the $\pi^0 n$ channel evaluated at the $\pi^- p$ threshold. The scattering length (14.19), unlike (14.15), does not depend upon the range. Inserting (14.19) in (14.18) and separating the real and the imaginary part, we end up with two real equations for the two unknowns $a_1$ and $a_3$. To more than sufficient accuracy, the explicit solutions, are

$$a_1 = \left[x \pm y(1 - 2p_t y)/\sqrt{2p_t y}\right]/(1 - p_t y); \tag{14.20a}$$

$$a_3 = \left[x \mp y(2 - p_t y)/\sqrt{2p_t y}\right]/(1 - p_t y), \tag{14.20b}$$

where $x = \epsilon/2\mu_2^2\alpha^3, y = \frac{1}{2}\gamma/2\mu_2^2\alpha^3$ and the double sign in (14.20) stems the fact that eq. (14.19) is quadratic in $a_{2I}$. If ($\epsilon$,$\gamma$) have been obtained in a model independent way then the results (14.20) are also model independent. As seen from Table 14.1 the uncertainty on $a_1$ and $a_3$ (3% and 9%, respectively) induced by experimental errors on $(\epsilon, \gamma)$ is much bigger than the uncertainty caused by varying $\beta$ (about 1%). Under these circumstances it is perfectly justified to infer the $\pi$N scattering lengths via Deser–Trueman formula and their numerical values obtained from (14.20) are displayed in Table 14.1 whereas the corresponding $b_0$ and $b_1$ are presented in Fig. 14.2.

## 14.2 Kaonic Hydrogen

The study of kaon-nucleon reactions and the experimental status of the kaonic hydrogen are far from matching the degree of completeness achieved in the pion-nucleon case. The first three independent measurements of kaonic hydrogen X-rays were carried out at CERN and the Rutherford Laboratory in the late 1970 through the early 1980 (cf. [Bird *et al.* (1983); Davies *et al.* (1979); Izycki *et al.* (1980)]). These pioneering experiments all used a liquid hydrogen target together with Si(Li) detectors. The observed spectra suffered from a large background and low statistics because the X-ray signals were very difficult to identify, having been strongly attenuated by the Stark mixing taking place in liquid target. The results obtained for the strong interaction 1s level shift and width are presented in Fig. 14.3. Despite the evident inconsistencies, the derived strong interaction shift values were found to be in qualitative agreement with each other in that the three experiments uniformly showed a *positive* value for the shift. However, this observation caused a serious embarrassment when it was realized that such experimental finding could not be reconciled with the analyses of the low-energy $N\bar{K}$ scattering data which invariably predicted a *negative* value for the strong interaction 1s level shift in kaonic hydrogen. This difficulty sustained for the following 15 years and the apparent conflict between the X-ray measurements and the low-energy scattering data was referred to as the "kaonic hydrogen puzzle". The attempts to find an explanation on a theoretical ground turned out to be unsuccessful and the "puzzle" remained unresolved only until recently when the kaonic hydrogen X-ray experiment was repeated at KEK. The first convincing measurement of the X-ray spectrum from KEK (cf. [Iwasaki *et al.* (1997); Ito *et al.* (1998)]) revealed a negative value for the shift in good accord with the scattering data (cf. Fig. 14.3). The KEK experiment introduced two novel experimental techniques to improve on the signal-to-noise ratio, and these were the use of a gaseous target combined with the two-pion tagging method. The negative kaon beam of 600 MeV/c was slowed down in a Carbon degrader and subsequently brought to rest in a cryogenic hydrogen target. A low-temperature pressurized gaseous hydrogen target was selected as a compromise between good kaon stopping rate and the unavoidable loss in the X-ray yield due to Stark mixing. In the previous experiments the major source of background was due to the $\gamma$ rays produced by $K^-p$ absorption reactions, directly from the $\pi^0$ decays, or resulting from $\Lambda$ and $\Sigma^+$ decay. Nevertheless, there are reaction channels where such processes giving rise to $\gamma$ rays are absent: these are reactions of the type $K^-p \to \Sigma^\pm\pi^\mp$ followed by $\Sigma^\pm \to n\pi^\pm$, which are characterized by the presence of two charged pions whose momenta in the final state exceed 150 MeV. It was precisely this latter property that was used to eliminate all reactions contaminating $\gamma$ ray background by tagging on two charged pions with momenta higher than 150 MeV. The KEK experiment result naturally requires confirmation and we can expect further much-improved data on kaonic hydrogen both from KEK and from the DAΦNE storage ring at Frascati [Bianco *et al.* (1999)].

In the kaonic hydrogen problem, it is crucial to have a reliable description of the characteristics of the N$\bar{K}$ interaction at energies close to threshold. The N$\bar{K}$ interaction at low energies is known to have a rich and complex structure and even at the very pK$^-$ threshold, at 1432 MeV of the total energy in the center of mass, there are several inelastic reaction channels opened, i.e. the hyperonic channels

$$pK^- \to \Lambda\pi^0, \quad \Sigma^0\pi^0, \quad \Sigma^\pm\pi^\mp,$$

but for incident momenta below 200 Mev/c there is no significant contribution from three-particle states (such as $\Lambda\pi\pi$). In addition to that there is the charge exchange channel $n\bar{K}^0$ whose threshold is located 5 MeV above that of the $pK^-$ channel. The N$\bar{K}$ interaction is strong, of comparable strength with NN interaction. Furthermore, not far from threshold there are prominent resonances present strongly coupled to the N$\bar{K}$ channel: in the s-wave the $\Lambda(1405)$ and the $\Sigma(1385)$ in the p-wave, both located below the pK$^-$ threshold, and, finally, the $\Lambda(1520)$ in the d-wave.

Our knowledge of the threshold properties of the N$\bar{K}$ interaction is based on the following data

- (i) *In-flight data.* The elastic channel, the charge-exchange and the hyperonic channels listed above make a total of six measured cross sections.
- (ii) *Capture at rest.* Three branching ratios for the pK$^-$ capture

$$\frac{\sigma(\pi^0\Lambda)}{\sigma(\pi^0\Lambda)+\sigma(\pi^0\Sigma^0)}, \qquad \frac{\sigma(\pi^+\Sigma^-)+\sigma(\pi^-\Sigma^+)}{\sigma(\pi^0\Lambda)+\sigma(\pi^0\Sigma^0)}, \qquad \frac{\sigma(\pi^+\Sigma^-)}{\sigma(\pi^-\Sigma^+)}.$$

- (iii) *Below threshold data.* The $\pi\Sigma$ effective mass distribution measured for values corresponding to imaginary pK$^-$ c.m. momenta ranging from zero to about 200 MeV [Hemingway (1985)].
- (iv) *Neutral kaons reactions.* Total cross section $\sigma_T(pK_L^0)$ for kaon momenta between 170 and 260 MeV together with the ratio

$$\frac{\sigma(Y\pi)}{\sigma(pK_S^0)}.$$

- (v) *Kaonic hydrogen X-ray data.* Strong interaction 1s level shift and width.

The N$\bar{K}$ interaction involves two isospin states I=0 and I=1 and assuming isospin invariance for I=0 we have two relevant channels $(N\bar{K},\Sigma\pi)$ whereas for I=1 there are three channels $(N\bar{K},\Sigma\pi,\Lambda\pi)$. Unitarity requires the s-wave multi-channel amplitudes T$^I$ be related to the K-matrix, or to its inverse $M=K^{-1}$

$$(T^I)^{-1} = M^I + \mathrm{i}\,Q$$

where the $M^I$ are real symmetric matrices and Q is a diagonal matrix of the channel centre-of-mass momenta. The Coulomb effects and mass differences have to be allowed for before confronting with the data as first proposed by [Dalitz *et al.* (1960)]. The longstanding difficulty associated with the determination of the low-energy $N\bar{K}$ amplitudes is that stand alone, the experimentally available data on $pK^-$ and $p\bar{K}^0_L$ initiated reactions do not contain enough information for a unique multi-channel analysis. The crucial point is that for obvious experimental reasons we are missing data for the $Y\pi \to Y'\pi$ channels where Y, Y' is $\Lambda$ or $\Sigma$. The simplest parameterization respecting multi-channel unitarity is to take the M matrices to be constant, which gives a total of 3+6=9 adjustable parameters. However, the data yield only six constraints so that only six combinations of the M-matrix elements will be determined. To make it more transparent we write the elastic and the inelastic amplitudes in the form

$$T^I(N\bar{K} \to N\bar{K}) = \frac{A_I}{1 - \mathrm{i}\,k\,A_I}; \tag{14.21a}$$

$$T^I(N\bar{K} \to \Sigma\pi) = \frac{M_I}{1 - \mathrm{i}\,k\,A_I}; \tag{14.21b}$$

$$T^1(N\bar{K} \to \Lambda\pi) = \frac{N_1}{1 - \mathrm{i}\,k\,A_1}; \tag{14.21c}$$

where the scattering lengths $A_I$ and the amplitudes $M_I$ and $N_1$ are all complex quantities which can be expressed in terms of the constant M-matrix elements. In (14.21) k denotes the $N\bar{K}$ c.m. momentum and the energy dependence enters the amplitudes (14.21) via the dependence on the channel momenta. The data on $N\bar{K}$ initiated processes are determined when we know six quantities: $A_I = a_I + \mathrm{i}\,b_I$ (with real $a_I, b_I$), the fraction of $\Lambda$ production relative to total I=1 hyperon production $\epsilon = |N_1|^2/(|N_1|^2 + |M_1|^2)$, and the relative phase $\phi = \arg M_0 - \arg M_1$. The problem is over parameterized because we have six constraints and need to determine nine parameters. Therefore, it is not possible to obtain a unique solution and e.g. the transformation $M_I \to M_I\,\mathrm{e}^{\mathrm{i}\,\mu}, N \to N\,\mathrm{e}^{\mathrm{i}\,\nu}$ with arbitrary $\mu$ and $\nu$ modifies the solution and yet leaves the fit intact. To improve the situation it is not enough to have better data – a different data are needed. To meet this requirement dispersion relations have been included into the fitting procedure but they imposed a rather weak constraint and required extra four adjustable parameters.

The inclusion of the below threshold data (iii) into the fitting procedure provides a new constraint but is not enough to make the solution unique. Actually, with a constant M-matrix it was not possible to reproduce the the shape of $\Lambda(1405)$ resonance and eventually a decent fit was obtained at the expense of an effective range term added to the I=0 M-matrix. This raised the total number of parameters to 12.

Although the M-matrix solutions are not unique they still fit the same data and we might expect that they should yield similar values for $A_0$, $A_1$, $\epsilon$ and $\phi$. In view of the difficulties with the M-matrix approach the latter quantities may be even regarded as constants. Such constant scattering lengths (CSL) fit involves then only six parameters and is unique. At $pK^-$ threshold the M-matrix fits are indeed quite close to the CSL fits. Nevertheless, it has to be kept in mind that the constant-scattering-length assumption involves a major over-simplification: the six parameters are not constant, for they must contain branch-point singularities associated with the different thresholds. In fact, they are not even analytic functions of the total energy.

The results from a variety of phenomenological analyses are summarized in Table 14.2 with reference to the method used.

The $N\bar{K}$ scattering lengths may be used to calculate the strong interaction 1s level shift $\epsilon$ and the width $\Gamma$. The latter are obtained from the Deser–Trueman formula

$$\epsilon + \mathrm{i}\,\tfrac{1}{2}\Gamma = 2\mu^2\,\alpha^3\,A^c$$

where $\mu$ is p-$K^-$ reduced mass and $A^c$ is the Coulomb corrected p-$K^-$ complex scattering length. To give allowance to the mass differences, similarly as in the pionium and pionic hydrogen case, we consider a two-channel situation in particle basis. In the present case, however, the charged p-$K^-$ channel threshold falls below the neutral channel threshold n-$\bar{K}^0$. Assuming that the mass splitting needs to be accounted for only for computing the channel momenta but the two-channel K-matrix respects isospin invariance, the s-wave scattering amplitude T at p-$K^-$ threshold, is obtained from the standard expression

$$T^{-1} = K^{-1} - \mathrm{i}\,Q,$$

where the diagonal Q matrix contains the channel momenta, i.e. zero, and, respectively, the n-$\bar{K}^0$ momentum $k_0$ evaluated at the p-$K^-$ threshold. Since the neutral channel is closed there we have to use the analytic continuation setting $k_0 \to i|k_0|$. Isospin invariance implies

$$K = \begin{pmatrix} \frac{1}{2}A_1 + \frac{1}{2}A_0 & \frac{1}{2}A_1 - \frac{1}{2}A_0 \\ \frac{1}{2}A_1 - \frac{1}{2}A_0 & \frac{1}{2}A_1 + \frac{1}{2}A_0 \end{pmatrix},$$

and this yields a formula for the p-$K^-$ scattering length A

$$A = \frac{\frac{1}{2}(A_1 + A_0) + |k_0|\, A_0\, A_1}{1 + |k_0|\frac{1}{2}(A_1 + A_0)},$$

identical to that originally obtained by [Dalitz *et al.* (1960)]. The inclusion of the Coulomb correction requires a model and to reduce the necessary algebra to a minimum we are going to use again the delta-shell potential. Within this model the Coulomb corrected scattering length $A^c$ can be very simply expressed in terms of A

$$\frac{1}{A^c} = \frac{1}{A}\frac{R^2}{\phi_0(0,R)^2} + \frac{1}{\phi_0(0,R)}\left[\frac{R}{\phi_0(0,R)} - \theta_0(0,R)\right]$$

where R is the range of the N$\bar{K}$ forces. Assuming that R is much smaller than the Bohr radius $r_B$, the above expression may be simplified, and we have

$$\phi_0(0,R) \approx R; \qquad \theta_0(0,R) \approx 1 - x(2\gamma + \log x),$$

where $\gamma$ is Euler constant and $x = 2R/r_B$. With this approximation the final formula, takes the form

$$\frac{1}{A^c} = \frac{1}{A} + \frac{2}{r_B}(2\gamma + \log\frac{2R}{r_B}),$$

which checks with the expression obtained by [Dalitz *et al.* (1960)]. Using the scattering lengths from Table 14.2 and applying the procedure outlined above we can predict the strong interaction 1s level shift and width in kaonic hydrogen and the corresponding values are presented in Fig. 14.3. Numerically, the correction due to mass difference goes in the opposite directions that the Coulomb correction and there is a good deal of cancellation. The results are largely insensitive to the adopted value of R, as long as R remains small in comparison with the Bohr radius. In our computations we have set R=0.5 fm as this value would be roughly expected if a rather plausible vector-meson-exchange mechanism generates the N$\bar{K}$ forces.

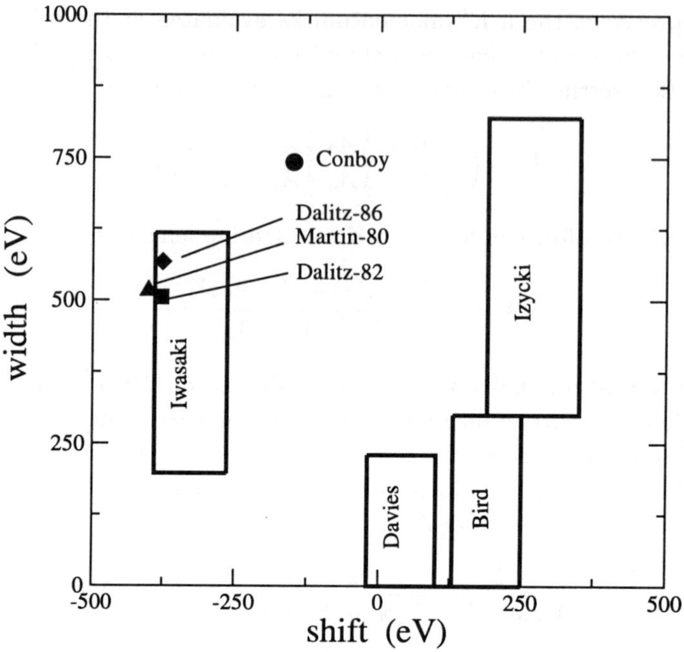


Fig. 14.3 Strong interaction 1s level shift and width in kaonic hydrogen. The rectangles give the region within one standard deviation of the measurements on kaonic hydrogen. They are compared with theoretical predictions deduced in the text from the phenomenological values of the low-energy scattering parameters.

It should be explained here that the different fits presented in Table 14.2 are not based on the same data. Nevertheless, the scattering lengths deduced by [Martin (1981)] and [Dalitz *et al.* (1982); Dalitz *et al.* (1986)] are quite close to each other. In fact, they are not far from the old values obtained by [Kim (1967)] from his own bubble chamber data. The CSL results obtained by [Conboy *et al.* (1986)] are significantly different from the fits obtained by other authors and presented in Table 14.2. In the I=0 $N\bar{K} \rightarrow N\bar{K}$ scattering amplitude reconstructed with the aid of the scattering length proposed by Conboy et al. the $\Lambda(1405)$ resonance has migrated to 1429 MeV and resides only 3 MeV away from the threshold. The main reason why the CSL fit by Conboy et al. differs from the CSL fit by [Dalitz *et al.* (1986)] is that the data sample used in these two fits were not the same. The data set collected by Durham-London-Warsaw collaboration (summarized in [Conboy *et al.* (1986)]) had so much higher precision as compared with the previous data on the low-energy $N\bar{K}$ initiated reactions

Table 14.2 The values of $N\bar{K}$ scattering length $A_I$ for I=0,1 at $pK^-$ threshold deduced from different fits.

| ref. | $A_0$(fm) | $A_1$(fm) | method |
|---|---|---|---|
| [Kim (1967)] | -1.67 +i 0.71 | -0.07+i 0.68 | CSL |
| [Martin (1981)] | -1.70 +i 0.68 | 0.37+i 0.60 | M-mat |
| [Dalitz *et al.* (1982)] | -1.57 +i 0.70 | 0.11+i 0.57 | M-mat |
| [Conboy *et al.* (1986)] | -1.03 +i 0.95 | 0.94+i 0.72 | CSL |
| [Dalitz *et al.* (1986)] | -1.57 +i 0.78 | 0.32+i 0.75 | CSL |
| [Dalitz *et al.* (1986)] | -1.59 +i 0.76 | 0.26+i 0.57 | M-mat |

that the authors felt that their data were on their own sufficient for making an independent determination of the scattering parameters. This resulted in the fit obtained by [Conboy *et al.* (1986)]. By contrast, the CSL fit by [Dalitz *et al.* (1986)] employed the same data set as that used by Conboy et al. *plus* the neutral kaon reaction data left out by Conboy et al. The M-matrix fit by [Dalitz *et al.* (1986)] includes additionally the below threshold data [Hemingway (1985)] and the scattering lengths emerging from that analysis are very close to those obtained from their CSL fit. The largest difference between the [Conboy *et al.* (1986)] CSL solution and [Dalitz *et al.* (1986)] CSL solution occurs in the real part of $\frac{1}{2}(A_0 + A_1)$ but this quantity is very weakly constrained by the scattering data. Complementary evidence, however, is obtained from the kaonic hydrogen X-ray data where the real part of $\frac{1}{2}(A_0 + A_1)$ may be directly related to strong interaction 1s level shift in kaonic hydrogen. As seen from Fig. 14.3, the kaonic hydrogen X-ray data do not seem to favor the Conboy et al. solution.

After so many years of research the low-energy $N\bar{K}$ parameters are still not well determined. Experimentally, we have two regimes: first, above the $pK^-$ threshold, where measurements scarcely exist for incident momenta smaller than 100 MeV/c, and, a second, below $pK^-$ threshold where the interesting physics can be studied only via final state interaction effects. The atomic $pK^-$ system of kaonic hydrogen standing between these two regimes links them, testing them both. In this situation, it is difficult to overestimate the importance of the imminent kaonic hydrogen X-ray experiment, especially that no other experiment bearing on the low-energy characteristics of the $N\bar{K}$ system will be conducted in the foreseeable future.

## 14.3 Antiprotonic Hydrogen

Antiprotonic hydrogen, owing to the similarity to positronium $e^+$-$e^-$ system, has been sometimes referred to as protonium. Notwithstanding, the $\bar{p}p$ system may be deposited in states with total isospin equal one, i.e. carrying quantum numbers not attainable to the $e^+ - e^-$ system. In consequence, the investigation of protonium states offers a new tool in meson spectroscopy capable of including "exotic" (i.e. non-$\bar{q}$-$q$) mesons. Furthermore, it has to be noted that in the $\bar{p}p$ system, not only the scale of the binding energies goes up by three orders of magnitude, but also the ordering of levels will be different than in the case of positronium. This is caused by the large anomalous magnetic moment of the proton increasing the role of the hyperfine interaction in the $\bar{p}p$ system. In consequence, the hyperfine level splitting is of the same size as the fine structure splitting. The relevant energy scale and the size of protonium are obtained by replacing the mass of the electron in the expression for the Rydberg energy by $\frac{1}{2}m$ where m is the proton mass. Describing the atomic orbits classically in terms of their principal quantum number n, the binding energies ($E_n$) and the expectation values of the radii associated with protonium levels, are

$$E_n = \tfrac{1}{2}m\frac{\alpha^2}{2n^2} = \frac{12.491}{n^2}\,keV; \tag{14.22a}$$

$$\langle r\rangle = r_B\,\tfrac{1}{2}[3n^2 - \ell(\ell+1)], \tag{14.22b}$$

where $r_B$=57.64 fm is the classical Bohr radius. The precise calculation of the binding energies includes a number of corrections to the above simple point-charge formula. These, account for effects due to finite size of the proton and the antiproton, relativistic corrections, recoil terms, hyperfine interaction, spin-orbit coupling and various vacuum polarization effects. All such corrections are small and can be treated using perturbation theory.

The earliest attempts to observe antiproton-hydrogen atoms used electrostatically separated, secondary antiproton beams at CERN and also at Brookhaven. Therefore, these experiments, similarly as in the kaonic hydrogen case, were severely limited by insufficient intensity, poor momentum resolution, large beam spot size and high contamination. However, in contrast with kaons, antiproton is a stable particle and today antiproton beams of several $10^{11}$ antiprotons can be stored in the Antiproton Accumulator at CERN. The antiproton beam can be extracted and further accelerated or decelerated. Experimental studies of antiproton atoms reached a new era following the commissioning of the Low-Energy-Antiproton-Ring (LEAR) at CERN in 1983. This facility remained in operation until 1996 and allowed to store $10^{10}$ antiprotons with the potential to accelerate or decelerate antiprotons to all momenta between 70 MeV/c and 2000 MeV/c, to cool antiprotons in phase space, and to extract the particles into external beam lines. This provided intense, high-quality antiproton beams and in the initial phase of the LEAR program three experiments were involved in searching for the characteristic lines of antiprotonic hydrogen atoms. Antiprotonic hydrogen atoms were formed by stopping antiprotons in $H_2$ and these experiments were conducted by using Si(Li) semiconductor detectors, gas scintillation counters, or X-ray drift chambers with which the strong-interaction energy shifts and broadenings of the Lyman $\alpha$ transition in hydrogen were measured directly. The last experiment in this series, used for the first time Charge-Coupled Devices for X-ray detection and this was combined with the cyclotron trap for degrading and stopping of the antiproton beam (a review of the experimental situation prior to 1989 can be found in [Batty (1989)], while more recent developments are summarized in [Gotta *et al.* (1999)])

The hadronic broadening of the protonium p states was found to be about two orders of magnitude larger than the radiative decay width. Therefore, the Balmer transitions are fed almost completely from that part of the cascade which does not involve s or p states and the measured line intensities are insensitive to the hadronic 2p-level width. From the intensity balance of the Lyman $\alpha$ transition and the total Balmer series only a spin-averaged 2p-level width ($\bar{\Gamma}_{2p}^{bal}$) was inferred. The measured world-average values for the spin-averaged strong-interaction parameters in protonium are

$$\bar{\epsilon}_{1s} = -721 \pm 14\ eV; \tag{14.23a}$$

$$\bar{\Gamma}_{1s} = 1097 \pm 42\ eV; \tag{14.23b}$$

$$\bar{\Gamma}_{2p}^{bal} = 32.5 \pm 2.1\ mV. \tag{14.23c}$$

It was also possible to extract the shifts and widths of the individual hyperfine components, ${}^1S_0$ and ${}^3S_1$, but this required additional assumptions (for details, c.f. [Gotta *et al.* (1999)]).

Theoretical interest in the properties of protonium arose a long time before the first measurements were conducted. It was very early noticed that the $\bar{N}N$ interaction is closely related to the NN interaction: the exchange of the same mesons generates forces in both systems. In a one-meson-exchange picture the only difference is a G factor equal to the G-parity of the exchanged meson leading to a sign change of the interaction potential when the G-parity of the exchanged meson is negative. Therefore, any meson exchange NN potential by G-parity transformation may be converted into a $\bar{\rm N}$N potential. This change of sign significantly modifies the total potential, and, e.g. the soft repulsive core originating in the NN potential mainly from $\omega$ exchange, in the $\bar{\rm N}$N case gives rise to a short-range attraction. Spin-orbit forces in the $\bar{\rm N}$N system are rather weak, while the spin-spin interaction is strong – in contrast to the well-known features of the NN interaction. Naturally, the $\bar{\rm N}$N potential obtained that way is not the whole story, as it contributes only to the real part of the interaction. The derivation of the annihilation part of the $\bar{\rm N}$N interaction is still a major challenge. Attempts have been made to construct the annihilation potential from a quark model but the best results are still obtained by resorting to a phenomenological treatment. The traditional methods of treating absorption are to either introduce an absorptive boundary condition or use a complex potential simulating the leakage of the flux into annihilation channels. Perhaps the simplest possibility is adding ad hoc an imaginary part to the meson-exchange potential. This imaginary part contains some parameters which are then adjusted by fitting the $\bar{\rm N}$N scattering data. It turns out that taking a simple short-ranged form, the same in all states, gives already a decent account of the data.

Perhaps the most spectacular result of $\bar{\rm N}$N potential models is the prediction of a rich spectrum of deeply bound states. The possible existence of narrow "baryonium" states has been a topic of great controversy but the initial experimental evidence did not receive confirmation from the high-precision experiments with antiprotons, both at rest and in flight which were conducted at LEAR. One method that has been used to search for baryonium states is to measure the energy distribution of annihilation products from antiprotonic-hydrogen atoms. The formation of a deeply bound baryonium state would be signaled by a monoenergetic $\pi$ or $\gamma$ accompanying the transition from the atomic $\bar{\rm p}$p level to the quasi-nuclear level.

The various $\bar{\rm N}$N potentials have been also used to predict the strong interaction level shifts and widths in antiprotonic hydrogen. Since the neutron proton mass difference is only 1.293 MeV, such calculation requires that the adjacent $\bar{\rm n}$n channel is explicitly included. At large separation the $\bar{\rm p}$p wave function is determined by the Coulomb interaction between the two constituents preserving their identity. At smaller distances, the strong interaction sets in and since the strong interaction potential mixes the $\bar{\rm p}$p and $\bar{\rm n}$n components, the protonium wave function will be a superposition of both of them. Assuming that the strong interaction is isospin invariant the strong interaction potential matrix appropriate for the above two-channel situation is

$$V = \tfrac{1}{2} \begin{pmatrix} V_0 + V_1 & V_0 - V_1 \\ V_0 - V_1 & V_0 + V_1 \end{pmatrix},$$

where $V_i$ are the isospin i=0,1 components of the antinucleon-nucleon interaction. The predictions from a representative sample of theoretical potentials are displayed in Fig. 14.4. The potentials derived by Bryan-Phillips [Bryan *et al.* (1968)], the Paris group [Ĉoté *et al.*(1982)]) Dover-Richard [Dover *et al.* (1980)] and Kaufmann [Kaufmann (1979)] are all based on a meson-exchange model. For comparison we present also results from

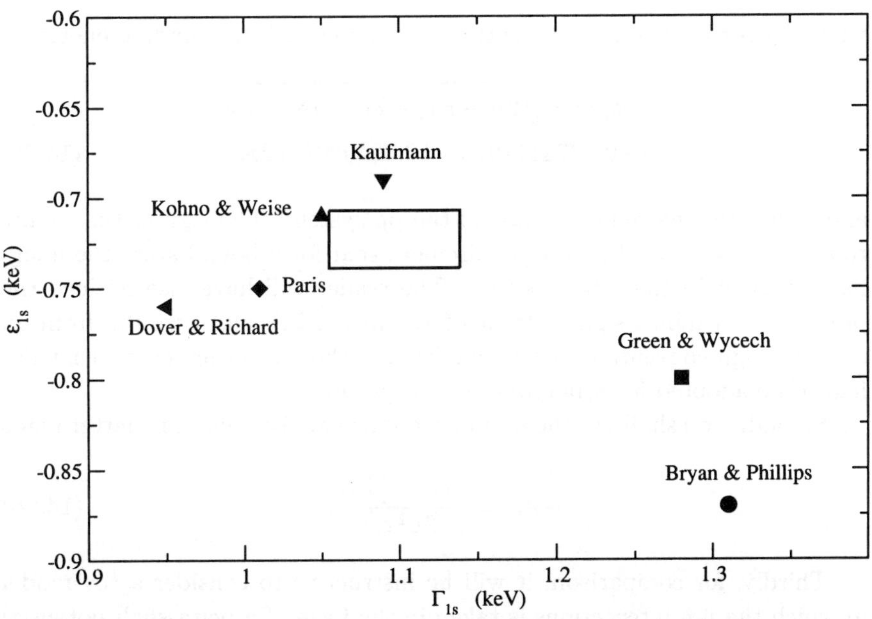


Fig. 14.4 The strong-interaction 1s level shift and width in protonium calculated from different p̄p potentials: The rectangle represents the region within one standard deviation of the measurements reported by [Gotta *et al.* (1999)]

data is some cases is better than in the other but this seems to be rather fortuitous and it is perhaps fair to say that all these models are capable of reproducing the measured values (14.23) in a semi-quantitative way.

Disregarding spin-isospin degrees of freedom, the measured strong interaction 1s level shift $\epsilon_{1s}$ and the width $\Gamma_{1s}$ in protonium may be used to deduce the antiproton-proton Coulomb-corrected complex scattering length $A_C$. The important quantity is here the expansion parameter in the small-shift approximation, i.e. the ratio of the complex energy shift to the unperturbed Coulomb energy. The non-relativistic binding energy corresponding to point Coulomb interaction is $E_1s$=12.491 keV so that this ratio $|\epsilon_{1s} + \mathrm{i}\frac{1}{2}\Gamma_{1s}|/E_{1s}$ is about 0.1 indicating that the first-order Trueman formula might not be sufficiently accurate. Therefore, the scattering length will be obtained by using three methods. Firstly, by application of the master equation in which we replace the K-matrix by the scattering length

$$-A_C \approx \mathcal{K}_0(k_0) = \frac{1}{g_0^+(k_0)}, \tag{14.24}$$

where $k_0$ is the complex momentum associated with the binding energy

$$\begin{aligned} k_0 &= \pm\sqrt{2\mu(-E_{1s} - \epsilon_{1s} - \mathrm{i}\tfrac{1}{2}\Gamma_{1s})} = \\ &= (-0.0774111976 + \mathrm{i}\,3.32408097)\,MeV. \end{aligned} \tag{14.25}$$

and $\mu$ denotes the reduced mass of the $\bar{\rm p}$p system. The sign of the square root in (14.24)is fixed be the requirement that for a bound state the imaginary part of $k_0$ must be positive. The reader will have also noticed the presence of a minus sign in front of $A_C$ in (14.24) which results from an opposite sign convention in the definition of the scattering length that has had to be adopted for a baryon-baryon system.

Secondly, we shall use the first order Trueman formula. The latter reads

$$-A_C = \frac{\epsilon_{1s} + \mathrm{i}\frac{1}{2}\Gamma_{1s}}{2\mu^2\,\alpha^3}. \tag{14.26}$$

Thirdly, for comparison, it will be instructive to consider a toy-model in which the $\bar{p}$-p interactions is taken in the form of a delta-shell potential V(r)

$$2\mu\,V(r) = -s\,\delta(R-r)/R, \tag{14.27}$$

where R is a range parameter and s is a dimensionless complex strength. For assigned value of R, the parameter s may be immediately obtained from the bound state condition, which takes the form of a difference of two logarithmic derivatives, i.e

$$s = R\frac{\phi_0(k_0, R)'}{\phi_0(k_0, R)} - R\frac{f_0^+(k_0, R)'}{f_0^+(k_0, R)}, \tag{14.28}$$

where the right-hand-side may be easily computed by continued fractions summation. With s in hand, the scattering length is obtained from the exact formula

$$-A_C = R\frac{s\,\phi_0(0, R)^2/R^2}{1 - s\,\theta_0(0, R)\,\phi_0(0, R)/R}, \tag{14.29}$$

and we may also calculate the Coulomb-free scattering length

$$-A = R\frac{s}{1 - s}. \tag{14.30}$$

In the above scheme the range R is not known but taking different values of R corresponds to a different model and this way we may study model dependence. The results of our computations conducted for the central values of $\epsilon_{1s}$ and $\Gamma_{1s}$ are presented in table Table 14.3. In accordance

Table 14.3 $A_C$ and A deduced from strong interaction 1s level shift and width in protonium; R is the range of the potential (14.27). All entries are in fermi. The two last rows are computed respectively, from (14.24) and (14.26).

| R | Re($A_C$) | Im($A_C$) | Re(A) | Im(A) |
|---|---|---|---|---|
| 0.25 | 0.8437 | -0.6984 | -0.8658 | 0.8974 |
| 0.50 | 0.8437 | -0.6985 | -0.8739 | 0.8682 |
| 0.75 | 0.8437 | -0.6985 | -0.8811 | 0.8544 |
| 1.00 | 0.8437 | -0.6985 | -0.8882 | 0.8467 |
| 1.25 | 0.8437 | -0.6986 | -0.8953 | 0.8424 |
| 1.50 | 0.8436 | -0.6986 | -0.9025 | 0.8401 |
| 1.75 | 0.8436 | -0.6986 | -0.9097 | 0.8393 |
| 2.00 | 0.8435 | -0.6986 | -0.9170 | 0.8395 |
| master | 0.8437 | -0.6984 | | |
| Trueman | 0.8318 | -0.6328 | | |

with expectations the Coulomb corrected scattering length is indeed almost identical with the value obtained from the master equation (14.24) whereas the accuracy of Trueman formula is at the level of about 10 %. The error in the master formula originates from neglecting the effective range term and is of the order $(R/r_B)^2$, so that for as large as R=2 fm is still at the level of 0.1 %. The Coulomb-free scattering length A is model dependent and varies quite a bit with R.

Taking into account the eperimental errors, given in (14.23), and using the master equation, the Coulomb corrected antiproton-proton scattering length is

$$A_C = [(0.848 \pm 0.018) - (0.699 \pm 0.028)]\, fm. \tag{14.31}$$

In principle, the Coulomb-modified $\bar{\mathrm{p}}\mathrm{p}$ scattering length $\mathrm{A}_C$ might be also determined experimentally by measuring the $\bar{\mathrm{p}}\mathrm{p}$ annihilation cross section close to threshold (cf. [Carbonell *et al.* (1997)]). In presence of absorption the Coulomb-distorted phase shifts $\delta_\ell^{sc}$ will be complex quantities and the angle integrated absorption cross section $\sigma_{abs}$ will be a difference between the total and the elastic cross section

$$\sigma_{abs}(k) = \frac{4\pi}{k^2} \sum_{\ell=0}^{\infty} (2\ell+1) \left[ \mathrm{Im}\, f_\ell^{sc}(k) - |f_\ell^{sc}(k)|^2 \right], \tag{14.32}$$

where k is the c.m. momentum. The Coulomb-distorted strong interaction partial wave amplitude $f_\ell^{sc}(k)$ is related to the K-matrix by the formula $\mathcal{K}_\ell(k)$

$$f_\ell^{sc}(k) = \frac{1}{\cot \delta_\ell^{sc} - \mathrm{i}} = \frac{C_\ell(\eta)^2\, k^{2\ell+1}\, \mathcal{K}_\ell(k)}{1 - g_\ell^+(k)\, \mathcal{K}_\ell(k)}. \tag{14.33}$$

By comparing (14.24) with (14.33) we can see that the $\ell$=0 partial wave amplitude (14.33) exhibits a pole at k=k$_0$, i.e. at the protonium bound state energy. For k sufficiently small, the dominant contribution to the absorption cross section comes from the s-wave and $\ell$=0 K-matrix may be approximated by the scattering length -$\mathrm{A}_C$. Keeping only the s-wave contribution, for small momenta we have

$$\lim_{k\to 0} k^2\, \sigma_{abs}(k) = \frac{8\pi^2 \mathrm{Im}(-A_C)/r_B}{|1 + 2\pi\, A_C/r_B|^2}, \tag{14.34}$$

i.e. the absorption cross section exhibits $1/\mathrm{v}^2$ behavior near threshold where v is the velocity of the incident particle. This change of the usual 1/v law well established for exothermic reactions is caused by the Coulomb interaction, as first pointed out by [Wigner (1948)]. Furthermore, the absorption cross section near threshold is determined by the scattering length $\mathrm{A}_C$. Therefore, knowing the latter from the antiprotonic X-ray experiment, we are in the the position to predict the $\bar{\mathrm{p}}\mathrm{p}$ absorption cross section close to threshold. This has been done in Fig. 14.5 where in order to see the $1/\mathrm{v}^2$ behavior we have plotted the theoretical absorption cross section multiplied by the square of the velocity of the incident antiproton $\beta^2$. The latter cross

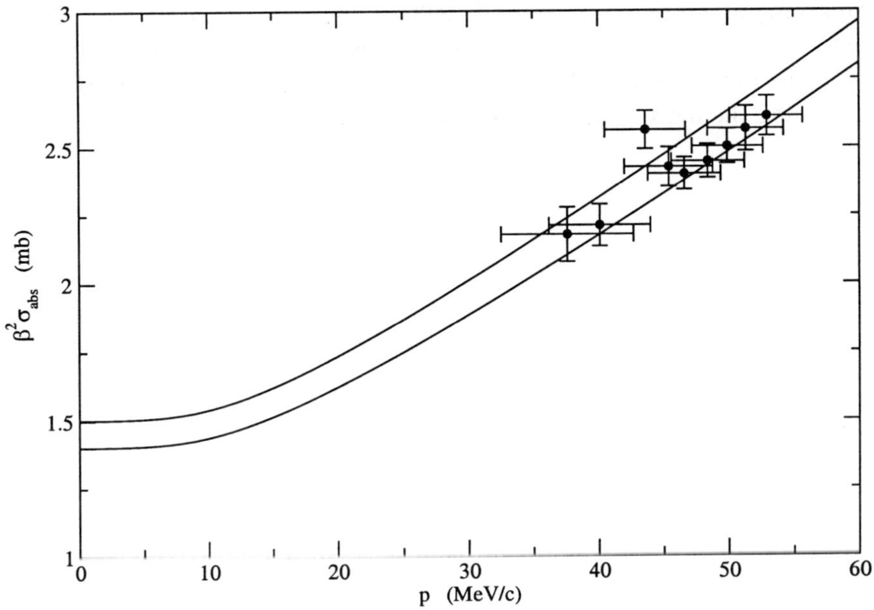


Fig. 14.5 Angle integrated p̄p absorption section multiplied by the square of of the beam velocity $\beta^2$ versus p̄p laboratory momentum.

section times $\beta^2$ indeed settles at a constant value but to see this happen one has to go down with the incident momenta below 10 MeV/c. The experimental data are available only at momenta considerably higher than that and they have been presented in Fig. 14.5. The theoretical curves were obtained by including the contribution from both, the s-wave and the p-wave. The area between the curves corresponds to a one standard deviation region associated with the experimental uncertainties on the scattering lengths (14.31). In the $\ell$=0 partial wave amplitude we used the complete K-matrix instead of taking the scattering length approximation. This K-matrix $\mathcal{K}_0(k)$ was obtained with the aid of the potential (14.27) putting somewhat arbitrarily R=1 fm and subsequently calculating s from (14.28). The explicit form of $\mathcal{K}_0(k)$ for this case is given in (11.17). The theoretical prediction shown in Fig. 14.5 based on protonium X-ray data is in very good agreement with the the experimentally measured absorption cross section. The cross section data are from [Berin *et al.* (1996); Zenoni *et al.* (1999)] and the errors bars represent quadratic addition of the statistical error and the systematical error divided by $\sqrt{12}$. In the above discussion spin and isospin degrees of freedom have been ignored so that an averaging procedure is implied. It has to be noted that close to threshold $k^2 \sigma_{abs}$ is almost linear in $\mathrm{Im}\, A_C$ so that averaging the cross sections with the appropriate statistical weights has essentially the same effect as averaging the individual scattering lengths.

# Chapter 15

# Hadronic Deuterium

## 15.1 Pionic Deuterium

The central issue we wish to address in this section is how to construct a theoretical framework in which we can use the pionic deuterium data to gain information on the $\pi$N scattering lengths. The measurement of the shift and the width of the 1s level in pionic deuterium presents us with the value of $\pi$d scattering length $a_{\pi d}$. The latter quantity is defined as the elastic $\pi$d scattering amplitude evaluated at zero kinetic energy of the incident pion. This amplitude is necessarily complex because absorption reaction channels are open even at the very threshold. The most important of them is the $\pi^- d \to nn$ reaction, and to a lesser extend the radiative absorption $\pi^- d \to \gamma nn$ channel. In principle, there would be also the charge-exchange break-up channel $\pi^- d \to \pi^0 nn$ that is open at threshold but this process is strongly suppressed by the centrifugal barrier. Indeed, with s-wave $\pi$N interaction there is no spin-flip possible so that for the two neutrons the ${}^1S_0$ state is not available, whereas the ${}^3S_1$ state is forbidden and they have to be produced in higher partial waves. On the whole, however, the absorptive effects are not large at threshold, judging from the magnitude of the imaginary part of the $\pi$d scattering length which empirically constitutes only about a quarter of the real part of $a_{\pi d}$. Strictly speaking, the absorptive processes contribute to both, the real and the imaginary part of $a_{\pi d}$ but in the following we are going to ignore the absorptive corrections to the real part of $a_{\pi d}$. Disregarding the absorptive processes, we shall concentrate our attention on a microscopic calculation of $a_{\pi d}$ and in order to be able to solve the ensuing three-body problem we introduce a potential description of the $\pi$N interaction to be used in the appropriate Faddeev equations.

In order to facilitate the discussion of the Faddeev approach, it is instructive to take the static model as our point of departure. The attractive feature of the static model is that it is much easier to develop and to compute since the final result for pion-deuteron scattering length takes the form of a single analytic formula that does not require off-shell information. Moreover, in our case the latter model also happens to be extremely good approximation to the full solution of the three-body problem. The earliest version of a static model, due to [Brueckner (1953)], was based on the fixed scatterer concept and ignored all isospin complications. Here, we wish to make it somewhat more realistic introducing as our dynamical framework a set of appropriate boundary conditions, but on the other hand, we are prepared to content ourselves with a theory that has isospin invariant point like interactions. Labeling the pion as 1 and the nucleons as 2 and 3, the boundary conditions representing the zero-range $\pi$-N interaction taking place on nucleon $i$ where $i = 2, 3$, may be written as

$$\begin{aligned} &\lim_{x_1 \to x_i} \overline{|\mathbf{x}_1 - \mathbf{x}_i| \Psi(\mathbf{x}_1, \mathbf{x}_2, \mathbf{x}_3)} = \\ &= \frac{\mu}{m}(b_0 + b_1 \mathbf{I} \cdot \boldsymbol{\tau}_i) \lim_{x_1 \to x_i} \frac{\mathrm{d}}{\mathrm{d}x_1} \overline{|\mathbf{x}_1 - \mathbf{x}_i| \Psi(\mathbf{x}_1, \mathbf{x}_2, \mathbf{x}_3)} \end{aligned} \tag{15.1}$$

where the bar denotes an average over directions $\mathbf{x}_1 - \mathbf{x}_i$ what is equivalent to projecting out the s-wave component of the wave function $\Psi$, and the boundary condition (15.1) is to be imposed for each of the two nucleons. The vectors $\mathbf{I}$ and $\boldsymbol{\tau}$ are, respectively, the pion and the nucleon isospin operators, whereas $b_0$ and $b_1$ denote the isoscalar and isovector $\pi$-N scattering lengths, $\mu$ is the $\pi$-N reduced mass and $m$ is the pion mass. In the following we choose the c.m. of the two nucleons as the origin of the coordinate system, i.e. we set $\mathbf{x}_2 = \frac{1}{2}\mathbf{r}$ and $\mathbf{x}_3 = -\frac{1}{2}\mathbf{r}$ with $\mathbf{r}$ being the nucleon-nucleon separation vector. The pion vector in this Jacobi coordinate system will be denoted as $\boldsymbol{\rho}$. When the wave function $\Psi(\mathbf{r}, \boldsymbol{\rho})$ describing the $\pi$NN system for the case of $\pi^-$ scattered off the deuteron is known, the amplitude leading to the final state with asymptotic wave function $\Phi_f$ is $-\langle \Phi_f | V | \Psi \rangle$ where $V$ denotes the potentials that have been taken out in the derivation of $\Phi_f$. For elastic scattering $\Phi_f(\boldsymbol{\rho}, \mathbf{r}) = \exp(\mathrm{i}\mathbf{p}' \cdot \boldsymbol{\rho})\, \psi_d(\mathbf{r})$ where $\mathbf{p}'$ is the momentum of the outgoing pion, $\psi_d$ is the deuteron wave function and $V$ is the sum of the two $\pi$N potentials as asymptotically there is no $\pi$-deuteron interaction. Although in our formalism we never needed $\pi$N potentials and the $\pi$N interaction is represented by the boundary condition (15.1), it is in fact possible to give a formal expression for such potential (cf. [Huang

(1963)]) and for the operator $V$ we take

$$V\Psi(\boldsymbol{\rho},\mathbf{r}) = -\frac{2\pi}{\mu}\Big\{(b_0 + b_1\,\mathbf{I}\cdot\boldsymbol{\tau}_2)\,\delta(\boldsymbol{\rho}-\tfrac{1}{2}\mathbf{r})\frac{\mathrm{d}}{\mathrm{d}\rho}|\boldsymbol{\rho}-\tfrac{1}{2}\mathbf{r}| + \\ + (b_0 + b_1\,\mathbf{I}\cdot\boldsymbol{\tau}_3)\,\delta(\boldsymbol{\rho}+\tfrac{1}{2}\mathbf{r})\frac{\mathrm{d}}{\mathrm{d}\rho}|\boldsymbol{\rho}+\tfrac{1}{2}\mathbf{r}|\Big\}\,\Psi(\boldsymbol{\rho},\mathbf{r}). \tag{15.2}$$

Denoting the incident pion momentum as $\mathbf{p}$ and making use of the boundary conditions (15.1) in (15.2), the $\pi$-d elastic scattering amplitude $f(\mathbf{p}',\mathbf{p})$ takes the form

$$f(\mathbf{p}',\mathbf{p}) = \frac{\nu}{m}\int \mathrm{e}^{-\mathrm{i}\mathbf{p}'\cdot\boldsymbol{\rho}}\,\psi_d^{\dagger}(\mathbf{r})\,\{\delta(\boldsymbol{\rho}-\tfrac{1}{2}\mathbf{r})|\boldsymbol{\rho}-\tfrac{1}{2}\mathbf{r}| + \\ + \delta(\boldsymbol{\rho}+\tfrac{1}{2}\mathbf{r})|\boldsymbol{\rho}+\tfrac{1}{2}\mathbf{r}|\}\,\Psi(\boldsymbol{\rho},\mathbf{r})\mathrm{d}^3\rho\,\mathrm{d}^3r, \tag{15.3}$$

where $\nu$ is $\pi$-d reduced mass. Given the elastic $\pi$-d scattering amplitude (15.3), the $\pi$-d scattering length follows immediately from

$$a_{\pi d} = f(0,0). \tag{15.4}$$

With the $\pi$-N interaction assumed to be isospin invariant, it will be convenient for us to adopt an isospin notation. For the initial $\pi^-$-d system, the isotopic spin wave function has the form

$$\chi_a = \pi^- \tfrac{1}{\sqrt{2}}(p_2n_3 - n_2p_3), \tag{15.5}$$

where the symbols $p, n, \pi^-$ in (15.5) stand for the isospin wave functions of the corresponding particles. The wave function (15.5) is antisymmetric in the nucleon labels, as appropriate for the state where the isospin of the two-nucleon subsystem $I_{23}$ equals zero. As a result of the interaction, the two nucleons can undergo a transition to a symmetric configuration corresponding to $I_{23} = 1$ and we shall need also a function that is symmetric under two-nucleon permutation

$$\chi_s = \tfrac{1}{2}\pi^-(p_2n_3 + n_2p_3) - \tfrac{1}{\sqrt{2}}\pi^0 n_2n_3. \tag{15.6}$$

Since our interest here is confined to s-wave interactions, no spin flip is possible and therefore the spin part of the wave function does not change. Regarding the nucleons as fixed scattering centers, we may anticipate that the wave function $\Psi(\boldsymbol{\rho},\mathbf{r})$ for the full system of the target nucleons and the meson will take the approximate form

$$\Psi(\boldsymbol{\rho}, \mathbf{r}) = e^{i\mathbf{p}\cdot\boldsymbol{\rho}}\, u_d(r)\, \chi_a + A(\mathbf{r}) \left[ \frac{\exp(ip|\boldsymbol{\rho} - \frac{1}{2}\mathbf{r}|)}{|\boldsymbol{\rho} - \frac{1}{2}\mathbf{r}|} + \frac{\exp(ip|\boldsymbol{\rho} + \frac{1}{2}\mathbf{r}|)}{|\boldsymbol{\rho} + \frac{1}{2}\mathbf{r}|} \right] \chi_a +$$
$$+ X(\mathbf{r}) \left[ \frac{\exp(ip|\boldsymbol{\rho} - \frac{1}{2}\mathbf{r}|)}{|\boldsymbol{\rho} - \frac{1}{2}\mathbf{r}|} - \frac{\exp(ip|\boldsymbol{\rho} + \frac{1}{2}\mathbf{r}|)}{|\boldsymbol{\rho} + \frac{1}{2}\mathbf{r}|} \right] \chi_s, \qquad (15.7)$$

where $u_d$ is the spatial part of deuteron wave function that includes also the deuteron spin and in particular may contain also the D-component. The projectile enters with momentum $\mathbf{p}$ and in the initial asymptotic region the pion and the target have separate wave functions (a plane wave and $u_d(r)$, respectively) and the propagation from one scattering center to another is described by a superposition of spherical waves. The hitherto unknown amplitudes denoted in (15.7), respectively, as $A(\mathbf{r})$ and $X(\mathbf{r})$ multiplying these outgoing waves emitted by the two centers account for the multiple scattering phenomena. They will be determined from the boundary conditions (15.1). To satisfy Pauli principle the wave function (15.7) must be antisymmetric in the two nucleon variables. This implies that we have to stipulate that the coefficients $A(\mathbf{r})$ and $X(\mathbf{r})$ are even under permutation of the nucleons, i.e. they must be invariant under the reflections $\mathbf{r} \to -\mathbf{r}$. For zero-energy scattering considered in this work, however, this is always the case because $A(r)$ and $X(r)$ depend then only upon the magnitude of $\mathbf{r}$. It is worth noting that the wave function (15.7) includes explicitly virtual charge exchange amplitude $X(r)$. Since our interest here is confined to zero-energy scattering, in the following we take $p = 0$ in (15.7). Equations for the functions $A(r)$ and $X(r)$ may be obtained by substituting (15.7) in (15.1) for $i = 2$ and equating the coefficients multiplying the same isospin functions. With two different isospin functions we obtain two equations and this procedure determines uniquely $A(r)$ and $X(r)$. Owing to the proper antisymmetrization of our wave function the boundary condition for $i = 3$ will be then automatically satisfied. The equations obtained from (15.1), are

$$A(r) = \tilde{b}_0 u_d(r) + \frac{\tilde{b}_0}{r} A(r) + \sqrt{2}\,\frac{\tilde{b}_1}{r} X(r), \qquad (15.8a)$$

$$-X(r) = \sqrt{2}\,\tilde{b}_1 u_d(r) + \sqrt{2}\frac{\tilde{b}_1}{r} A(r) + \frac{\tilde{b}_0 + \tilde{b}_1}{r} X(r). \qquad (15.8b)$$

In (15.8) we introduced the abbreviation $\tilde{b}_j = (1 + m/M) b_j$ where $M$ is the nucleon mass. The $\pi$-d scattering length is given by the overlap integral

$$a_{\pi d} = (2\nu/m) \int u_d(r)^\dagger A(r)\, \mathrm{d}^3 r \tag{15.9}$$

where $A(r)$ is the solution of (15.8)

$$A(r) = \frac{\tilde{b}_0 + (\tilde{b}_0 + \tilde{b}_1)(\tilde{b}_0 - 2\tilde{b}_1)/r}{1 - \tilde{b}_1/r - (\tilde{b}_0 + \tilde{b}_1)(\tilde{b}_0 - 2\tilde{b}_1)/r^2}\, u_d(r). \tag{15.10}$$

Using (15.10) in (15.9) and expanding $A(r)$ in powers of the $\pi$N scattering lengths, we retrieve the well known second order formula for the $\pi$-d scattering length (cf. [Ericson *et al.* (1988)])

$$a_{\pi d}^{(2)} = \frac{2\nu}{m}\left[\tilde{b}_0 + (\tilde{b}_0^2 - 2\,\tilde{b}_1^2)\left\langle \frac{1}{r} \right\rangle\right], \tag{15.11}$$

where the expectation value is taken with respect to the deuteron wave function. As advertised at the beginning of this section, formula (15.10) provides a complete solution of the problem. To examine the accuracy of the static formula we have to compare it with the exact solution of the three-body problem. The latter will be obtained by solving the Faddeev equations on which we now embark.

To solve the Faddeev equations it will be convenient for us to work in momentum space. Introducing the Faddeev partitions, we write the three-body wave function as

$$\Psi = \psi^{(1)}(\mathbf{q}_1, \mathbf{k}_1) + \psi^{(2)}(\mathbf{q}_2, \mathbf{k}_2) + \psi^{(3)}(\mathbf{q}_3, \mathbf{k}_3), \tag{15.12}$$

where $\mathbf{q}_1$ denotes the relative momentum of the (23) pair whereas $\mathbf{k}_1$ is the c.m. momentum of particle 1 and cyclic permutations are implied. To obtain Faddeev equations for the amplitudes, the different partitions are written as (cf. [Peresypkin *et al.* (1974)])

$$\begin{aligned}\psi^{(1)}(\mathbf{q}, \mathbf{k}) &= (2\pi)^3 \phi(\mathbf{q}) \delta(\mathbf{k} - \mathbf{p}) \chi_a + \\ &\quad + \frac{F(\mathbf{q}, \mathbf{k})\chi_a + G(\mathbf{q}, \mathbf{k})\chi_s}{E - q^2/M - k^2/2\nu}; \end{aligned} \tag{15.13a}$$

$$\psi^{(2)}(\mathbf{q}, \mathbf{k}) = \frac{A(-\mathbf{q}, \mathbf{k})\chi_a - X(-\mathbf{q}, \mathbf{k})\chi_s}{E - q^2/2\mu - k^2/2\nu_N}; \tag{15.13b}$$

$$\psi^{(3)}(\mathbf{q}, \mathbf{k}) = \frac{A(\mathbf{q}, \mathbf{k})\chi_a + X(\mathbf{q}, \mathbf{k})\chi_s}{E - q^2/2\mu - k^2/2\nu_N}; \tag{15.13c}$$

where $\nu_N$ is the reduced mass of the nucleon and that of the $\pi$N pair, $E$ is the c.m. three particle kinetic energy and $\phi(\mathbf{q})$ is the deuteron wave function in the momentum space. In (15.13) we have introduced four scattering amplitudes $F(\mathbf{q},\mathbf{k}), G(\mathbf{q},\mathbf{k}), A(\mathbf{q},\mathbf{k})$ and $X(\mathbf{q},\mathbf{k})$. However, the amplitude $G(\mathbf{q},\mathbf{k})$ to be non-zero requires at least p-wave NN interaction and therefore will be excluded from our considerations, while the three remaining amplitudes will be determined from the Faddeev equations. It is evident from (15.13) that under the $P_{23}$ permutation $\psi^{(1)} \to -\psi^{(1)}$ and $\psi^{(2)} \leftrightarrow -\psi^{(3)}$, so that the total wave function is, as required, antisymmetric in the nucleon labels. Assuming exact isospin conservation, we can write the Faddeev equations

$$F(\mathbf{q},\mathbf{k}) = \int \frac{\mathrm{d}^3 k'}{(2\pi)^3} \frac{\left\langle \mathbf{q} \left| t(E - \frac{k^2}{2\nu}) \right| \frac{1}{2}\mathbf{k} + \mathbf{k}' \right\rangle + \left\langle \mathbf{q} \left| t(E - \frac{k^2}{2\nu}) \right| -\frac{1}{2}\mathbf{k} - \mathbf{k}' \right\rangle}{E - \frac{1}{2\mu}\left(\mathbf{k} + \frac{\mu}{M}\mathbf{k}'\right)^2 - \frac{k'^2}{2\nu_N}} \times A(\mathbf{k} + \mathbf{k}'\frac{\mu}{M}, \mathbf{k}'); \tag{15.14a}$$

$$\begin{aligned} A(\mathbf{q},\mathbf{k}) &= \left\langle \mathbf{q} \left| t_0(E - \frac{k^2}{2\nu_N}) \right| \frac{\mu}{M}\mathbf{k} + \mathbf{p} \right\rangle \phi(\mathbf{k} + \tfrac{1}{2}\mathbf{p}) + \\ &+ \int \frac{\mathrm{d}^3 k'}{(2\pi)^3} \frac{\left\langle \mathbf{q} \left| t_0(E - \frac{k^2}{2\nu_N}) \right| \frac{\mu}{M}\mathbf{k} + \mathbf{k}' \right\rangle}{E - \frac{1}{M}\left(\mathbf{k} + \frac{1}{2}\mathbf{k}'\right)^2 - \frac{k'^2}{2\nu}} F(-\mathbf{k} - \tfrac{1}{2}\mathbf{k}', \mathbf{k}') + \\ &+ \int \frac{\mathrm{d}^3 k'}{(2\pi)^3} \frac{\left\langle \mathbf{q} \left| t_0(E - \frac{k^2}{2\nu_N}) \right| -\frac{\mu}{m}\mathbf{k} - \mathbf{k}' \right\rangle}{E - \frac{1}{2\mu}\left(\mathbf{k} + \frac{\mu}{m}\mathbf{k}'\right)^2 - \frac{k'^2}{2\nu_N}} A(-\mathbf{k} - \frac{\mu}{m}\mathbf{k}', \mathbf{k}') + \\ &+ \sqrt{2} \int \frac{\mathrm{d}^3 k'}{(2\pi)^3} \frac{\left\langle \mathbf{q} \left| t_1(E - \frac{k^2}{2\nu_N}) \right| -\frac{\mu}{m}\mathbf{k} - \mathbf{k}' \right\rangle}{E - \frac{1}{2\mu}\left(\mathbf{k} + \frac{\mu}{m}\mathbf{k}'\right)^2 - \frac{k'^2}{2\nu_N}} X(-\mathbf{k} - \frac{\mu}{m}\mathbf{k}', \mathbf{k}'); \end{aligned} \tag{15.14b}$$

$$-X(\mathbf{q},\mathbf{k}) = \sqrt{2}\left\langle \mathbf{q}\left| t_1(E-\frac{k^2}{2\nu_N})\right| \frac{\mu}{M}\mathbf{k}+\mathbf{p}\right\rangle \phi(\mathbf{k}+\tfrac{1}{2}\mathbf{p})+$$

$$+\sqrt{2}\int\frac{\mathrm{d}^3\,k'}{(2\pi)^3}\,\frac{\left\langle \mathbf{q}\left| t_1(E-\frac{k^2}{2\nu_N})\right| \frac{\mu}{M}\mathbf{k}+\mathbf{k}'\right\rangle}{E-\frac{1}{M}\left(\mathbf{k}+\frac{1}{2}\mathbf{k}'\right)^2-\frac{k'^2}{2\nu}}\,F(-\mathbf{k}-\tfrac{1}{2}\mathbf{k}',\,\mathbf{k}')+$$

$$+\int\frac{\mathrm{d}^3\,k'}{(2\pi)^3}\,\frac{\left\langle \mathbf{q}\left| t_0(E-\frac{k^2}{2\nu_N})-t_1(E-\frac{k^2}{2\nu_N})\right| -\frac{\mu}{m}\mathbf{k}-\mathbf{k}'\right\rangle}{E-\frac{1}{2\mu}\left(\mathbf{k}+\frac{\mu}{m}\mathbf{k}'\right)^2-\frac{k'^2}{2\nu_N}}\times$$

$$\times\; X(-\mathbf{k}-\frac{\mu}{m}\mathbf{k}',\,\mathbf{k}')+$$

$$+\sqrt{2}\int\frac{\mathrm{d}^3\,k'}{(2\pi)^3}\,\frac{\left\langle \mathbf{q}\left| t_1(E-\frac{k^2}{2\nu_N})\right| -\frac{\mu}{m}\mathbf{k}-\mathbf{k}'\right\rangle}{E-\frac{1}{2\mu}\left(\mathbf{k}+\frac{\mu}{m}\mathbf{k}'\right)^2-\frac{k'^2}{2\nu_N}}\,A(-\mathbf{k}-\frac{\mu}{m}\mathbf{k}',\,\mathbf{k}'); \tag{15.14c}$$

where in (15.14) $\langle \mathbf{q}'|t(E)|\mathbf{q}\rangle$ is the NN scattering t-matrix for zero isospin and $\langle \mathbf{q}'|t_j(E)|\mathbf{q}\rangle$ are, respectively, the isoscalar ($j=0$) and isovector ($j=1$) $\pi$N scattering t-matrices. The elastic scattering amplitude is given by the expression

$$f(\mathbf{p}',\mathbf{p}) = \lim_{p'\to p}\frac{p'^2-p^2}{4\pi}\int\phi(\mathbf{q})^\dagger\frac{F(\mathbf{q},\mathbf{p}')}{E-q^2/M-p'^2/2\nu}\,\frac{\mathrm{d}^3q}{(2\pi)^3} \tag{15.15}$$

and the scattering length is obtained from (15.4). We can use (15.14a) to eliminate $F(\mathbf{q},\mathbf{k})$ in (15.15) in favor of the amplitude $A(\mathbf{q},\mathbf{k})$. In the NN scattering matrices occurring in (15.14), as a result of the limiting procedure, only the deuteron pole contributes and scattering length is given as an overlap integral

$$a_{\pi d} = -\frac{\nu}{\pi}\int\phi(k)^\dagger A(\mathbf{k}\,\frac{\mu}{M},\mathbf{k})\,\frac{\mathrm{d}^3k}{(2\pi)^3}. \tag{15.16}$$

The above formula is analogous to (15.9), and, in fact, the static approximation results (15.9)–(15.10) could have been derived from the Faddeev formalism. In order to demonstrate that (15.9)–(15.10) follow from (15.14) we note that when the nucleons are static they are not supposed to scatter

($t \to 0$) and the amplitude $F(\mathbf{q}, \mathbf{k})$ drops out in (15.14b) and (15.14c) so that we are left with only two coupled integral equations. When the underlying forces are of zero range, the off-shell $\pi$N scattering amplitudes can be simplified, and in that case

$$\langle \mathbf{q}'|t_j(E)|\mathbf{q}\rangle = -\frac{2\pi}{\mu}\frac{b_j}{1+\kappa b_j}, \qquad j = 0, 1;$$

where $\kappa^2 = 2\mu B$ and $B$ is the binding energy of the deuteron. The important consequence of the zero-range assumption, apparent from the above formula, is that the t-matrices become independent upon the off-shell momenta. Therefore, the amplitudes $A(\mathbf{q}, \mathbf{k})$ and $X(\mathbf{q}, \mathbf{k})$ will be functions of one variable only and it will be convenient for us to introduce a notation that emphasizes that fact, setting $A(\mathbf{q}, \mathbf{k}) = -(m/2\pi)\mathcal{A}(k)$ and $X(\mathbf{q}, \mathbf{k}) = -(m/2\pi)\mathcal{X}(k)$, where $\mathcal{A}(k)$ and $\mathcal{X}(k)$ are two, hitherto unknown amplitudes. With static nucleons, the energy denominators in (15.14b) and (15.14c) become all equal to $-B - (\mathbf{k}' + \mathbf{k})^2/2m$ and we end up with the following set of integral equations for the amplitudes $\mathcal{A}(k)$ and $X(k)$:

$$\begin{aligned}\mathcal{A}(k) = \hat{b}_0\phi(k) + 4\pi\,\hat{b}_0 \int \frac{\mathrm{d}^3k'}{(2\pi)^3}\frac{\mathcal{A}(k')}{\kappa^2 + (\mathbf{k}' + \mathbf{k})^2} +\\ + \sqrt{2}\,4\pi\hat{b}_1 \int \frac{\mathrm{d}^3k'}{(2\pi)^3}\frac{\mathcal{X}(k')}{\kappa^2 + (\mathbf{k}' + \mathbf{k})^2};\end{aligned} \tag{15.17a}$$

$$\begin{aligned}-\mathcal{X}(k) = \sqrt{2}\hat{b}_1\phi(k) + 4\pi(\hat{b}_0 - \hat{b}_1) \int \frac{\mathrm{d}^3k'}{(2\pi)^3}\frac{\mathcal{X}(k')}{\kappa^2 + (\mathbf{k}' + \mathbf{k})^2} +\\ + \sqrt{2}\,4\pi\hat{b}_1 \int \frac{\mathrm{d}^3k'}{(2\pi)^3}\frac{\mathcal{A}(k')}{\kappa^2 + (\mathbf{k}' + \mathbf{k})^2},\end{aligned} \tag{15.17b}$$

where

$$\hat{b}_j = b_j\,\frac{1 + m/M}{1 + \kappa b_j}; \qquad j = 0, 1. \tag{15.18}$$

The above set of integral equations can be immediately solved by introducing the Fourier transform

$$A(r) = \int e^{\mathbf{ikr}}\mathcal{A}(k)\mathrm{d}^3k \tag{15.19}$$

together with a similar relationship for $\mathcal{X}(k)$ and $\phi(k)$ and using the well known formula

$$\frac{4\pi}{\kappa^2 + (\mathbf{k}+\mathbf{k}')^2} = \int e^{-\mathrm{i}(\mathbf{k}+\mathbf{k}')\mathbf{r}} \frac{e^{-\kappa r}}{r} \mathrm{d}^3 r.$$

In order to solve (15.17) we multiply the latter equations by $e^{\mathbf{ikr}}$ and subsequently integrate them over $\mathbf{k}$. As a result, we obtain a set of two algebraic equations for $A(r)$ and $X(r)$ that differ from (15.8) only by $\exp(-\kappa r)/r$ replacing $1/r$ and $\hat{b}_j$ replacing $\tilde{b}_j$. Since (15.16) goes over into (15.9), we are led to the extension of the static formula (15.10)

$$A(r) = \frac{\hat{b}_0 + (\hat{b}_0 + \hat{b}_1)(\hat{b}_0 - 2\hat{b}_1)e^{-\kappa r}/r}{1 - \hat{b}_1 e^{-\kappa r}/r - (\hat{b}_0 + \hat{b}_1)(\hat{b}_0 - 2\hat{b}_1)e^{-2\kappa r}/r^2}\, u_d(r). \tag{15.20}$$

This formula is to be used in (15.9) but now accounts for the binding energy correction.

Concluding our discussion of the static model we wish to recall that a closed form expression for $\pi$d scattering length has been also obtained by effecting an explicit summation of Feynman diagrams and the most complete treatment can be found in [Kolybasow *et al.* (1973)]. The ultimate static formula for $a_{\pi d}$, that takes into account isospin degree of freedom, given in [Kolybasow *et al.* (1973)] is rather complicated and at first sight appears to be different from (15.20). However, a closer inspection reveals that the authors of [Kolybasow *et al.* (1973)] apparently did not realize that their fractional formula for $a_{\pi d}$ could have been significantly simplified because a common factor equal

$$1 + \tilde{b}_1 \frac{\mathrm{e}^{-\kappa r}}{r} - (\tilde{b}_0 + \tilde{b}_1)(\tilde{b}_0 - 2\tilde{b}_1)\frac{\mathrm{e}^{-2\kappa r}}{r^2}$$

may be pulled out both, from the numerator, as well as from the denominator and eventually drops out. Indeed, when the redundant factor has been canceled, the resulting expression is identical with (15.20). Therefore, when binding corrections are disregarded, this approach reproduces the static model result (15.10) and it is reassuring that in this case all three methods give the same answer.

To improve upon the static model one needs a numerical solution of the Faddeev equations and in the following, similarly as in the previous calculations [Peresypkin *et al.* (1974); Afnan *et al.* (1974); Mizutani *et al.* (1977)], in order to reduce the computational effort, all the pairwise interactions invoked will be represented by rank-one separable potentials.

The $\pi$N s-wave interaction is taken in the form of a standard Yamaguchi potential with the same form factor in both isospin states. Since the inverse range parameter $\beta$ that enters that form factor is not known, similarly as before, we consider the zero-range limit, i.e. $\beta \to \infty$. For an assigned value of $\beta$, the strength parameter of the potential may be eliminated in favor of the scattering length and the appropriate s-wave t-matrices, are

$$\begin{aligned}\langle k|t_j(E)|k'\rangle = &-\frac{2\pi}{\mu}\frac{1}{1+k^2/\beta^2}\times \\ &\times \frac{b_j}{1-\mathrm{i}pb_j\,(1-2\mathrm{i}p/\beta)(1-\mathrm{i}p/\beta)^{-2}}\,\frac{1}{1+k'^2/\beta^2}\end{aligned} \tag{15.21}$$

where $p = \sqrt{2\mu E}$ and $j = 0, 1$ and it is evident from (15.21) that the zero-range limit can be effected. When the nucleon motion is taken into account, the p-wave $\pi$N interaction gives contribution to the $\pi$d scattering amplitude even at threshold. Therefore, in addition to s-wave, we are going to include also the p-wave interaction, limiting ourselves only to the P33 wave as in that case both the strength and the statistical weight are dominant rendering the remaining p-waves negligible. The corresponding p-wave form factor of the form

$$g_\Delta(k) = \frac{k}{k^2+\beta_\Delta^2}$$

has been adopted from [Afnan *et al.* (1974)] with $\beta_\Delta = 5.33\,fm^{-1}$ where the depth of the separable potential can be adjusted to the experimentally known value of the P33 scattering volume taken to be $0.64\,fm^3$. It is well known that with the above form, the shape of the delta resonance cannot be well reproduced but this is less important here, the essential thing is to have the P33 amplitude at threshold correctly reproduced. Besides, the p-wave constitutes only a small correction and using a more complicated model does not seem to be currently justified. For the NN interaction we use two separable models: a simple Hulthen-Yamaguchi potential with inverse range parameter equal $\beta_N = 6.01162\,\sqrt{MB}$ whose strength is fixed by the deuteron binding energy, and the PEST potential constructed by [Zankel *et al.* (1983)] with a more sophisticated form factor of the form

$$g(k) = \sum_{i=1}^{6} \frac{C_i}{k^2+\beta_i^2}, \tag{15.22}$$

where the parameters $C_i$ and $\beta_i$ have been tabulated by [Zankel *et al.* (1983)]. This potential has been devised in such a way that the corresponding NN half-off-shell T-matrix has the same behavior as that of the Paris potential. [Lacombe *et al.* (1981)] This separable replica of the Paris potential takes into account the short range repulsion that is absent in the Yamaguchi potential yet retaining the simplicity of the latter.

Using standard partial wave projections the Faddeev equations (15.14) can be reduced to a system of four coupled inhomogeneous integral equations in a single variable that are amenable for numerical treatment. In the actual practice, in order to cross-check our numerical procedures, we used two independent methods of solving these equations. The direct method introduces an integration mesh what allows us to replace integrals by sums so that the integral equations take the form of a system of linear algebraic equations easily solvable by standard methods. The second method solves the system of integral equations by successive iterations. The iterative procedure is equivalent to a power expansion in $\pi$N scattering lengths what allows tracing down the contribution from the different orders. Since the scattering lengths are rather small, as compared with the deuteron size, the iterative sequence proves to be rapidly convergent.

The experimental $\pi$-d scattering length has been extracted form the 1s level shift in pionic deuterium by using the Deser–Trueman [Deser *et al.* (1954)] formula. Therefore, the extracted quantity is in fact the Coulomb corrected scattering length, denoted hereafter as $a^c_{\pi d}$, and before confronting the calculated pion-deuteron scattering length with experiment one needs the experimental value of $a_{\pi d}$, i.e. the purely nuclear scattering length. Of course, Coulomb correction could be anticipated to be very small but since the experimental errors are also small, it is of interest to give some quantitative estimate of the Coulomb correction. In principle, for calculating the latter one needs to know the pion-deuteron nuclear potential responsible for the level shift. This potential is not known but with the zero-range potential simulating the $\pi$N interaction in the first approximation it is reasonable to expect that the effective potential is proportional to the nuclear density, so that the shape of the nuclear potential is given by the square of the deuteron wave function $u_d(2r)^2$. Still, the depth is not known, but on the nuclear scale this potential must be rather weak because the experimental value of $a^c_{\pi d}$ is quite small. Therefore, to quantify the value of the ratio of $a^c_{\pi d}/a_{\pi d}$ it is sufficient to keep only the first order terms in the nuclear potential. Since in this case the potential depth drops out, we are led to the formula

$$\frac{a^{\mathrm{c}}_{\pi d}}{a_{\pi d}} = \frac{\int_0^\infty u_d(2r)^2\,\phi_0(0,r)^2\,\mathrm{d}r}{\int_0^\infty u_d(2r)^2\,r^2\,\mathrm{d}r}, \tag{15.23}$$

where $\phi_\ell(k,r)$ denotes the regular Coulomb wave function that for zero-momentum (k=0) and zero orbital momentum ($\ell$=0), simplifies to the form (cf. (3.43))

$$\phi_0(0,r) = r\,\frac{J_1(2\sqrt{2\nu\alpha r})}{\sqrt{2\nu\alpha r}}, \tag{15.24}$$

where $\alpha$ is the fine structure constant and $J_1(x)$ denotes the Bessel function. Expanding (15.23) in powers of $\alpha$, we obtain quite adequate first order formula $a^{\mathrm{c}}_{\pi d}/a_{\pi d} = 1 - \alpha\nu\,\langle r\rangle$ where the expectation value is with respect to the deuteron wave function. We have checked that for a variety of deuteron wave functions the calculated ratio (15.23) has been very stable and its numerical value is 0.985. Using this number together with the experimental value of $a^{\mathrm{c}}_{\pi d}$

$$a^{\mathrm{c}}_{\pi d} = [-(2.61 \pm 0.05) + \mathrm{i}\,(0.63 \pm 0.07)] \times 10^{-2}/m_\pi$$

taken from [Hauser *et al.* (1998)] where $m_\pi$ is the mass of the charged pion, we deduce the value of the purely nuclear $\pi$d scattering length

$$a_{\pi d} = (-2.65 \pm 0.05) \times 10^{-2}/m_\pi, \tag{15.25}$$

and hereafter the above number will be referred to as the "experimental" $\pi$d scattering length in which all absorptive effects have been neglected.

Adopting the zero-range model of the $\pi$N interaction, for calculating the $\pi$d scattering length one needs as input just the isoscalar and the isovector $\pi$N scattering scattering lengths. The values of $b_0$ and $b_1$ that have been extracted from the pionic hydrogen data given by [Schröder *et al.* (1999)], are

$$\begin{aligned} b_0 &= -(0.22 \pm 0.43) \times 10^{-2}/m_\pi; \\ b_1 &= -(9.05 \pm 0.42) \times 10^{-2}/m_\pi, \end{aligned} \tag{15.26}$$

where the quoted uncertainty comprises the experimental errors together with the uncertainty introduced by applying a specific procedure that allows to deduce $b_0$ and $b_1$ from the measured X-ray spectra. The theoretical uncertainty is quoted to be about twice as large as the experimental error. Besides, the errors on $b_0$ and on $b_1$ are strongly correlated.

Using (15.26) as our input, we have calculated the $\pi$d scattering length and the results are presented in Table 15.3. All entries are doubled because we employ two models of NN interaction: the numbers without brackets have been obtained using PEST wave function and, respectively, the bracketed quantities correspond to the Yamaguchi potential. For each set of input values of $(b_0, b_1)$ we computed $a_{\pi d}$ using five different methods discussed before, beginning from the simplest second order formula (15.11), through the static model (15.10) and (15.20), up to the full Faddeev calculation without, and, with $\Delta$, respectively. The results of the Faddeev calculation with s-wave interaction only (without $\Delta$) constitute a benchmark for the various approximations. Contrary to what has been often claimed in the literature, the second order formula is insufficient as the error incurred is roughly four times bigger than the present experimental uncertainty on $a_{\pi d}$. It is apparent from Table 15.3 that the closest to Faddeev result is in all cases the static model (15.10). The accuracy of the latter is very good, the error being always below 2%. By contrast, the performance of the implementation (15.20) of static model is rather disappointing, especially that from formula (15.20) containing binding energy correction, one might expect further improvement. Nevertheless, the numbers show just the opposite, that in fact the included corrections go in the wrong direction worsening the results so much that even the second order formula proves to be more accurate. Of course, it is not just the binding energy correction that is responsible for the difference between the static model and the Faddeev result, as only the latter properly accounts for the nucleon recoil. However, the lions share of the recoil correction seems to be canceled with the binding energy correction and this cancellation explains the success of the static formula (15.10) containing neither of these corrections. An explicit demonstration that, at least to the second order, such mechanism is at work can be found in [Fäldt (1977)]. Since the static model (15.10) proves to be so accurate for Yamaguchi and PEST models of the NN interaction, we took advantage of this fact, using it to examine more realistic NN potentials containing also the D-wave part. The results of our computations are displayed in Table 15.1 where we compare the two separable models (Hulthen-Yamaguchi and PEST), used in Faddeev calculations, with two popular local potentials (Paris [Lacombe *et al.* (1981)] and Bonn [Machleidt *et al.* (1987)]). As expected, the PEST wave function results are indeed very close to those obtained with Paris wave function despite the lack of the D-component in the PEST wave function. Therefore, neglect-

Table 15.1 The expectation values of r, 1/r and the values of $\pi$d scattering length calculated for different NN wave functions. For $\pi$N scattering lengths we have adopted their central values, i.e. $b_0 = -0.22$ and $b_1 = -9.05$. All scattering lengths are given in $10^{-2}/m_\pi$ units.

| | NN wave function | | | |
|---|---|---|---|---|
| | Hulthen | PEST | Paris | Bonn |
| $\langle r \rangle\, fm$ | 3.1345 | 3.2309 | 3.2685 | 3.2536 |
| $\langle 1/r \rangle\, fm^{-1}$ | 0.55501 | 0.45507 | 0.44864 | 0.46314 |
| 2-nd order $a_{\pi d}$ | -3.66 | -3.06 | -3.04 | -3.13 |
| static $a_{\pi d}$ | -3.09 | -2.78 | -2.78 | -2.82 |

Table 15.2 $\pi$d scattering lengths calculated from consecutive iterations of the Faddeev equations. All entries are in $10^{-2}/m_\pi$ units.

| order | PEST no $\Delta$ | PEST with $\Delta$ | Yamaguchi no $\Delta$ | Yamaguchi with $\Delta$ |
|---|---|---|---|---|
| 1 | -1.66 | -1.21 | -1.70 | -1.23 |
| 2 | -2.98 | -2.66 | -3.42 | -3.30 |
| 3 | -2.89 | -2.44 | -3.20 | -2.77 |
| 4 | -2.85 | -2.48 | -3.11 | -2.91 |
| 5 | -2.85 | -2.45 | -3.14 | -2.82 |
| 6 | | -2.46 | -3.15 | -2.87 |
| 7 | | -2.46 | -3.14 | -2.84 |
| 8 | | | -3.14 | -2.86 |
| 9 | | | | -2.85 |
| 10 | | | | -2.85 |

ing the D-wave in the Faddeev calculation does not appear to be a serious omission. It is also gratifying that PEST, Paris and Bonn models give very similar results. In Table 15.2 we present the values of $\pi$d scattering length obtained in result of iterative solution of the Faddeev equations. Since for zero-range $\pi$N interaction there is no additional suppression due to the $\pi$N form factor, the rate of convergence is somewhat slower but the converged result is obtained in less than 10 iterations. We give $a_{\pi d}$ values calculated with and without p-wave $\pi$N interaction what allows to evaluate the p-wave contribution in each order. For Yamaguchi NN interaction the p-wave correction in the first order is quite large and contributes $0.47 \times 10^{-2}/m_\pi$. The p-wave contribution to the second order (called sp-term by [Baru *et al.* (1997)]) has opposite sign and equals $-0.35 \times 10^{-2}/m_\pi$. In general, the net effect of the p-wave interaction on the converged result is reduced owing to the destructive interference between repulsive s-waves and attractive p-waves, amounting in total only $0.29 \times 10^{-2}/m_\pi$. Similar features are observed for the PEST model but since the convergence rate is faster, the higher order corrections are suppressed and the interference effects seem to be smaller, i.e. the first order p-wave correction is $0.45 \times 10^{-2}/m_\pi$ while the corresponding correction to the converged result is $0.39 \times 10^{-2}/m_\pi$.

Table 15.3 $\pi$d scattering length obtained from the static model and from a Faddeev calculation in the zero-range model for different $b_0$ and $b_1$. For the NN forces we used PEST and Yamaguchi potentials, the results for the latter case are presented here in brackets. All entries are in $10^{-2}/m_\pi$ units.

| | | | $b_1$ | |
|---|---|---|---|---|
| $b_0$ | model | -9.47 | -9.05 | -8.63 |
| | 2-nd order | -4.22 (-4.87) | -3.97 (-4.57) | -3.74 (-4.28) |
| | static (15.10) | -3.89 (-4.21) | -3.69 (-3.98) | -3.49 (-3.77) |
| -0.65 | static (15.20) | -3.44 (-3.77) | -3.29 (-3.58) | -3.10 (-3.39) |
| | Faddeev | -3.97 (-4.27) | -3.76 (-4.04) | -3.55 (-3.81) |
| | *ditto* with $\Delta$ | -3.59(-3.97) | -3.37 (-3.73) | -3.16 (-3.50) |
| | 2-nd order | -3.30 (-3.96) | -3.06 (-3.66) | -2.82 (-3.37) |
| | static (15.10) | -2.99 (-3.32) | -2.78 (-3.09) | -2.58 (-2.87) |
| -0.22 | static (15.20) | -2.53 (-2.87) | -2.36 (-2.68) | -2.19 (-2.49) |
| | Faddeev | -3.07 (-3.37) | -2.85 (-3.14) | -2.65 (-2.92) |
| | *ditto* with$\Delta$ | -2.68 (-3.08) | -2.46 (-2.85) | -2.25 (-2.62) |
| | 2-nd order | -2.38 (-3.04) | -2.14 (-2.74) | -1.90 (-2.45) |
| | static (15.10) | -2.08 (-2.42) | -1.87 (-2.19) | -1.68 (-1.97) |
| 0.21 | static (15.20) | -2.62 (-1.97) | -1.45 (-1.77) | -1.28 (-1.59) |
| | Faddeev | -2.16 (-2.47) | -1.95 (-2.24) | -1.74 (-2.02) |
| | *ditto* with $\Delta$ | -1.76 (-2.20) | -1.54 (-1.96) | -1.34 (-1.73) |

It is apparent form Table 15.3 that the calculated $\pi$d scattering length values are rather sensitive to the input values of $(b_0, b_1)$ and therefore it is not so easy to see when the calculation agrees with experiment. To facilitate the comparison with experiment the values of $a_{\pi d}$ resulting from Faddeev calculation (PEST with $\Delta$) and displayed in Table 15.3 have been represented analytically using bilinear interpolation on a grid in the $(b_0, b_1)$

plane. Then, given the interpolating polynomial, we equated it to the experimental value of $a_{\pi d}$, adding or subtracting the experimental error. This procedure gave us two constraints of algebraic form in the $(b_0, b_1)$ variables, readily solvable with respect to one of these variables. The two functions obtained this way may be plotted in the $(b_0, b_1)$ plane where, as shown in Fig. 15.1 they set the boundary of the tilted band representing the one standard deviation constraint imposed by the $\pi d$ scattering length deduced form pionic deuterium data. The rectangle in Fig. 15.1 represents the experimental values of $(b_0, b_1)$ to within one standard deviation inferred from pionic hydrogen data. The ultimate $(b_0, b_1)$ values that are consistent with both the pionic hydrogen and the pionic deuterium data fill the area of the black strip.

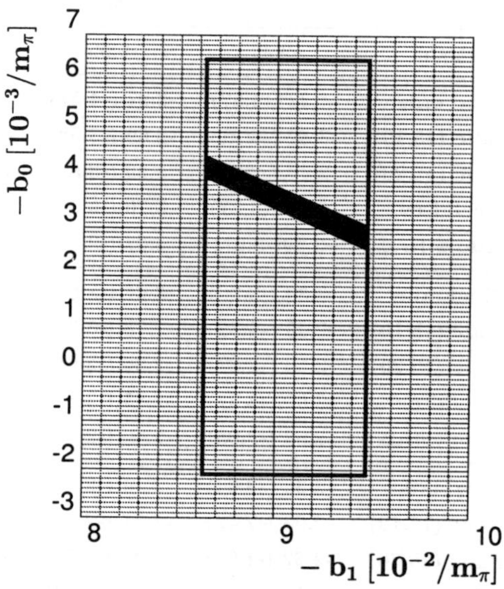


Fig. 15.1 Constrains on the isoscalar and isovector scattering lengths imposed by pionic deuterium data. The black strip corresponds the one standard deviation region. The rectangle corresponds to the values obtained from pionic hydrogen data in [Sigg *et al.* (1996)].

It is apparent from Table 14.1 that to improve upon Deser–Trueman formula we need some additional clue concerning $\beta$ and it becomes something of a challenge to find ways to ferret out more precisely what the value of $\beta$ might be. So far in our considerations we have not mentioned yet the pionic deuterium data and at this stage it is logical to ask whether this additional information might not help to pin down the range parameter of the $\pi$N potential. Therefore, in the next step, we use the values given in Table 14.1 as input for a three-body calculation, i.e. we use the separable potential (14.7) in the Faddeev equations for calculating the $\pi d$ scattering length. The results of our computations are presented in Fig. 15.2 where we have plotted the $\pi d$ scattering length versus $\beta$. The full circles represent the results obtained by the including the p-wave interaction (more precisely, just the P33 wave), while the open circles correspond to a situation where the delta has been left out. For reasons of clarity of the presentation these two sets of points have been given at different $\beta$ values. The indicated error bars reflect the uncertainty in the input values (cf. Table 14.1 ). For comparison, the experimental value of $\pi d$ scattering length to within one standard deviation is given in Fig 15.2 as the area between the two horizontal lines. The striking feature apparent from Fig 15.2 is that the results are almost independent of the range parameter $\beta$. Furthermore, the calculated scattering lengths are consistent with experiment for all $\beta$ no matter whether the delta has been included or not. This result may come as a disappointment since the deuteron data give no illumination how to bracket the value of $\beta$.

In order to understand how the above result comes about we shall invoke again the static model, taking advantage of the fact that with the Yamaguchi potential representing the $\pi$N interaction the static solution of the Faddeev equations may be readily obtained (cf. [Foldy *et al.* (1969)]). Thus, introducing the Yamaguchi form factors and going to the static limit we can repeat the procedure outlined in the preceding section. The static solution of the Faddeev equations may be then sought in the form

$$A(\mathbf{q}, \mathbf{k}) = -\frac{m}{2\pi} \frac{\beta^2}{q^2 + \beta^2} \mathcal{A}(k);$$
$$X(\mathbf{q}, \mathbf{k}) = -\frac{m}{2\pi} \frac{\beta^2}{q^2 + \beta^2} \mathcal{X}(k),$$

and the above ansatz used in the Faddeev equations yields a set of two integral equations that differ from (15.17) in that the appropriate kernels contain now an extra factor $1/[1 + (\mathbf{k} + \mathbf{k}')^2/\beta^2]^2$. Despite this additional complication, the Fourier transform of this extended kernel still can be

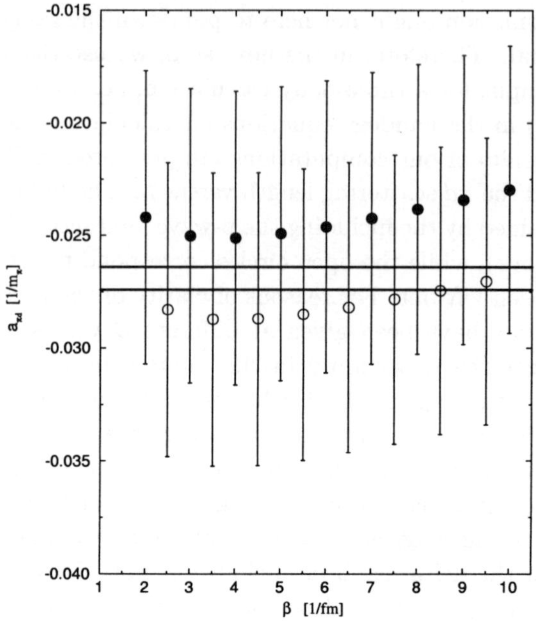


Fig. 15.2 $\pi$d scattering length vs. the inverse range parameter $\beta$ of the $\pi$N potential. Full (open) circles correspond to a Faddeev calculation with (without) p-wave $\pi$N interaction.

effected and leads to a simple analytic expression

$$\frac{4\pi}{\kappa^2 + (\mathbf{k}+\mathbf{k}')^2}\frac{\beta^4}{[\beta^2 + (\mathbf{k}+\mathbf{k}')^2]^2} =$$

$$= (1-\frac{\kappa^2}{\beta^2})^{-2}\int e^{-\mathrm{i}(\mathbf{k}+\mathbf{k}')\mathbf{r}}\frac{\mathrm{d}^3 r}{r}\left\{e^{-\kappa r} - e^{-\beta r}[1+\frac{\beta r}{2}(1-\frac{\kappa^2}{\beta^2})]\right\}.$$

Using the above formula, similarly as before, we end up with a system of two algebraic equations for $A(r)$ and $X(r)$. Neglecting the binding energy correction ($\kappa \to 0$), the resulting equations differ from (2.11) in that the zero-range pion propagator $1/r$ has to be multiplied by the function $g(r)$ given by the formula

$$g(r) = 1 - e^{-\beta r}(1 + \tfrac{1}{2}\beta r). \tag{15.27}$$

Therefore, the sought for solution for $A(r)$ follows from (2.13) after replacing $1/r$ by $g(r)/r$. Formula (15.27) proves to be quite useful for estimating the size of the $\beta$ dependent correction and to this end we need to evaluate

$g(r)$ at some average value of $r$ and a plausible candidate for such average value is the deuteron radius $r_d = \frac{1}{2}\sqrt{< r^2 >} \approx 2$ fm. Indeed, with this choice the second order formula (2.15) that provides for a major contribution to $a_{\pi d}$ will be little affected since by setting $r = r_d$, we get $\langle 1/r \rangle =$ 0.5 fm$^{-1}$, not far from the values listed in Table 15.1. When $\beta$ is varied in the range $2 - 10$ fm$^{-1}$, we have $r_d\beta > 4$ in the exponential damping factor in (15.27), so that the $\beta$ dependent terms make a contribution to $g(r)$ at the level of a few percent and the resulting $\pi$d scattering length is almost independent upon $\beta$. This feature, sustained in the full Faddeev solution, is a consequence of the fact that the adopted range of the $\pi$N forces was small as compared with the deuteron radius.

In conclusion, we have seen that the uncertainty in the calculated $a_1$ and $a_3$, as well as in $a_{\pi d}$, connected with the lack of knowledge of the range parameter constitutes only a small fraction of the uncertainty resulting from the experimental errors on the pionic hydrogen data. The above results may be viewed as an *a posteriori* justification of our zero-range model developed in the previous chapter introducing a finite range would be merely a fine tuning which is not yet affordable in the current state of affairs.

## 15.2 Kaonic Deuterium

Owing to the lack of sufficiently intensive low-energy antikaon beams kaonic deuterium X-ray experiment has not been carried out yet. There are hopes that this situation may change soon as such beams are now coming into operation, e.g. using the DAΦNE storage ring at Frascati [Bianco *et al.* (1999)]. Another kaonic atom experiment is under preparation at KEK. The measurement of the strong interaction level shift and width of the ground state in the kaonic deuterium would be of great value providing direct access to the $K^-$-d scattering length. The interest in the latter quantity is rather obvious: supplementing the $K^-$-p scattering length, known from the kaonic-hydrogen X-ray experiment, with the $K^-$-d scattering length, one would have two constraints from which it should be possible to deduce the two $\bar{K}$N scattering length $A_0$ and $A_1$. The problem, however, is more complicated because the $K^-$-d scattering length is not just a sum of $K^-$-p and $K^-$-n scattering lengths. In fact, impulse approximation fails and the multiple scattering series very likely diverges. The $K^-$-d scattering length does not depend upon $K^-$-p and $K^-$-n scattering lengths in a simple way and to find this relationship one has to solve a complicated multi-channel three-body problem. The underlying forces are not specified by the $\bar{K}N$ scattering lengths alone and a number of additional parameters must be supplied. The main difficulty here is not so much the computational aspect of the problem but the lack of knowledge concerning the two-body forces, and in particular hyperon-nucleon interaction. There have been very few Faddeev type calculations done (cf. [Toker *et al.* (1982); Torres *et al.* (1986); Bahaoui *et al.* (1990)]) for the $K^-$-d system and all of them consider the zero kinetic energy scattering case because that is the only energy where there are any significant data. The three-body calculations quoted above used the same Faddeev equations although the detailed treatments were slightly different. To extend the calculations below $K^-$-d threshold or above the deuteron break-up energy required very much greater effort than was at that time justified and in absence of new data there was little incentive to continue these efforts so that for nearly two decades the investigation of the subject became suspended.

Recently, there has been a renewed theoretical interest in the kaonic deuterium problem (cf. [Barrett *et al.* (1999); Kamalov *et al.* (2001); Bahaoui *et al.* (2002)]), partly because of the kaonic deuterium experiment under way at DAΦNE and at KEK, for which reliable theoretical predictions would be quite useful. The other reason to take on the $K^-d$ problem afresh is connected with the advances of the Chiral Perturbation Theory, especially in the meson-baryon sector. The advantageous feature of this approach is that for the first time a full many-channel calculation might be carried out introducing only a few adjustable parameters.

On a phenomenological level, the principal motivation behind the study of the $K^-d$ scattering length is that it provides an important experimental constraint imposed upon the elementary $\bar{K}N$ scattering lengths. The derivation of a concrete form of this constraint may involve various degrees of sophistication in handling the underlying three-body problem. Since the $\bar{K}N$ input information is rather limited, it seems natural that one should start with something simple introducing successively further refinements. A convenient starting point is the static model in which the nucleons do not move and the presence of the hyperonic channels is simulated by introducing absorptive $\bar{K}N$ interactions. Furthermore, the latter are taken to be operative in s-waves only in which case there is no spin-flip. Assuming that the $\bar{K}N$ scattering operators are isospin invariant, we introduce two isospin functions, respectively, symmetric ($\chi_s$) and antisymmetric ($\chi_a$) in the two-nucleon labels

$$\chi_s = \sqrt{\frac{1}{3}}\,K^-\frac{p_2n_3 + n_2p_3}{\sqrt{2}} + \sqrt{\frac{2}{3}}\bar{K}^0 n_2n_3; \tag{15.28}$$

$$\chi_a = -K^-\frac{p_2n_3 - n_2p_3}{\sqrt{2}}, \tag{15.29}$$

where the incident kaon is labeled as 1 and the nucleons are numbered as 2 and 3. The zero-range absorptive K̄N interactions are introduced in the form of two boundary conditions (cf. [Chand *et al.* (1962)] )

$$\begin{aligned}\lim_{x_1\to x_i} \overline{|\mathbf{x}_1 - \mathbf{x}_i|\Psi(\mathbf{x}_1,\mathbf{x}_2,\mathbf{x}_3)} &= \\ &= \frac{\mu}{m}(A_x + A_y\,\boldsymbol{\tau}_1\cdot\boldsymbol{\tau}_i)\lim_{x_1\to x_i}\frac{\mathrm{d}}{\mathrm{d}x_1}\overline{|\mathbf{x}_1 - \mathbf{x}_i|\Psi(\mathbf{x}_1,\mathbf{x}_2,\mathbf{x}_3)},\end{aligned} \tag{15.30}$$

where i=2,3, m denotes the kaon mass, $\mu$ is K̄N reduced mass and

$$A_x = \frac{3}{4}A_1 + \frac{1}{4}A_0, \quad A_y = \frac{1}{4}(A_1 - A_0), \tag{15.31}$$

with $A_0$ and $A_1$ denoting I=0 and I=1 K̄N scattering lengths. Formally, the wave function $\Psi$ takes the same form as in the $\pi$d case (cf. (15.7)), but now it involves different isospin functions given in (15.28) and (15.29). Repeating all steps of the procedure outlined in the preceding section, from (15.30) one obtains the appropriate amplitudes A(r) and X(r) occurring in the wave function and the ultimate formula for the $K^-$d scattering length is

$$A_{Kd} = \frac{2\nu}{m}\int u_d(r)^\dagger\,A(r)\,\mathrm{d}^3r, \tag{15.32}$$

where $\nu$ denotes K-d reduced mass. The amplitude A(r) occurring in the above formula, is

$$A(r) = \frac{\tilde{A}_x + (\tilde{A}_x + \tilde{A}_y)(\tilde{A}_x - 3\tilde{A}_y)/r}{1 - 2\tilde{A}_y/r - (\tilde{A}_x + \tilde{A}_y)(\tilde{A}_x - 3\tilde{A}_y)/r^2}\,u_d(r), \tag{15.33}$$

where $\tilde{A}_x = (1 + m/M)A_x$ with similarly defined $\tilde{A}_y$.

It has to be noted that in the derivation of the static formula for $K^-$d scattering length exact isospin conservation is not mandatory and may be relaxed. As input one needs then four complex K-matrix elements evaluated at the $K^-$p threshold, associated with the processes: $K^-p \to K^-p$, $K^-p \to \bar{K}^0n$, $\bar{K}^0n \to \bar{K}^0n$ and $K^-n \to K^-n$. The appropriate extension of (15.33) may be found in [Chand *et al.* (1962); Kamalov *et al.* (2001)].

In order to refine the static formula (15.33) by including properly the recoil and the binding energy corrections one has to resort to the Faddeev formalism, first applied to $K^-d$ scattering problem by [Hetherington *et al.* (1965); Hetherington *et al.* (1967)]. The latter approach, however, requires that both, the on-shell and the off-shell $\bar{K}N$ scattering amplitudes be supplied as input. Clearly, this can not be arranged without introducing further unknown parameters. One possibility to reduce the number of free parameters is to assume that the $\bar{K}N$ forces may be simulated by a Yamaguchi potential with a single range parameter $\beta$. In this case the system of Faddeev equations for $K^-d$ scattering below the break-up threshold may be obtained by taking the same ansatz for the Faddeev partitions as in the $\pi$-d case considered above, changing only the isospin functions in (15.13). At the same time the two-body $\pi N$ amplitudes must be replaced by the $\bar{K}N$ scattering amplitudes. As a result of these modifications, similarly as in $\pi d$ case, the three complex functions $F(\mathbf{q},\mathbf{k}), A(\mathbf{q},\mathbf{k}), X(\mathbf{q},\mathbf{k})$ appearing in (15.13) are bound to satisfy a system of Faddeev equations. This system of equations may be obtained from (15.14) by introducing the following replacements: $t_0 \to t_x$ and $\sqrt{2}t_1 \to -\sqrt{3}t_y$ in (15.14b) and, respectively, $\sqrt{2}t_1 \to -\sqrt{3}t_y$ and $(t_0 - t_1) \to (t_x - 2t_y)$ in (15.14c). Equation (15.14a) applies intact. Here, similarly as in (15.31), the just introduced amplitudes $t_x$ and $t_y$ denote, respectively, the isospin no-flip and isospin flip $\bar{K}N$ amplitudes. Since the range parameter $\beta$ is not known, it makes sense taking the limit $\beta \to \infty$ which corresponds to a zero-range interaction and in this regime a Faddeev-type calculation may be compared with the static model result. This gives a clue how accurate is the static formula. In Table 15.4 we have compared the $K^-$-d scattering lengths obtained from different multichannel three-body calculations with the results based upon the zero-range model. The $K^-d$ scattering lengths predicted by the zero-range model are by about 30% bigger than those resulting from a multi-channel approach. Such bigger values have been also favored in a recent multi-channel three-body calculation [Bahaoui *et al.* (2002)] in which the two-body input is derived from a chiral Lagrangian. The average value $A_{Kd} \approx (-1.8 + \mathrm{i}\,1.5)$ fm quoted by [Bahaoui *et al.* (2002)] agrees well with the zero-range model values given in Table 15.4.

Table 15.4 Multi-channel calculation of $K^-d$ scattering length is compared with a Faddeev calculation using zero-range absorptive interaction and with the static formula result, for three different choices of the $\bar{K}N$ scattering lengths. The multi-channel calculations are from [Toker *et al.* (1982); Torres *et al.* (1986); Bahaoui *et al.* (1990)], respectively. All entries are in fm.

| | Toker | Torres | Bahaoui |
|---|---|---|---|
| $A_0$ | -1.77+i 0.73 | -1.69+i 0.76 | -1.61+i 0.86 |
| $A_1$ | 0.41+i 0.54 | 0.28+i 0.67 | 0.45+i 0.62 |
| Multi-channel | -1.47+i 1.08 | -1.34+i 1.04 | -1.11+i 1.30 |
| Faddeev | -1.9 +i 1.3 | -1.9 +i 1.3 | -1.8 +i 1.4 |
| static | -1.7 +i 1.3 | -1.6 +i 1.2 | -1.5 +i 1.4 |

## 15.3 Antiprotonic Deuterium

Antiprotonic deuterium has been observed in a LEAR experiment at CERN and the measured X-ray spectrum revealed the values of the strong interaction shift and width of the ground-state level (cf. [Gotta *et al.* (1999)]). In a deuterium target the annihilation rate from the p-levels is bigger, as compared with that in protonium, which gives rise to a further reduction of the yields of the Lyman transitions. Therefore, only scarce information is available on the hadronic ground-state transition based on a limited statistics. On the other hand, the strong interaction 2p level shift and width were directly measured with about 10% accuracy. The deduced spin-averaged values for the strong interaction 1s and 2p level shifts and widths, given by [Gotta *et al.* (1999)] are

$$\bar{\epsilon}_{1s} = -1.05 \pm 0.25 \quad \text{keV}; \tag{15.34a}$$

$$\bar{\Gamma}_{1s} = 1.10 \pm 0.75 \quad \text{keV}. \tag{15.34b}$$

$$\bar{\epsilon}_{2p} = -243 \pm 26 \quad \text{meV}; \tag{15.34c}$$

$$\bar{\Gamma}_{2p} = 489 \pm 30 \quad \text{meV}. \tag{15.34d}$$

Using the above values in the master equation, one obtains spin-averaged Coulomb-modified s-wave $\bar{p}$-d scattering length

$$A_0^{cs} = [(0.7 \pm 0.2) - i(0.4 \pm 0.3)] \text{ fm};. \tag{15.35}$$

and the corresponding p-wave scattering volume

$$A_1^{cs} = [(3.3 \pm 0.3) - \mathrm{i}(3.18 \pm 0.18)]\ \mathrm{fm}^3. \qquad (15.36)$$

The imaginary part of the above s-wave scattering length plays also a crucial role in the p̄-d annihilation cross section for small incident momenta. The latter cross section has been measured but only at three different momenta of the incident antiproton. Recently, [Protasov *et al.* (2000)] took advantage of the fact that the imaginary part of the scattering volume (15.36) is quite well determined and used the p̄-d annihilation in-flight data to pin down the imaginary part of $A_0^{cs}$. As a result of their fit, the precision of the determination of the imaginary part of $A_0^{cs}$ has been improved and reads

$$\mathrm{Im}\, A_0^{cs} = -[0.62 \pm 0.02(\mathrm{stat.}) \pm 0.05(\mathrm{sys.})]\ \mathrm{fm} \qquad (15.37)$$

with $\chi^2$ per point equal 8.3. The different attempts to determine the s-wave p̄-d scattering length have been illustrated in Fig. 15.3 where the X-ray experiment result to within one standard deviation is represented by the empty rectangle and the values of the scattering length consistent with the in-flight data (15.37) are displayed as the black strip.

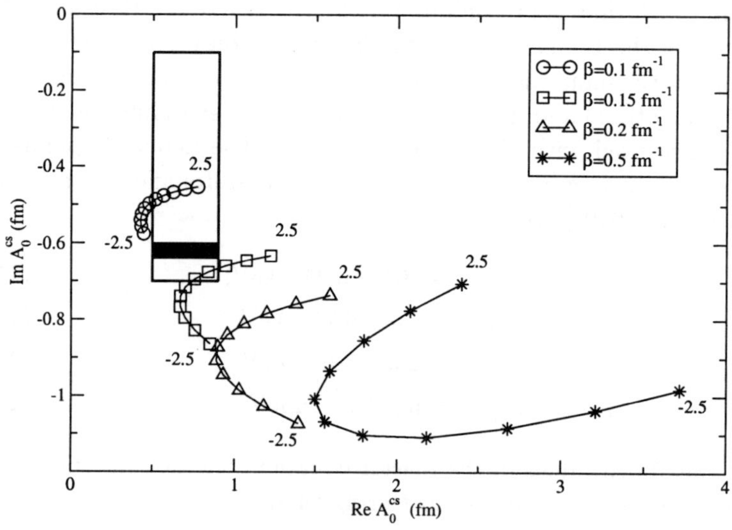


Fig. 15.3 Coulomb-modified s-wave p̄-d scattering length.

Standing on the ground of a naive geometrical picture of the annihilation process, the imaginary part of the $\bar{p}$-d scattering length might be expected to be approximately equal to the sum of the imaginary parts of the scattering length corresponding to scattering off each of the two constituents of the deuteron. Therefore, the experimental finding that the absolute value of $\mathrm{Im}A_0^{cs}$ is in fact smaller than $|\mathrm{Im\ a}(\bar{p}n)|$ alone came as a surprise to some people. The latter quantity is of course not directly measurable but we have $\mathrm{Im\ a}(\bar{p}n)= \mathrm{Im\ a}(\bar{n}p)$ and the value $\mathrm{Im\ a}(\bar{n}p)$=-[83± 0.07] fm has been inferred from $\bar{n}p$ annihilation data by [Mutchler *et al.* (1988)].

The fact that $|\mathrm{Im\ A}_0^{cs}|$ is smaller than expected indicates merely that simple impulse approximation is not valid which is quite conceivable because both $\bar{p}p$ and $\bar{p}n$ scattering lengths are rather large, comparable with the deuteron size. Having made this observation, the problem we wish to address to is whether the $\bar{p}$-d scattering length could be explained by a microscopic calculation. To meet this challenge one should be able to express the $\bar{p}$-d scattering length in terms of $\bar{p}p$ and $\bar{p}n$ scattering lengths. Clearly, the ensuing $\bar{p}$-d zero-energy scattering problem in its full generality is very difficult owing to its multi-channel nature, the presence of Coulomb interaction, spin complications, *etc.* Therefore, to simplify matters, we are going to consider spin-averaged phenomenological two-body forces where the annihilation channels will be simulated by introducing absorptive interactions. As our input we will use the experimental values of the $\bar{p}p$ and $\bar{p}n$ scattering lengths. The three-body dynamics will be solved without Coulomb interaction and we will content ourselves by adding the Coulomb correction at the very last stage. Finally, to reduce the calculational effort the two-body interactions will be taken in a separable form with Yamaguchi form-factors. Despite all these simplifications our input is still incomplete and the first difficulty is connected with the fact that we do know only the $\bar{p}p$ scattering length, and, additionally, the imaginary part of $\bar{p}n$ scattering length. Since we are missing the real part of $\bar{p}n$ scattering length the latter quantity has to be regarded as a free parameter whose value may be varied in some reasonable limits. The other adjustable parameter is associated with the range of the forces and we take this parameter to be the same for both $\bar{p}p$ and $\bar{p}n$ interactions. The three-body problem is now well posed but involves two real adjustable parameters.

With spin-averaged forces, the $\bar{p}$-d scattering problem is formally very similar to that of $K^-$-d scattering and in particular the isospin structure is identical in both cases, with the $\bar{p}p \rightarrow \bar{n}n$ process being the counterpart of charge-exchange. In particular, this means that the static model formula

(15.33) is applicable also for $\bar{p}d$ scattering. The same is true for Faddeev equations in which absorptive interactions simulate the presence of the annihilation channels. We used such system of Faddeev equations to compute the $\bar{p}d$ scattering length. Introducing the inverse range parameter $\beta$, following (15.21), the explicit expression for the $\bar{N}N$ scattering t-matrices with isospin I=0,1, takes the form

$$\begin{aligned} \langle k|t_I(E)|k'\rangle &= -\frac{2\pi}{\mu}\frac{1}{1+k^2/\beta^2}\times \\ &\times\frac{a_I}{1-\mathrm{i}pa_I\,(1-2\mathrm{i}p/\beta)(1-\mathrm{i}p/\beta)^{-2}}\,\frac{1}{1+k'^2/\beta^2}, \end{aligned} \tag{15.38}$$

with $\mu=\frac{1}{2}M$ where M is nucleon mass and $p=\sqrt{2\mu E}$. The $\bar{N}N$ scattering lengths $a_I$ I=0,1 occurring in (15.38) are related to the $\bar{p}p$ and the $\bar{p}n$ scattering lengths: $a_1$=a($\bar{p}$n) and $a_0$=2a($\bar{p}$p)-a($\bar{p}$n). Actually, the $\bar{p}p$ scattering length (14.31) includes the Coulomb correction which needs to be removed as in the above isospin relations we have the $\bar{p}p$ scattering length that involves only the strong interaction. For a Yamaguchi potential this step is straight-forward and (14.16) gives the exact relationship between these two scattering lengths. Since the the real part of $\bar{p}n$ scattering length a($\bar{p}$n) is not known we set a($\bar{p}$n)=$\xi-\mathrm{i}0.83$ where the parameter $\xi$ will be varied in some physically sensible limits. The value of $\beta$ is not known but the results of our computations turn out to be insensitive when $\beta$ is pushed above 5 fm$^{-1}$ and where essentially the zero-range limit is attained. Since we are dealing with effective interactions, the lowest value of $\beta$, corresponding to the longest range of the $\bar{N}N$ forces, is harder to estimate but in the present case there will be no need of taking $\beta$ smaller than 0.1-0.2 fm$^{-1}$.

For assigned values of $\beta$ and $\xi$ the solution of the Faddeev equations presents us with the $\bar{p}d$ scattering length. However, in order to compare it with experiment Coulomb correction must be included and for that purpose we used the approximate formula (15.23). The ultimate values of $\bar{p}d$ scattering length are presented in Fig. 15.3 where by varying the parameters $\beta$ and $\xi$ the resulting scattering lengths can be arranged as a set of trajectories. To expose these trajectories the points with the same $\beta$ value have been connected by a continuous curve, to guide the eye. On the plot each trajectory contains 11 points corresponding to different $\xi$, varied from -2.5 fm up to 2.5 fm with a step size equal 0.5 fm. The points with the largest and the smallest value of $\xi$ are marked at each trajectory. It is apparent from Fig. 15.3 that for very short range of the $\bar{N}N$ interaction the associated trajectories all miss the rectangle representing the data and the calculated $\bar{\text{p}}$d scattering length cannot be brought to agreement with experiment be varying $\xi$. In particular, the static model formula (15.33) and the zero-range Faddeev calculation predict $\bar{\text{p}}$d scattering lengths which cannot be reconciled with the available data. However, if we admit smaller values of $\beta$ the situation changes: the calculated $\bar{\text{p}}$d scattering length becomes much less sensitive to $\xi$ and the calculated values fall very close to experiment. It is not clear whether such small values of $\beta$ are physically plausible but the above results demonstrate that the relationship between the $\bar{\text{p}}$d scattering length and the elementary $\bar{\text{p}}$p and $\bar{\text{p}}$n scattering lengths can be very different from that based on a simple geometrical picture.

# Chapter 16

# Hadronic Atoms with A $\geq$ 4

The use of strongly interacting negatively charged probe particles such as $\pi^-$, $K^-$, $\bar{p}$, $\Sigma^-$, *etc.* at zero kinetic energy opens interesting prospects for nuclear physics studies. In nuclear targets these particles can be captured into atomic orbits and the X-ray or Auger transitions can be observed as the particle cascades toward the nucleus. This permits to establish the spectroscopic state from which the absorption finally occurs. The data on strong interaction level shifts, widths and yields in hadronic atoms contain valuable information concerning zero-energy hadron-nucleus interaction. Experimentally, $\pi^-$, $K^-$, $\bar{p}$ and $\Sigma^-$ hadronic atoms have been observed so far. In total, they span a wide range of interaction strengths providing means to probe a broad range of nuclear densities, from very low values up to almost nuclear matter density. For very strongly absorbed hadrons, the radial distance from the nuclear center at which absorption occurs is localized outside the nuclear radius, i.e. at the nuclear periphery (cf. Fig. 16.1). This remarkable specificity to the nuclear density in the peripheral region gives a new insight into our understanding of the nuclear structure because the conventional probes such as electrons or protons are sensitive merely to the global properties such as the rms radius and surface thickness.

The extensive hadronic atom data have been analyzed with the aid of the hadron-nucleus optical potential which is usually obtained from multiple-scattering equations for the low-energy hadron propagating in the nuclear medium. Equivalently, it can be also derived from the study of the hadron's Green function in nuclear matter by expanding the hadron self-energy in powers of nuclear density. Both approaches have their virtues in so far as they stress different aspects of the underlying assumptions and indicate a logical sequence of approximations leading to a manageable calculational scheme. One basic premise of the microscopic theory is that the range of the elementary hadron-nucleon interaction is short compared with the typical inter-nucleon separation. Since the interaction takes place mainly in the nuclear surface region where the local density is significantly reduced, the product of the average hadron-nucleon scattering amplitude and of $\sqrt{\rho}$ has been regarded as the basic expansion parameter of the multiple-scattering theory. However, the progress in deriving a reliable optical potential has been slow which hindered the extraction of much reliable nuclear structure information from the hadronic atom data. Hitherto, the nuclear structure information has had to be supplied into the hadronic atom analyzes rather than being extracted from such an analysis.

Since the range of the hadron-nucleon force is small in comparison with the nuclear size, the range of the hadron-nucleus potential is essentially given by the nuclear radius. At a qualitative level, the absorption rate $\Gamma$ for a hadron in a $n\ell$ orbital state on a nucleus is proportional to the overlap

$$\Gamma \sim \int_0^\infty u_{n\ell}(r)^2\,\rho(r)\,r^2\,\mathrm{d}r, \tag{16.1}$$

where $u_{n\ell}(r)$ is the unperturbed (coulombic) radial wave function and $\rho(r)$ is the nuclear density. In Fig. 16.1, we display the radial dependence of the density, the square of the wave function and their product for a typical case of kaonic sulphur. The absorption probability is seen to peak at the nuclear periphery where the density has fallen to 20% of its value at the origin. Since $u_{n\ell}(r) \sim r^\ell$ the absorption rate in (16.1) is in practice a probe of the $2\ell$-th moment of the density and for typical values of $\ell$ ~3-6 we encounter rather high moments. Consequently, small changes in the radii are magnified in importance.

For fixed Z, the capture rate varies rapidly with $\ell$. Indeed, in the first approximation we can take uniform density with a nuclear radius $R \sim A^{\frac{1}{3}} \sim Z^{\frac{1}{3}}$. In this case, (16.1) yields

$$\Gamma \sim (Z\,R)^{2\ell+3} \sim Z^{4+\frac{8}{3}\ell} \tag{16.2}$$

and this simple estimate explains immediately two general features. Firstly, in the outer part of the cascade the hadron runs predominantly through circular orbits ($n=\ell+1$) since for fixed principal quantum number n, they have the smallest overlap with the nuclear density. The inner orbits have bigger overlap and as a result the hadron is absorbed by the nucleus. Secondly, the absorption rate is a slowly varying function of n but exhibits a very rapid variation with $\ell$. Typically, passing from the upper circular level to the lower circular level the capture rate increases by a factor $10^2$-$10^3$.

For comparison, juxtaposing the same levels the radiative width $\Gamma_{rad}$ for a dipole transition increases only by a factor 2-3. This implies that the atomic cascade terminates abruptly at a critical level when $\Gamma$ becomes bigger that $\Gamma_{rad}$. The absorption probability rapidly increases and the hadron has no chance to experience yet another radiative transition. Actually, such transition would lead to an extremely broad level so that the identification the X-ray leading to it might not have been possible.

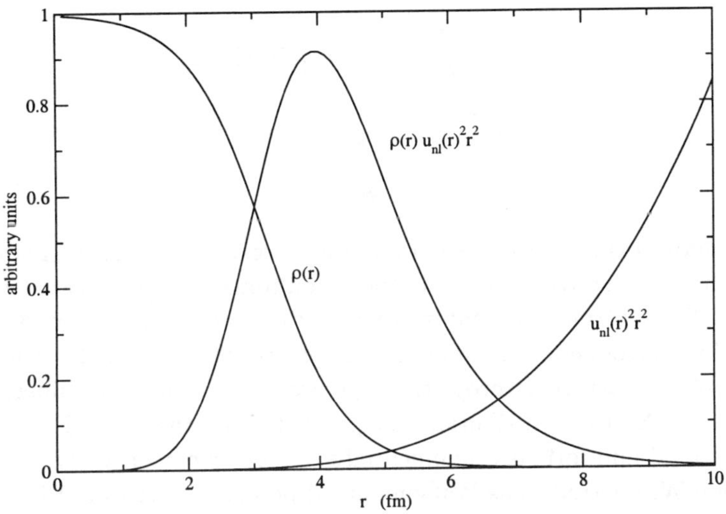


Fig. 16.1 Radial overlap for kaonic sulphur $^{32}S$. In the purely coulombic wave function we set $\ell$=n-1=2. For better presentation the functions have arbitrary normalization.

## 16.1 Hadron-Nucleus Effective Potential

In the standard treatment the effective potential simulating hadron-nucleus interaction is derived from multiple scattering theory developed by Watson and collaborators [Watson (1953)] or from the completely equivalent formulation presented by Kerman, McManus and Thaler [Kerman *et al.* (1951)] (abbreviated KMT hereafter). These two approaches are identical in content and the resulting effective potentials may be readily transformed one into the other [Nagarajan *et al.* (1975)]. We begin the discussion of the various approximations by re-deriving briefly the basic results of both of the above mentioned theories using formal scattering theory [Eisenberg *et al.* (1980)].

To avoid inessential complications, we consider the scattering of a hadron h from a collection of A identical fermions in a given state (nuclear ground state). In order to make the above restriction explicit, following KMT, we shall work in a limited Hilbert space in which only totally anti-symmetric target wave functions are admitted. Formally, one introduces a projection operator $\mathfrak{A}$ onto this limited Hilbert space and the Lippmann-Schwinger equation for the scattering T operator is

$$T(E) = \sum_{\alpha=1}^{A} v_\alpha [1 + G(E)\, \mathfrak{A}\, T(E)], \qquad (16.3)$$

where $v_\alpha$ are hadron-nucleon potentials and G(E) is the Green's operator

$$G(E) = (E - K - H_A + i\epsilon)^{-1}. \qquad (16.4)$$

In the above expressions E is parametric energy, K denotes the hadron kinetic energy operator and $H_A$ is the Hamiltonian of the isolated nuclear target. The intermediate states in which the target is not excited are extracted by introducing the ground state projection operator P. Defining Q as complementary to P projection operator in the limited Hilbert space (i.e. $P+Q=\mathfrak{A}$), the contribution in eq. (16.3) from the excited states may be separated and formally summed into an effective energy dependent interaction W, referred to as Watson optical potential operator. Eliminating the Q part, the Lippmann-Schwinger equation (16.3) breaks into a pair of equations

$$T(E) = W[1 + G(E)\, P\, T(E)]; \qquad (16.5)$$

$$W = \sum_{\alpha=1}^{A} [1 + G(E)\, Q\, W]. \qquad (16.6)$$

The one-body effective potential $V_W$ is obtained from the many-body operator W by taking the ground state expectation value

$$V_W = \langle 0 | W | 0 \rangle, \qquad (16.7)$$

where $V_W$ as defined above is still an operator in the incident hadron space. By construction, the effective potential $V_W$ when inserted into a one-body Lippmann-Schwinger equation yields exactly the elastic scattering amplitude $\langle 0|T|0\rangle$.

The next step consists in expressing W in terms of hadron-nucleon scattering amplitudes. To accomplish this most conveniently, we introduce the scattering operators $\theta_\alpha$ describing the hadron-nucleon scattering in the nuclear medium. They are obtained as solutions of the following equation

$$\theta_\alpha = v_\alpha \,(1 + G\,Q\,\theta_\alpha) \tag{16.8}$$

and it will prove useful to introduce a set of "potentials" $W_\alpha$ defined as

$$W_\alpha = v_\alpha \,(1 + G\,Q\,W). \tag{16.9}$$

In view of (16.6), we have $W = \sum_\alpha W_\alpha$. Eliminating $v_\alpha$ between (16.8) and (16.9), we arrive at the equation

$$W_\alpha = \theta_\alpha \,(1 + G\,Q \sum_{\beta \neq \alpha} W_\beta), \tag{16.10}$$

which iterated yields a series representation for W

$$W = \sum_\alpha \theta_\alpha + \sum \sum_{\alpha \neq \beta} \theta_\alpha\, G\, Q\, \theta_\beta + \cdots . \tag{16.11}$$

In (16.11) W is given as an expansion in terms of the number of nucleons the projectile encounters at the same time. Needless to say, the operator W obtained by formal manipulations contains all the many-body aspects of the problem. The excited states enter twice: explicitly as seen from the presence of Q in formula (16.11), and implicitly, via the scattering operators $\theta_\alpha$ (cf. (16.8)). Since the latter are very complicated many-body operators, all the approximation schemes seek to relate them to the free space scattering operators $t_\alpha$ which obey the Lippmann-Schwinger equation

$$t_\alpha(\omega) = v_\alpha \,[1 + g_\alpha(\omega)\, t_\alpha(\omega)]. \tag{16.12}$$

In (16.12) $\omega$ denotes parametric energy and the free space hadron-nucleon Green's function is given as

$$g_\alpha(\omega) = (\omega - K - K_\alpha + \mathrm{i}\,\epsilon)^{-1} \tag{16.13}$$

where $K_\alpha$ is the kinetic energy operator of the nucleon $\alpha$. Eliminating the potentials $v_\alpha$ between (16.12) and (16.8), we arrive at an operator relation connecting the scattering in medium to free space scattering

$$\theta_\alpha(E) = t_\alpha(\omega)\; \{1 + [G(E)\,Q - g_\alpha(\omega)]\theta_\alpha(E)\}\,. \tag{16.14}$$

Formally, eq. (16.14) together with (16.11) completes the Watson scheme and the standard procedure is to expand (16.14) and (16.11) in powers of $t_\alpha$, usually keeping only the leading term. This first order expression in $t_\alpha$ gives the familiar impulse approximation form of the optical potential

$$V_W \approx \langle 0 | \sum_\alpha t_\alpha | 0 \rangle. \tag{16.15}$$

We turn now to KMT formalism [Kerman *et al.* (1951)]. These authors took slightly different point of departure by noting that the requirement of antisymmetrization may be employed to simplify the underlying algebra. Accordingly, for identical target constituents, (16.3) may be cast to the form

$$T(E) = A\bar{v}[1 + G(E)\mathfrak{A}\, T(E)], \tag{16.16}$$

where

$$\bar{v} = \frac{1}{A}\sum_\alpha v_\alpha. \tag{16.17}$$

Defining the scattering operators in medium $\tau_\alpha$ as solutions of the Lippmann-Schwinger equation

$$\tau_\alpha = v_\alpha\,(1 + G\,\mathfrak{A}\,\tau_\alpha), \tag{16.18}$$

it is convenient to define the operator $\bar{\tau}$ similar to (16.17), i.e.

$$\bar{\tau} = \frac{1}{A}\sum_\alpha \tau_\alpha. \tag{16.19}$$

Now, the potential $\bar{\mathrm{v}}$ in (16.16) may be eliminated in favor of $\bar{\tau}$, and the resulting equation for T reads

$$T = A\bar{\tau} + (A-1)\,\bar{\tau}\,G\,\mathfrak{A}\,T, \tag{16.20}$$

which upon scaling T'=[(A-1)/A]T may be cast to the standard Lippmann-Schwinger form

$$T' = (A-1)\,\bar{\tau}\,(1 + G\,\mathfrak{A}\,T'). \tag{16.21}$$

It is understood that after solving eq. (16.21) for T' one has to return to T. Similarly as before, the Q part in (16.21) may be separated from the P part which allows to break (16.21) into two equations

$$T' = U' (1 + G P T'); \tag{16.22a}$$

$$U' = (A-1)\,\bar{\tau}\,(1 + G Q U'), \tag{16.22b}$$

where the operator U' is identified as the effective pseudo-potential. Equation (16.22b) can be formally solved by iteration which yields the ultimate expression for the KMT effective pseudo-potential

$$U' = \left(\frac{A-1}{A}\right) \sum_{\alpha} \tau_\alpha + \left(\frac{A-1}{A}\right)^2 \sum_{\alpha} \sum_{\beta} \tau_\alpha\, G\, Q\, \tau_\beta + \cdots . \tag{16.23}$$

The relation between the effective scattering operators $\bar{\tau}_\alpha$ entering (16.23) and the free-space scattering operators $t_\alpha$ is again rather complicated. Eliminating $v_\alpha$ between (16.12) and (16.18), one obtains

$$\tau_\alpha(E) = t_\alpha(\omega)\,\{1 + [G(E)\,\mathfrak{A} - g_\alpha(\omega)]\,\tau_\alpha(E)\}\,. \tag{16.24}$$

This expression together with (16.23) completes the KMT scheme. Similarly as in the Watson formulation, practical solutions may be obtained by expanding (16.23)–(16.24) in powers of $t_\alpha$ setting somewhat arbitrarily $G \approx g_\alpha$. This procedure leads to the correlation expansion of the optical potential investigated extensively by Feshbach and his collaborators (for a review cf [Feshbach (1993)])

The explicit relation between the Watson optical potential W and the KMT pseudo-potential U' may be readily derived upon eliminating T between (16.5) and (16.22a). This yields the following relation between these two potentials [Nagarajan *et al.* (1975)]

$$(A-1)\,W = A\,U' - U'\,G\,P\,W, \tag{16.25}$$

and (16.25) transforms formally W into U' and vice versa.

For completeness we shall also re-derive the Watson series solution for the scattering operator T. This can be accomplished by introducing auxiliary operators $T_\alpha$, defined as

$$T_\alpha = v_\alpha\,(1 + G\,\mathfrak{A}\,T), \tag{16.26}$$

so that

$$T = \sum_{\alpha} T_\alpha. \tag{16.27}$$

Eliminating $v_\alpha$ between (16.26) and (16.18), we obtain

$$T_\alpha = \tau_\alpha \,(1 + G\,\mathfrak{A} \sum_{\beta \neq \alpha} T_\beta), \tag{16.28}$$

and iteration of (16.28) yields the Watson series in the form

$$T = \sum_\alpha \tau_\alpha + \sum \sum_{\alpha \neq \beta} \tau_\alpha \, G\,\mathfrak{A}\, \tau_\beta + \cdots . \tag{16.29}$$

The above formal considerations have been exact and now, having assembled all the necessary tools, we are ready to embark at a discussion of the various approximations.

It has been perpetually emphasized in the literature that the hadron-nucleus interaction in a hadronic atom is of peripheral nature, i.e. the hadron fails to penetrate deeply inside the nucleus. Since the hadron-nucleon interaction is confined to a small volume and the hadron encounters a low density of the nucleus it may be expected that in this situation very few degrees of freedom may be excited Formally, this means that the nucleus can be regarded as a dilute system so that in the correlation expansion (16.11), or (16.23) the dominant contribution comes from the single scattering term. Although, the single scattering approximation appears to be justified and results in a major simplification, this is not enough to bring the multiple scattering scheme into a manageable form. The main difficulty stems from fact that both the $\theta_\alpha$ entering W and the $\tau_\alpha$ entering U' are still very complicated many-body operators. In this situation the prevailing tendency has been to identify the scattering in medium with free-space scattering taking the optical potential as $\langle 0 | \sum_\alpha t_\alpha | 0 \rangle$ known as "t$\rho$" form. Such procedure may be expected to be reliable for sufficiently high-energy scattering where the nucleon binding may be regarded as a small effect. By contrast, in a hadronic atom typical hadron binding energies are of the order of several hundred keV, being by an order of magnitude smaller than the average binding energy per nucleon in a nucleus. Even if allowance is given to the fact that in the nuclear periphery the nucleon binding is slightly reduced, the hadron energy still constitutes a negligibly small fraction of the total binding energy of the system. Under such circumstances the scattering from a bound nucleon could be significantly different from the free-space scattering. Formally, this means that the expansion of $\theta_\alpha$ or $\tau_\alpha$ in powers of $t_\alpha$ may be very slowly convergent (or even divergent) so that truncating the series after the first term is completely unjustified.

Actually, in the low-energy regime the optical potential of the form $\langle 0 | \sum_\alpha t_\alpha | 0 \rangle$ raises a number of disconcerting formal questions. To begin with, we wish to note that the optical potential as such should be applicable for an arbitrary hadron-nucleon interaction. Let us consider then a hadron-nucleon interaction in which close to threshold there are no inelastic channels open. When the incident hadron energy is smaller than the nucleus excitation energy, only elastic scattering is possible and the optical potential should be real. By contrast, the $\langle 0 | \sum_\alpha t_\alpha | 0 \rangle$ optical potential is in general complex because the amplitudes $t_\alpha$ are complex quantities. One may try to find a way out arguing that the $t_\alpha$ operators ought to be taken at threshold where they are real and in the s-wave case proportional to the hadron-nucleon scattering length. This variant is not satisfactory either as the scattering length is allowed to take any value and in particular may even become infinite. A potential whose depth might become infinite is not acceptable and this indicates that the optical potential construct simply proportional to the scattering length involves a serious over-simplification.

When we admit hadron-nucleon interactions in which the absorptive channels are open at threshold (e.g. kaonic and antiprotonic atoms) additional difficulties emerge. To get further insight, we assume that the underlying multi-channel hadron-nucleon interaction respects exactly isospin invariance. In a hadron-nucleon system the inelastic channels may be conveniently eliminated at the expense of introducing complex potentials. This is a highly non-linear operation and in the resulting hadron-nucleon complex potential isospin invariance is lost. As an illustration we can take the $K^-p$ system where isospin invariance tells us that the $K^-p$ complex scattering length $a(K^-p)$ is a superposition of I=0 and I=1 scattering lengths $a(K^-p) = \frac{1}{2}(a_0 + a_1)$. We may construct a complex $K^-p$ potential $v(K^-p)$ that reproduces $a(K^-p)$ as well as complex potentials $v_0$ and $v_1$ reproducing $a_0$ and $a_1$, respectively. Since for effective potentials there is no such thing as isospin invariance there is no reason why $v(K^-p)$ should be equal to $\frac{1}{2}(v_0 + v_1)$. A problem appears when we pass from a single nucleon situation to a collection of nucleons. Since the effective potential between a hadron and a single nucleon does not respect isospin invariance it is difficult to see how by taking many nucleons isospin invariance could be restored in the hadron-nucleus optical potential. Notwithstanding, the optical potential proportional to the free-space hadron-nucleon scattering amplitudes would be isospin invariant.

A possible resolution of the above difficulties offers a prescription where the "t$\rho$" form is replaced by the "v$\rho$" form. This approximation is based on the assumption that in the zero-energy hadron-nucleus interaction the dominant contribution comes from the nuclear ground state and in the first approximation the excited nuclear states may be disregarded as if the nucleus were to stay in the ground state at all times. Such procedure has been well known in atomic and nuclear physics and is sometimes called rigid or static approximation because one ignores the nuclear polarization caused by hadron-nucleus interaction. Several arguments may be advanced in favor of this approximation. Firstly, as has been already mentioned, the interaction takes place in a relatively diluted region of the nuclear medium where single scattering processes dominate. In these circumstances a single nucleon excitation is unlikely to be strongly coupled with excited nuclear states. Secondly, for hadron energies close to threshold most of the inelastic channels may be regarded as sufficiently distant and therefore effectively decoupled from the ground state. Thirdly, the presence of inelastic channels would lead to a correction to the effective interaction which is of the second order in the off-diagonal potential. This may be seen by iterating eq. (16.6). The resulting correction may be expected to be of a shorter range and should be eventually suppressed by the centrifugal barrier. Finally, the ground state dominance ought to be viewed as a first approximation which, when need arises, might be improved be including excited states.

The rigid nucleus assumption brings major simplifications into multiple scattering theory. Formally, the suppression of nuclear excitations is achieved by setting Q→0 and in the following we put a superfix nought indicating that we consider a non-excitable nucleus situation. In the limit Q→0, we have

$$\theta_\alpha \to \theta_\alpha^{(0)} = v_\alpha; \tag{16.30}$$

$$\tau_\alpha \to \tau_\alpha^{(0)} = v_\alpha\,(1 + G\,P\,\tau_\alpha^{(0)}); \tag{16.31}$$

$$W \to W^{(0)} = \sum_\alpha \theta_\alpha^{(0)}; \tag{16.32}$$

$$U' \to U^{(0)\prime} = (1 - A^{-1}) \sum_\alpha \tau_\alpha^{(0)}; \tag{16.33}$$

$$T \to T^{(0} = \sum_\alpha \tau_\alpha^{(0)} + \sum \sum_{\alpha \neq \beta} \tau_\alpha^{(0)}\,G\,P\,\tau_\beta^{(0)} + \cdots\,. \tag{16.34}$$

As seen from (16.32) and (16.33), in absence of nuclear excitations the single scattering term is the sole contributor to the optical potential. Furthermore, the potentials $W^{(0)}$ and $U^{(0)\prime}$ are then completely equivalent, and as may be directly checked, eq (16.25) will be satisfied with W and U' given by (16.32) and (16.33), respectively. The formal solution of Lippmann-Schwinger equation (16.5) with the Watson potential (16.32), takes the form

$$T^{(0)} = \sum_{\alpha} v_\alpha + \sum_{\alpha} v_\alpha \, G \, P \sum_{\beta} v_\beta + \cdots \,, \tag{16.35}$$

which, after making use of (16.31), reduces correctly into the Watson series expression (16.34). The KMT pseudo-potential (16.33) used in the Lippmann-Schwinger equation (16.22a), yields

$$\begin{aligned} T^{(0)} &= \sum_{\alpha} \tau_\alpha^{(0)} + \left(\frac{A-1}{A}\right) \sum_{\alpha} \tau_\alpha^{(0)} \, G \, P \sum_{\beta} \tau_\beta^{(0)} + \\ &+ \left(\frac{A-1}{A}\right)^2 \sum_{\alpha} \tau_\alpha^{(0)} \, G \, P \sum_{\beta} \tau_\beta^{(0)} \, G \, P \sum_{\gamma} \tau_\gamma^{(0)} + \cdots \,. \end{aligned} \tag{16.36}$$

At first sight the above expression might appear different from (16.34) but a closer examination shows that the terms 1/A exactly cancel in (16.36) the redundant terms with repeated summation indices and the outcome is identical with the Watson series (16.34). This reduction is a consequence of the requirement of antisymmetry of the nuclear ground state.

Even though the optical potentials $W^{(0)}$ and $U^{(0)\prime}$ are formally equivalent but KMT pseudo-potential still renders an impractical scheme because the operator $\tau_\alpha^{(0)}$ can not be simply evaluated. This is not surprising since the KMT scheme has been originally devised for high-energy scattering. By contrast, the Watson operator $W^{(0)}$ leads to a more manageable expression for the optical potential. As seen from (16.30), the effective scattering operator in the nuclear medium is just equal to the hadron-nucleon potential and the corresponding optical potential may be written as

$$V = \langle 0 | W^{(0)} | 0 \rangle = \langle 0 | \sum_{\alpha} v_\alpha | 0 \rangle , \tag{16.37}$$

which is the "v$\rho$" form. To compare (16.37) with the "t$\rho$" potential, the hadron-nucleon potentials $v_\alpha$ in (16.37) are eliminated in favor of the free-space scattering operators $t_\alpha$. This allows to cast V to the form

$$V = \langle 0 | \sum_{\alpha} t_\alpha \, (1 + g_\alpha \, t_\alpha)^{-1} | 0 \rangle , \tag{16.38}$$

in which the non-linear dependence upon $t_\alpha$ has been made explicit. It is only in the high-energy scattering case that the optical potential (16.38) may be expanded in powers of $t_\alpha$ and eventually approximated by the "$t\rho$" form. In the low-energy regime this would be equivalent to Born approximation which is certainly inadequate for interactions whose strength, as inferred from free-space hadron-nucleon scattering, is comparable with that of the nucleon-nucleon interaction. The non-linear dependence upon $t_\alpha$ immediately alleviates the difficulties inherent in the "$t\rho$" form discussed above. Clearly, for real hadron-nucleon potentials the optical potential must be also real and its depth always stays finite. For absorptive interactions even if the amplitudes $t_\alpha$ respect isospin invariance the complex hadron-nucleon potentials do not and isospin invariance will not be transmitted to the optical potential (16.37).

The first order optical potential may be immediately brought to a manageable form. For simplicity, we consider isospin-zero nuclei with infinitely heavy nucleons, i.e. there will be no recoil. In momentum space, (16.15) may be written as

$$\langle \mathbf{k}' | V | \mathbf{k} \rangle = \langle 0 | \sum_\alpha \mathrm{e}^{-\mathrm{i}\mathbf{k}' \cdot \mathbf{x}_\alpha}\, t(\mathbf{k}', \mathbf{k})\, \mathrm{e}^{\mathrm{i}\mathbf{k} \cdot \mathbf{x}_\alpha} | 0 \rangle, \tag{16.39}$$

where we have invoked translational invariance of the t matrix. Now, the optical potential can be expressed in terms of Fourier transform of the nuclear density $\tilde{\rho}(\mathbf{k})$

$$\begin{aligned} \langle \mathbf{k}' | V | \mathbf{k} \rangle = t(\mathbf{k}', \mathbf{k}) \int \sum_\alpha \mathrm{e}^{-\mathrm{i}\mathbf{k}' \cdot \mathbf{x}} \langle 0 | \delta(\mathbf{x} - \mathbf{x}_\alpha) | 0 \rangle \mathrm{e}^{\mathrm{i}\mathbf{k} \cdot \mathbf{x}}\, \mathrm{d}^3 x = \\ = -\frac{4\pi}{2\mu_h} f(\mathbf{k}', \mathbf{k})\, A \tilde{\rho}(\mathbf{k} - \mathbf{k}'), \end{aligned} \tag{16.40}$$

where $\mu - h$ is hadron-nucleon reduced mass and in (16.40), we used the conventional expression for the nuclear density (normalized to unity)

$$A\, \rho(x) = \sum_\alpha \langle 0 | \delta(\mathbf{x} - \mathbf{x}_\alpha) | 0 \rangle. \tag{16.41}$$

At low energies the hadron-nucleon scattering amplitude is dominated just by the s-waves and possibly by the p-waves, so that the isospin averaged amplitude may be taken as

$$f(\mathbf{k}', \mathbf{k}) = b_0 + c_0\, \mathbf{k}' \cdot \mathbf{k}, \tag{16.42}$$

where $\mathbf{k}(\mathbf{k}')$ is the incident (final) hadron c.m. momentum. Close to threshold $b_0$ and $c_0$ are the scattering length and the scattering volume, respectively. In order to specify completely the optical potential (16.40) we need to know the amplitude both off and on the energy shell but the off-shell extrapolation unavoidably introduces some arbitrariness. Indeed, it will be noted that when (16.42) is taken literary, it stipulates the off-shell continuation. Using (16.42) in (16.40), we have

$$\begin{aligned}2\mu\langle\mathbf{k}'|V|\mathbf{k}\rangle = -4\pi\,(1+\frac{m}{M})\,Ab_0 \int \mathrm{e}^{-\mathrm{i}\mathbf{k}'\cdot\mathbf{r}}\,\rho(r)\,\mathrm{e}^{\mathrm{i}\mathbf{k}\cdot\mathbf{r}}\mathrm{d}^3r\, - \\ -4\pi c_0 \int (\boldsymbol{\nabla}\mathrm{e}^{-\mathrm{i}\mathbf{k}'\cdot\mathbf{r}})\cdot\rho(r)\,(\boldsymbol{\nabla}\mathrm{e}^{\mathrm{i}\mathbf{k}\cdot\mathbf{r}})\mathrm{d}^3r.\end{aligned} \tag{16.43}$$

where $\mu$ is hadron-nucleus reduced mass and the ratio $\mu/\mu_h$ has been approximated by 1+m/M where m and M denote the masses of the hadron and the nucleon, respectively. Integration by parts in the last term, yields

$$\begin{aligned}2\mu\langle\mathbf{k}'|V|\mathbf{k}\rangle = -4\pi\,A\,(1+\frac{m}{M})\,\times \\ \times \int \mathrm{e}^{-\mathrm{i}\mathbf{k}'\cdot\mathbf{r}}\,[b_0\,\rho(\mathbf{r}) - c_0\,\boldsymbol{\nabla}\cdot\rho(\mathbf{r})\boldsymbol{\nabla}]\,\mathrm{e}^{\mathrm{i}\mathbf{k}\cdot\mathbf{r}}\mathrm{d}^3r,\end{aligned} \tag{16.44}$$

where each nabla operator is understood to to act on all functions to the right of it. The optical potential in configuration space is obtained from (16.44) by a Fourier transform which in practice boils down to stripping off the plane wave factors from the integrand. The final result is

$$2\mu\langle\mathbf{r}'|V|\mathbf{r}\rangle = -4\pi\,A\,(1+\frac{m}{M})\,\delta(\mathbf{r}'-\mathbf{r})\,\left[b_0\,\rho(\mathbf{r}) - c_0\,\boldsymbol{\nabla}\cdot\rho(\mathbf{r})\boldsymbol{\nabla}\right], \tag{16.45}$$

and this velocity-dependent, non-local optical potential, derived originally for pions, is known as Kisslinger potential [Kisslinger (1955)].

The obvious alternative to the off-shell continuation in (16.42) is to insist that k and k' should be taken on-shell, i.e. both equal to the on-shell momentum p. Introducing momentum transfer vector $\mathbf{q} = \mathbf{k}' - \mathbf{k}$, formula (16.42) may be written as

$$f(\mathbf{k}',\mathbf{k}) = b_0 + c_0\,p^2 - \tfrac{1}{2}\,c_0\,q^2. \tag{16.46}$$

Using this form in (16.40) the optical potential depends on $\mathbf{k}'$ and $\mathbf{k}$ only through their difference and henceforth will be local. Instead of (16.45), we have

$$\begin{aligned}2\mu\langle\mathbf{r}'|V|\mathbf{r}\rangle = -4\pi\,A\,(1+\frac{m}{M})\delta(\mathbf{r}'-\mathbf{r})\,\times \\ \times\,\left[b_0\,\rho(\mathbf{r}) + c_0\,(p^2 + \tfrac{1}{2}\nabla^2)\rho(r)\right],\end{aligned} \tag{16.47}$$

where the nabla operator acts only on the density. The form (16.47) is known as the Laplacian potential.

The optical potential of the first-order in t is however a high energy approximation. Consequently, when free-space scattering parameters are inserted in t, not surprisingly the first-order potential fails to reproduce the hadronic atoms data. Going to higher orders, not to mention a summation of the whole Watson series, in general presents a very difficult problem. As a result of inclusion of higher order terms, the optical potential not only becomes a more complicated function of the nuclear density but also acquires a dependence upon nucleon-nucleon correlations. This was demonstrated quite early by [Ericson *et al.* (1966)] who pointed out to the analogy between pion-nucleus scattering near threshold and the scattering of light in a medium composed of non-polar molecules. In both cases we encounter strong p-wave dominance: for pions this is a consequence of the proximity of the delta resonance, while in optics because dipole interaction prevails in the long wave-length limit. In electrodynamics, under the influence of light each molecule becomes a source of electric field. A test body placed in the medium would measure an average electric field which is a coherent sum of the fields generated by all the molecules plus the field representing a plane wave propagating through the medium. However, a particular molecule in the medium experiences a local field which is different from the average field by the contribution generated by the molecule itself. This difference gives rise to a non-linear relationship between the index of refraction and the medium density known as the Lorentz-Lorenz effect. In the case of pions propagating in nuclear matter, an analogous effect occurs but it is the optical potential that becomes a non-linear function of the nuclear density (Lorentz-Lorenz-Ericson-Ericson effect). As a result, the p-wave part in the Kisslinger potential (16.45) has had to be changed

$$\nabla \cdot \rho(\mathbf{r}) \nabla \to \nabla \cdot \frac{\rho(\mathbf{r})}{1 + (4\pi/3)\, c_0\, (A-1)\, \rho(r)} \nabla \tag{16.48}$$

and to derive the modified potential it was necessary to assume that $\pi$-N interaction is of zero-range and to introduce hard-core nucleon-nucleon repulsion keeping the nucleons apart from each other. Subsequently, this subject was taken on by [ Eisenberg *et al.* (1973)] who demonstrated that the Lorentz-Lorenz modified optical potential may be obtained as a result of a partial summation of the Watson series. However, with more realistic nucleon-nucleon correlations or with $\pi$-N forces not necessarily of zero-range, partial summation of the Watson series was no longer possible.

It is desirable to express also the potential (16.36) in terms of the nuclear density. In momentum space (16.36) can be written as

$$\langle \mathbf{k}'|V|\mathbf{k}\rangle = \langle 0|\int \mathrm{e}^{-i\mathbf{k}'\cdot\mathbf{y}}\textstyle\sum_\alpha v(|\mathbf{y}-\mathbf{x}_\alpha|)\mathrm{e}^{i\mathbf{k}\cdot\mathbf{y}}\,\mathrm{d}^3y|0\rangle =$$

$$= \int \mathrm{e}^{-i\mathbf{q}'\cdot\mathbf{y}}\, v(|\mathbf{y}-\mathbf{x}|)\langle 0|\textstyle\sum_\alpha \delta(\mathbf{x}-\mathbf{x}_\alpha)|0\rangle\,\mathrm{d}^3x\,\mathrm{d}^3\Delta y = \tag{16.49}$$

$$= A\int \mathrm{e}^{-i\mathbf{q}\cdot\mathbf{y}}\,\rho(x)\,v(|\mathbf{y}-\mathbf{x}|)\,\mathrm{d}^3x\,\mathrm{d}^3y,$$

and this potential is manifestly local. In configuration space the optical potential (16.49) takes the form of a convolution of the density with the hadron-nucleon potential

$$\langle \mathbf{r}'|V|\mathbf{r}\rangle = \delta(\mathbf{r}'-\mathbf{r})\,V(r); \tag{16.50a}$$

$$V(r) = A\int v(x)\,\rho(|\mathbf{r}-\mathbf{x}|)\,\mathrm{d}^3x, \tag{16.50b}$$

and is sometimes referred to as the folding model potential. In order to examine the shape of this potential the density may be expanded in a Taylor series about a fixed position r, large comparing with the range of the potential v. In the integral (16.50b) only even powers of x will give non-vanishing contribution and the optical potential takes the form

$$V(r) = A\int v(x)\,\mathrm{d}^3x\left[1+\sum_{n=1}^{\infty}\frac{\langle x^{2n}\rangle}{(2n+1)!}\Delta^n\right]\rho(r), \tag{16.51}$$

where $\langle x^{2n}\rangle$ denotes (2n)-th moment of v(x) and $\Delta$ is a Laplacian acting on $\rho(r)$. For short-ranged v, the dominant term in the potential (16.51) has the shape of the nuclear density while the correction terms involving the powers of the Laplacian represent a typical surface effect. For spherically symmetric density $\rho$(r), formula (16.50b) may be cast to the form

$$V(r) = A\,4\pi\int_0^\infty v(x)\tilde{\rho}(x,r)\,x^2\,\mathrm{d}x, \tag{16.52}$$

where $\tilde{\rho}(x,r) = \tilde{\rho}(r,x)$ is angle averaged density

$$\tilde{\rho}(x,r) = \tfrac{1}{2}\int_{-1}^{1}\rho(\sqrt{x^2+r^2-2\,r\,x\,t})\,\mathrm{d}t. \tag{16.53}$$

It is evident from (16.53) that $\tilde{\rho}(0,r) = \rho(r)$ and $\tilde{\rho}(x,r) \to \rho(r)$ for r≫x.

To examine the departure from the $\rho(r)$ shape caused in (16.50b) by the finite range of v, for simplicity the latter may be taken as a delta-shell potential, setting v(r)=$v_0\,R\,\delta(r-R)$ where R and $v_0$ are the range and the depth of the hadron-nucleon interaction. The folding integral in (16.52) can be in this case written as

$$V(r) = A \int \rho(x)\, v(|\mathbf{r} - \mathbf{x}|)\, \mathrm{d}^3x = A\, 4\pi\, v_0\, R^3\, \tilde{\rho}(R, r). \tag{16.54}$$

When the hadron-nucleon interaction is of a short-range, the departure of V(r) from the density shape may be anticipated to be minor, hence we put

$$\tilde{\rho}(R, r) = \rho(r) + \delta\rho(r). \tag{16.55}$$

In Fig. 16.2 we compare $\rho(r)$ and $\delta\rho(r)$ for $^{32}S$ taking the range of the hadron-nucleon potential to be R=1 fm. The surface nature of the correction $\delta\rho$ is apparent and if this term is treated perturbatively the change of sign in $\delta\rho(r)$ may result in some cancellations. The finite-range correction

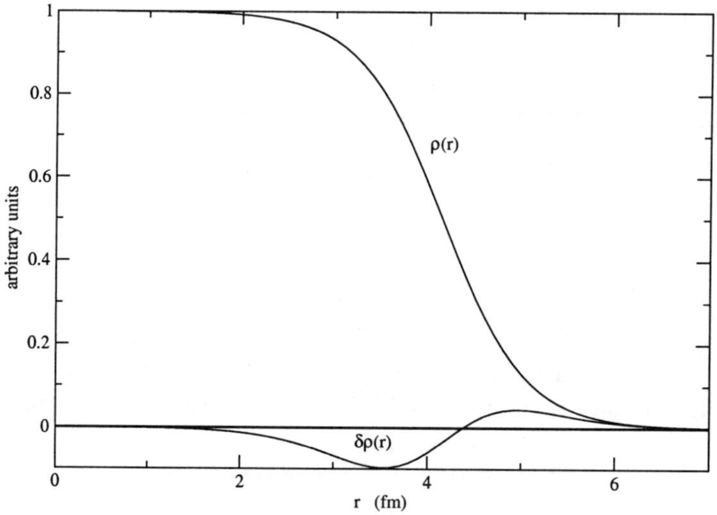


Fig. 16.2 Correction due to the finite range in the shape of the optical potential. The presented density $\rho$ corresponds to $^{32}S$.

contains both attractive and repulsive component and their relative contributions depend upon the shape of the density and the magnitude of the range parameter.

The size of the correction due to the $\delta\rho$ term in the optical potential rapidly increases in importance with the orbital momentum. This follows quite generally from the the fact that the measurement of the width essentially probes the $2\ell$-th moment of the potential given by (16.50b). Expanding the Fourier transform of this potential in a Taylor series about zero, we have

$$\int e^{i\mathbf{k}\cdot\mathbf{r}}\, V(r)\, d^3r = \int V(r)\, d^3r \left\{ 1 + \sum_{n=1}^{\infty} \frac{(-1)^n k^{2n}}{(2n+1)!} \left\langle r^{2n} \right\rangle_V \right\}, \quad (16.56)$$

where $\left\langle r^{2n} \right\rangle_V$ is a 2n-th moment of the potential V(r) given in (16.50b). However, the Fourier transform of the folding model potential equals also to a product of the Fourier transforms of $\rho$ and of v. Expanding these in powers of k, we can relate the moments of the potential V with the moments of the density and the moments of v. For short-ranged hadron-nucleon forces, it is sufficient to keep only the second moment of v (i.e. $\left\langle x^2 \right\rangle_v$), and we obtain

$$\left\langle r^{2\ell} \right\rangle_V \approx A \left\langle r^{2\ell} \right\rangle_\rho + \tfrac{1}{3}\ell(2\ell+1)\, A \left\langle r^{2\ell-2} \right\rangle_\rho \left\langle x^2 \right\rangle_v, \quad (16.57)$$

so that the coefficient multiplying the finite-range correction term increases as $\ell^2$.

In order to understand the difference between the optical potential based on the "t$\rho$" prescription and the folding model potential (16.50) it is best to turn to a specific example in which the solution of the hadron-nucleon inverse scattering problem becomes a trivial matter. This is the delta-shell potential model discussed above (16.54), and, for simplicity, we assume that only s-wave are of importance. The advantageous feature of the delta-shell potential is that the scattering length is very simply related to depth $v_0$ of the potential and this relation may me easily inverted whereby $v_0$ may be expressed in terms of the scattering length. Thus, eliminating $v_0$ in (16.54) in favor of the scattering length $b_0$, the folding model potential takes the form

$$2\mu\, V(r) = -4\pi\, A\, (1 + \frac{m}{M}) \frac{b_0}{1 + b_0/R} \left[\rho(r) + \delta\rho(r)\right], \quad (16.58)$$

and this expression has to be compared with the first order optical potential (16.45), setting $c_0$=0. In contrast with the potential (16.45), just proportional to $b_0$, in (16.58) the dependence upon $b_0$ enters in a non-linear way. Thus, these two forms will be equivalent only when $|b_0/R| \ll 1$ but this inequality in neither type of hadronic atoms known today is likely to be satisfied. In the low-energy regime the first order optical potential makes a poor approximation which is borne out by the fact that such potential fails to reproduce hadronic atoms data. It has to be stated clearly that this non-linear dependence of the folding model potential upon the scattering length, demonstrated here for the delta-shell potential, is quite a general feature. Indeed, for a hadron-nucleon potential of an unspecified shape the scattering length will be a rational function of the potential strength. To obtain the potential strength as a function of the scattering length this formula has to be inverted which in general must be performed numerically. Nevertheless, for potentials not too strong that can support no more than one bound state, in the scattering length the pole in the potential strength closest to zero dominates. Keeping only the pole term, the inversion is trivial and we arrive at (16.58) times a slowly varying function of $b_0$.

The medium correction to the first order optical potential must therefore be very important. The next level of sophistication consists in starting with the free-space hadron-nucleon amplitude and computing an effective interaction appropriate to nucleons embedded in the medium. This program has been taken on by many authors. In the independent pair approximation the optical potential may be sought in the form

$$\langle \mathbf{r}|V(E)|\mathbf{r}'\rangle = \sum_\alpha \int \mathrm{d}^3x\, \mathrm{d}^3x'\, \varphi_\alpha^*(\mathbf{x})\, t(\mathbf{r},\mathbf{x},\mathbf{r}',\mathbf{x}';E+E_\alpha)\, \varphi_\alpha(\mathbf{x}'), \quad (16.59)$$

where the $\varphi_\alpha(\mathbf{x})$ are a set of single nucleon optical model wave functions and t is an effective hadron-nucleon t-matrix in the medium. The effective t-matrix is related to the free-space amplitude $t_0$ by an equation of the form

$$t = t_0 + t_0\,(G - G_0)\,t, \quad (16.60)$$

where $G$-$G_0$ is the modification of the hadron-nucleon propagator due to the presence of the medium. Since this difference involves the optical potential V which enters G, (16.59) and (16.60) must be solved simultaneously in a self-consistent way for the optical potential [Lutz *et al.* (2002)]. This procedure also generates the desired non-linear dependence upon the free-space scattering length, via eq. (16.60), and in addition to that, the requirement of self-consistency induces a non-linear dependence upon the density. In the limit of zero-density all medium corrections must go away and the in-medium amplitude reduces to free-space scattering amplitude.

The above scheme provided some general guidelines quite useful for constructing phenomenological forms of the optical potential. One such form that received favor in certain quarters generalizes the "t$\rho$" optical potential by introducing a more complicated density dependence

$$2\mu V(r) = -4\pi A\,(1+\frac{\mu}{M})\left\{b_0 + B_0\left[\frac{\rho(r)}{\rho(0)}\right]^{\alpha}\right\}\rho(r). \qquad (16.61)$$

This flexible formula allows to examine the departures of the optical potential from the low-density limit, in which case $b_0$ is identified with the threshold value of the free-space hadron-nucleon scattering amplitude. Since the potential (16.61) is justified only by plausibility arguments, it is not clear what is the physical interpretation of the parameters $\alpha$ and $B_0$.

## 16.2 Pionic Atoms

Historically, pionic atoms were the first hadronic atoms observed experimentally, cf. [Camac *et al.* (1952)]. Today, the data base housing pionic atoms strong interaction level shifts and widths covers a large number of nuclear species ranging from 1s level in pionic hydrogen up to 4f level in $^{237}$Np. Until very recently, the number of entries in this data base was larger than the number of measured level shifts and widths in all the remaining hadronic atoms taken together. With the advent of the new LEAR measurements, however, the data base for $\bar{p}$-atoms and that for pionic atoms are of about the same size. The early measurements (for a review, cf. [Backenstoss (1970)]) were confined to the peripheral states while all lower-lying states turned out to be too broad which precluded their identification from the measured X-rays. The situation has improved with the advent of intense pion beams at meson factories and thanks to the recent advances in detection technique. High-quality data obtained with the use high-resolution Compton-suppressed solid state $\gamma$-ray detectors provided access to the previously unobserved deeper bound and much broadened levels for many targets. The numerical values for the pionic atoms level shifts and widths taken from [Konijn *et al.* (1990)] are presented in Table 16.1 and in Table 16.2. However, it is important to keep in mind that [Konijn *et al.* (1990)] use an alternative definition of the level shifts $\epsilon$, where the shift is defined as the difference beteween the measured transition energy and the calculated transition energy for *point-like* Coulomb potential, *excluding* finite-size effects but taking into account all other electromagnetic corrections. The shifts presented in Table 16.1 and Table 16.2 have not been corrected for for the effects of the finite charge distribution.

Pionic atoms were also the first hadronic atoms in which the strong interaction effects were investigated theoretically. At zero-energy the nuclear interaction of the pion is relatively weak and therefore the "t-$\rho$" prescription seems here to be better justified than in the other types of hadronic atoms. Accordingly, the optical potential of the Kisslinger type has been the prototype of the various theoretical potentials, or at least provided a theoretical guidance in obtaining the form of the potential. By and large, the starting point in most investigations is the optical potential put forward by Ericson and Ericson (cf. [Ericson *et al.* (1966); Krell *et al.* (1969)]). This potential has two components: the theoretically motivated part associated with multiple scattering of the s- and p-wave pions in the nuclear medium and another part purely phenomenological simulating absorptive effects. This "standard" optical potential takes the form

$$2\mu\, V(\mathbf{r}) = -4\pi \left[ q(r) - \boldsymbol{\nabla} \cdot \frac{\alpha(r)}{1 + (4\pi/3)\xi\, \alpha(r)} \boldsymbol{\nabla} \right], \tag{16.62}$$

where $\mu$ is $\pi$-nucleus reduced mass and

$$q(r) = (1 + \frac{m}{M})\, [b_0\, \rho(r) + b_1\, \delta\rho(r)] + (1 + \frac{m}{2M}) B_0\, \rho(r)^2; \tag{16.63}$$

$$\alpha(r) = (1 + \frac{m}{M})^{-1}\, [c_0\, \rho(r) + c_1\, \delta\rho(r)] + (1 + \frac{m}{2M})^{-1} C_0\, \rho(r)^2; \tag{16.64}$$

where m and M denote the masses of the pion and the nucleon, respectively. We have introduced the densities

$$\rho(r) = \rho_n(r) + \rho_p(r); \qquad \delta\rho(r) = \rho_n(r) - \rho_p(r); \tag{16.65}$$

with $\rho_n$ and $\rho_p$ denoting the neutron and proton densities, normalized to N and Z, respectively. The parameter $\xi$ in (16.62) in the original formulation of [Ericson *et al.* (1966)] was set equal to one and this corresponds to zero-range forces and repulsive hard-core NN correlations but now values not only between 0 and 1 but also bigger than one have been also considered. The q(r) dependent term in (16.62) is due to the elementary $\pi$-N s-wave interaction and $b_0$ and $b_1$ represent, respectively, the isoscalar and isovector scattering lengths. The $\alpha(r)$ term is associated with p-wave interaction and depends upon the isoscalar and isovector scattering volumes $c_0$ and $c_1$, respectively. The values of the scattering lengths and scattering volumes are expected to be modified by the nuclear medium.

The terms proportional to $B_0$ and $C_0$ in (16.63) and (16.64) have been put in by hand and they are intended to simulate the effects of true pion absorption. Hence, both $B_0$ and $C_0$ are assumed to be complex quantities, and their imaginary parts are directly connected with pion absorption, i.e. the only open channel for the pion at this low an energy. The non-zero real parts of these parameters are very much needed and in fact play an important role although their interpretation is less clear. Certainly they partly represent higher order terms in the optical potential but may be also understood as a manifestation of dispersive effects through which the absorption mechanism can change the real part of the optical potential. The quadratic dependence upon the density is motivated by the fact that the pion absorption reaction predominantly leads to the ejection of two nucleons. In this approximate form the absorption mechanism is assumed to be of zero-range and the two nucleons involved are uncorrelated. It is perhaps worth recalling that the ejection of a single nucleon is highly improbable since the emitted nucleon absorbing then all of the pion's rest energy emerges with a momentum equal to $\sqrt{2mM} \sim 500$ MeV. Since the average Fermi momentum reaches only half of that value, it is difficult to maintain energy and momentum conservation. Actually, the p-wave absorption enters the potential (16.62) in a rather complicated way since the quadratic term, for purely traditional reasons, has been also included in the Lorentz-Lorenz denominator. It seems fair to say that the absorptive con-

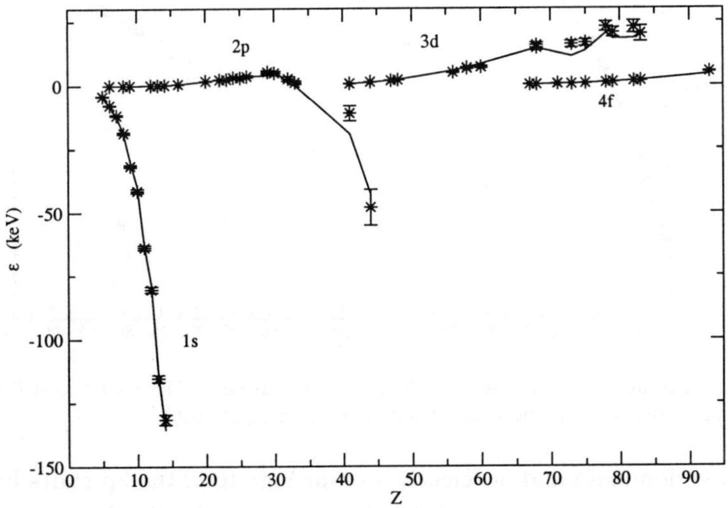


Fig. 16.3 Strong interaction level shifts in pionic atoms. The continuous lines join points calculated with the best-fit conventional optical potential.

tribution still remains the least justified part of the standard potential and the variations thereof. Experimentally, very little information is available that would provide a clue how to improve the phenomenological description of pion absorption.

For an assigned $\xi$ value, the standard potential contains four real and two complex parameters to be adjusted to the experimentally measured strong interaction level shifts and widths throughout the whole periodic table. The results of such a fitting procedure conducted by [Konijn *et al.* (1990)], in which a $\chi^2$/data-point of about 3 was obtained, are presented in Fig. 16.3 and in Fig. 16.4. It is quite impressive that the widths spanning over five orders of magnitude may be reproduced by a single potential. The s-wave part of the optical potential is repulsive and the importance of this component for a given state in pionic atom will be magnified with the increasing binding energy, due to the increased overlap of the atomic wave function with the nuclear density. Eventually, the s-wave repulsion

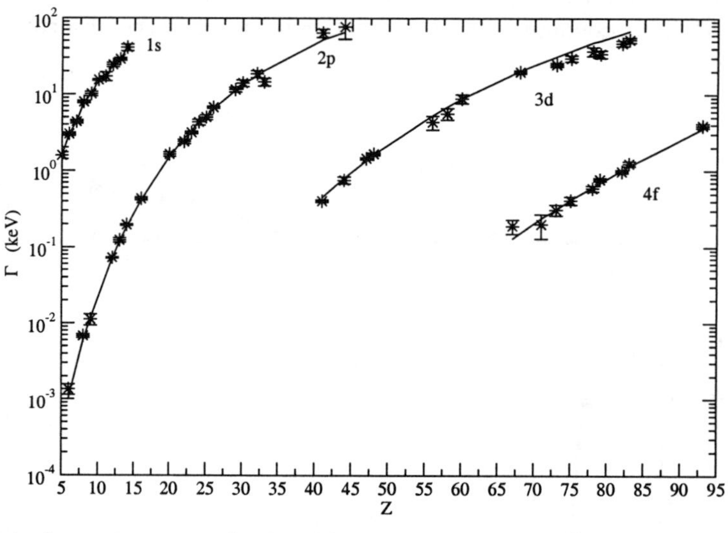


Fig. 16.4 Strong interaction level widths in pionic atoms. The continuous lines join points calculated with the best-fit conventional optical potential.

may even dominate and, as clearly seen in Fig. 16.3, the 2p shifts become repulsive. This effect was predicted long time ago by [Krell *et al.* (1969)] and received full experimental confirmation.

Table 16.1 Compilation of pionic atom data adopted from [Konijn *et al.* (1990)] where the references to the original papers may be found.

| Nucl | level | $\epsilon$ (keV) | $\Gamma$ (keV) |
|---|---|---|---|
| $^{10}$B | 1s | -3.25 ± 0.08 | 1.58 ± 0.16 |
| $^{11}$B | 1s | -4.17 ± 0.09 | 1.77 ± 0.16 |
| $^{12}$C | 1s | -6.79 ± 0.15 | 2.99 ± 0.14 |
| $^{13}$C | 1s | -7.76 ± 0.13 | 2.59 ± 0.14 |
| $^{14}$N | 1s | -11.72 ± 0.22 | 4.34 ± 0.24 |
| $^{16}$O | 1s | -18.7 ± 0.4 | 7.9 ± 0.3 |
| $^{18}$O | 1s | -23.3 ± 0.3 | 6.3 ± 0.4 |
| $^{19}$F | 1s | -31.6 ± 0.5 | 10.1 ± 0.7 |
| $^{20}$Ne | 1s | -41.4 ± 0.8 | 15.4 ± 0.4 |
| $^{22}$Ne | 1s | -49.7 ± 0.5 | 10.8 ± 0.7 |
| $^{23}$Na | 1s | -63.9 ± 0.9 | 17.1 ± 2.3 |
| $^{24}$Mg | 1s | -80.5 ± 1.2 | 24.3 ± 1.6 |
| $^{27}$Al | 1s | -115.5 ± 1.4 | 28.8 ± 1.2 |
| $^{28}$Si | 1s | -131.6 ± 2. | 41. ± 4. |
| $^{12}$C | 2p | $(3.06 \pm 0.16)10^{-3}$ | $(1.36 \pm 0.22)10^{-3}$ |
| $^{16}$O | 2p | $(1.48 \pm 0.03)10^{-2}$ | $(0.68 \pm 0.04)10^{-2}$ |
| $^{18}$O | 2p | $(1.51 \pm 0.04)10^{-2}$ | $(0.75 \pm 0.04)10^{-2}$ |
| $^{19}$F | 2p | 0.03 ± 0.003 | 0.0112 ± 0.0019 |
| $^{24}$Mg | 2p | 0.125 ± 0.004 | 0.0725 ± 0.0018 |
| $^{26}$Mg | 2p | 0.122 ± 0.004 | 0.0811 ± 0.0019 |
| $^{27}$Al | 2p | 0.19 ± 0.007 | 0.124 ± 0.008 |
| $^{28}$Si | 2p | 0.275 ± 0.01 | 0.196 ± 0.005 |
| $^{30}$Si | 2p | 0.271 ± 0.01 | 0.196 ± 0.008 |
| $^{32}$S | 2p | 0.59 ± 0.022 | 0.43 ± 0.021 |
| $^{40}$Ca | 2p | 1.77 ± 0.08 | 1.64 ± 0.11 |
| $^{42}$Ca | 2p | 1.53 ± 0.08 | 1.65 ± 0.15 |
| $^{44}$Ca | 2p | 1.41 ± 0.08 | 1.64 ± 0.12 |
| $^{48}$Ca | 2p | 1.18 ± 0.08 | 1.64 ± 0.11 |
| $^{46}$Ti | 2p | 2.27 ± 0.14 | 2.39 ± 0.15 |
| $^{48}$Ti | 2p | 2.07 ± 0.13 | 2.62 ± 0.15 |
| $^{50}$Ti | 2p | 1.72 ± 0.11 | 2.15 ± 0.27 |
| $^{51}$V | 2p | 2.41 ± 0.16 | 3.13 ± 0.13 |
| $^{50}$Cr | 2p | 3.17 ± 0.21 | 4.3 ± 0.4 |
| $^{52}$Cr | 2p | 2.74 ± 0.19 | 3.85 ± 0.21 |
| $^{54}$Cr | 2p | 2.43 ± 0.19 | 3.84 ± 0.29 |

Table 16.2 Compilation of pionic atom data, continuation of Table 16.1.

| Nucl | level | $\epsilon$ (keV) | $\Gamma$ (keV) |
|---|---|---|---|
| $^{55}$Mn | 2p | 3.04 ± 0.25 | 5. ± 0.4 |
| $^{56}$Fe | 2p | 3.6 ± 0.4 | 6.75 ± 0.21 |
| $^{nat}$Cu | 2p | 5.3 ± 0.6 | 11.5 ± 0.8 |
| $^{nat}$Zn | 2p | 4.9 ± 0.8 | 14. ± 1.4 |
| $^{nat}$Ge | 2p | 2.7 ± 0.9 | 18.5 ± 2. |
| $^{75}$As | 2p | 1.2 ± 0.7 | 14.5 ± 1.7 |
| $^{93}$Nb | 2p | -11. ± 3. | 64. ± 8. |
| $^{nat}$Ru | 2p | -48. ± 7. | 77. ± 24. |
| $^{93}$Nb | 3d | 0.74 ± 0.02 | 0.402 ± 0.016 |
| $^{nat}$Ru | 3d | 1.39 ± 0.08 | 0.75 ± 0.08 |
| $^{nat}$Ag | 3d | 1.94 ± 0.07 | 1.44 ± 0.05 |
| $^{nat}$Cd | 3d | 2.14 ± 0.09 | 1.65 ± 0.07 |
| $^{138}$Ba | 3d | 5.1 ± 0.3 | 4.3 ± 0.9 |
| $^{140}$Ce | 3d | 6.6 ± 0.3 | 5.6 ± 1. |
| $^{142}$Ce | 3d | 6.8 ± 0.3 | 6.5 ± 0.9 |
| $^{148}$Nd | 3d | 7.2 ± 0.4 | 8.8 ± 1.2 |
| $^{150}$Nd | 3d | 7.1 ± 0.4 | 9.2 ± 1.1 |
| $^{166}$Er | 3d | 15.6 ± 1. | 19.7 ± 0.9 |
| $^{168}$Er | 3d | 14.6 ± 1. | 19.4 ± 1. |
| $^{181}$Ta | 3d | 16. ± 1.3 | 24.6 ± 1.5 |
| $^{nat}$Re | 3d | 16.5 ± 1.3 | 29.7 ± 2.7 |
| $^{nat}$Pt | 3d | 22.8 ± 1.9 | 37. ± 5. |
| $^{197}$Au | 3d | 20.6 ± 1.9 | 34. ± 4. |
| $^{208}$Pb | 3d | 22.7 ± 2.4 | 47. ± 4. |
| $^{209}$Bi | 3d | 20.1 ± 3. | 52. ± 4. |
| $^{165}$Ho | 4f | 0.3 ± 0.08 | 0.19 ± 0.04 |
| $^{166}$Er | 4f | 0.3 ± 0.03 | 0. ± 0. |
| $^{168}$Er | 4f | 0.3 ± 0.03 | 0. ± 0. |
| $^{175}$Lu | 4f | 0.6 ± 0.07 | 0.2 ± 0.07 |
| $^{181}$Ta | 4f | 0.56 ± 0.04 | 0.31 ± 0.05 |
| $^{nat}$Re | 4f | 0.76 ± 0.04 | 0.41 ± 0.05 |
| $^{nat}$Pt | 4f | 1.09 ± 0.04 | 0.59 ± 0.05 |
| $^{197}$Au | 4f | 1.25 ± 0.07 | 0.77 ± 0.04 |
| $^{208}$Pb | 4f | 1.68 ± 0.04 | 0.98 ± 0.05 |
| $^{209}$Bi | 4f | 1.78 ± 0.06 | 1.24 ± 0.09 |
| $^{237}$Np | 4f | 5.26 ± 0.19 | 3.88 ± 0.26 |

## 16.3 Kaonic Atoms

The data base for strong interaction level shifts and widths in kaonic atoms is presented in Table 16.3. Early calculations revealed two facts, firstly, that a simple optical potential proportional to the nuclear density reproduces quite well the data, and, secondly, that the best fit depth of the optical potential violates the low-density limit. Indeed, taking the optical potential as

$$2\mu V(r) = -4\pi\, A\,(1+\frac{\mu}{M})\, b_0\, \rho(r). \tag{16.66}$$

and using e.g. the values of the $\bar{\mathrm{K}}\mathrm{N}$ scattering lengths from [Martin (1981)], we have

$$\begin{aligned} b_0 = \tfrac{1}{2}[a(K^-p)+a(K^-n)] &= \tfrac{1}{2}[\tfrac{1}{2}(a_0+a_1)+a_1] = \\ &= (-0.15+\mathrm{i}\,0.62)\,\mathrm{fm}. \end{aligned} \tag{16.67}$$

Not much different $\mathrm{b}_0$ is obtained adopting for $\bar{\mathrm{K}}\mathrm{N}$ scattering lengths the more recent values inferred from a chirally inspired model of [Oset *et al.* (2000)], namely $a(K^-p) = (-1.00+\mathrm{i}\,0.94)\,\mathrm{fm}$ and $a(K^-n) = (0.53+\mathrm{i}\,0.62)\,\mathrm{fm}$, which yields

$$b_0 = (-0.24+\mathrm{i}\,0.78)\,\mathrm{fm}. \tag{16.68}$$

However, when $\mathrm{b}_0$ in (16.66) is regarded as an adjustable parameter, a fit to the kaonic atom data shows a big departure from the above free-space value. The most recent values are

$$b_0 = [(0.62\pm 0.05)+\mathrm{i}\,(0.92\pm 0.05)]\,\mathrm{fm}, \tag{16.69}$$

as obtained by [Batty *et al.* (1997)], or

$$b_0 = [(0.52\pm 0.03)+\mathrm{i}\,(0.80\pm 0.03)]\,\mathrm{fm}, \tag{16.70}$$

as quoted by [Baca *et al.* (2000)]. The difference between (16.69) and (16.70) has to be attributed to slightly different nuclear densities used by these two groups. The quality of the fit may be appreciated from Fig. 16.5 and Fig. 16.6 where the level shifts and widths calculated by [Baca *et al.* (2000)] are compared with experiment (the quoted value of $\chi^2$ per No. of degree of freedom is 2.1).

We note that the real part of the phenomenological potential (16.66) has been invariably attractive, whereas the "t$\rho$" potential employing the free-space scattering length (16.67) would be repulsive. Clearly, a simple "t$\rho$"

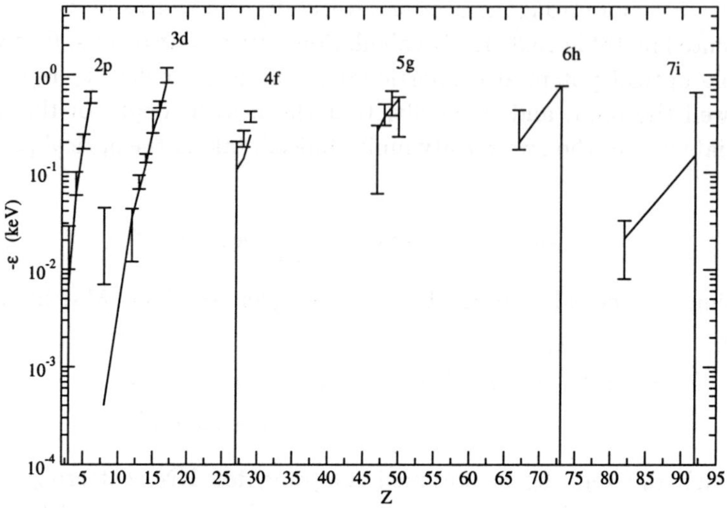


Fig. 16.5 Strong interaction level shifts in kaonic atoms. The continuous lines join points calculated with the best-fit conventional optical potential.

potential is inadequate and the medium corrections are indispensable. As a matter of fact, this difference in sign is hardly surprising in view of the fact that $\bar{\mathrm{K}}$N interaction is rather strong and the first order Born approximation is insufficient. This point may be further amplified by considering the "v$\rho$" form of the optical potential resulting from a partial summation of the multiple scattering series. Taking the elementary $\bar{\mathrm{K}}$N potential in the delta-shell form, and neglecting the surface correction $\delta\rho$, the potential (16.58) follows from (16.66) in result of a substitution

$$b_0 \to \frac{b_0}{1 + b_0/R},$$

and the presence of this denominator is crucial for explaining the change of sign. Admittedly, the range parameter R is not known, but it may be varied in some reasonable limits. Using for $b_0$ the value of [Oset *et al.* (2000)] given in (16.68), for R=1.29 fm, we obtain

$$\frac{b_0}{1 + b_0/R} \approx (0.63 + \mathrm{i}\,0.94)\,\mathrm{fm},$$

which is remarkably close to the best fit value (16.69) of [Batty *et al.*

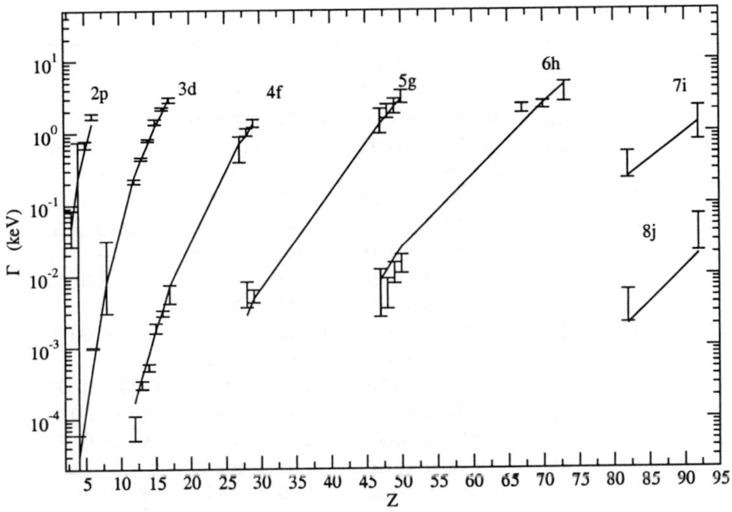


Fig. 16.6 Strong interaction level widths in kaonic atoms. The continuous lines join points calculated with the best-fit conventional optical potential.

(1997)]. The essential point here is that the depth of the optical potential depends in a non-linear way upon the free-space scattering length. Recently, such a non-linear dependence has been also generated within the "t$\rho$" model by introducing the medium corrections in a self-consistent way. Self-consistency is stipulated by including the optical potential into the in-medium $K^-$ propagator entering the appropriate integral equation for the scattering operator in medium (cf. (16.60)). For details the Reader is referred to the original papers, cf. [Baca *et al.* (2000); Cieplý *et al.* (2001); Lutz *et al.* (2002)].

Table 16.3 Compilation of kaonic atom data, references to the experimental papers may be traced down from [Friedman *et al.* (1994)]

| Nucl | level | $\epsilon$ (keV) | $\Gamma$ (keV) | $\Gamma_u$ (eV) |
|---|---|---|---|---|
| He | 2p | -0.04 ± 0.03 | – | – |
| He | 2p | -0.035 ± 0.012 | 0.03 ± 0.03 | – |
| Li | 2p | 0.002 ± 0.026 | 0.055 ± 0.029 | – |
| Be | 2p | -0.079 ± 0.021 | 0.172 ± 0.58 | 0.04 ± 0.02 |
| $^{10}$B | 2p | -0.208 ± 0.035 | 0.81 ± 0.1 | – |
| $^{11}$B | 2p | -0.167 ± 0.035 | 0.7 ± 0.08 | – |
| C | 2p | -0.59 ± 0.08 | 1.73 ± 0.15 | 0.99 ± 0.2 |
| O | 3d | -0.025 ± 0.018 | 0.017 ± 0.014 | – |
| Mg | 3d | -0.027 ± 0.015 | 0.214 ± 0.015 | 0.08 ± 0.03 |
| Al | 3d | -0.13 ± 0.05 | 0.49 ± 0.16 | – |
| Al | 3d | -0.076 ± 0.014 | 0.442 ± 0.022 | 0.3 ± 0.04 |
| Si | 3d | -0.24 ± 0.05 | 0.81 ± 0.12 | – |
| Si | 3d | -0.13 ± 0.015 | 0.8 ± 0.033 | 0.53 ± 0.06 |
| P | 3d | -0.33 ± 0.08 | 1.44 ± 0.12 | 1.89 ± 0.3 |
| S | 3d | -0.55 ± 0.06 | 2.33 ± 0.2 | 3.1 ± 0.36 |
| S | 3d | -0.43 ± 0.12 | 2.31 ± 0.17 | – |
| S | 3d | -0.462 ± 0.054 | 1.96 ± 0.17 | 2.9 ± 0.5 |
| Cl | 3d | -0.77 ± 0.4 | 3.8 ± 1. | 5.8 ± 1.7 |
| Cl | 3d | -0.94 ± 0.4 | 3.92 ± 0.99 | – |
| Cl | 3d | -1.08 ± 0.22 | 2.79 ± 0.25 | – |
| Co | 4f | -0.099 ± 0.106 | 0.64 ± 0.25 | – |
| Ni | 4f | -0.18 ± 0.07 | 0.59 ± 0.21 | 5.9 ± 2.3 |
| Ni | 4f | -0.246 ± 0.052 | 1.23 ± 0.14 | – |
| Cu | 4f | -0.24 ± 0.22 | 1.65 ± 0.72 | 7. ± 3.8 |
| Cu | 4f | -0.377 ± 0.048 | 1.35 ± 0.17 | 5.1 ± 1.1 |
| Ag | 5g | -0.18 ± 0.12 | 1.54 ± 0.58 | 7.3 ± 4.7 |
| Cd | 5g | -0.4 ± 0.1 | 2.01 ± 0.44 | 6.2 ± 2.8 |
| In | 5g | -0.53 ± 0.15 | 2.38 ± 0.57 | 11.4 ± 3.7 |
| Sn | 5g | -0.41 ± 0.18 | 3.18 ± 0.64 | 15.1 ± 4.4 |
| Ho | 6h | -0.3 ± 0.13 | 2.14 ± 0.31 | – |
| Yb | 6h | -0.12 ± 0.1 | 2.39 ± 0.3 | – |
| Ta | 6h | -0.27 ± 0.5 | 3.76 ± 1.15 | – |
| Pb | 7i | – | 0.37 ± 0.15 | 4.1 ± 2. |
| Pb | 7i | -0.02 ± 0.012 | – | – |
| U | 7i | -0.26 ± 0.4 | 1.5 ± 0.75 | 45. ± 24. |

## 16.4 Antiprotonic Atoms

Antiprotonic atoms provide insight into the antiproton-nucleus interaction and supply valuable information about the nuclear periphery where the antiproton annihilation takes place. The data base for antiprotonic atoms strong interaction level shifts and widths is presented in Table 16.4 containing data from earlier experiments and in Table 16.5 containing the most recent LEAR measurements reported by [Trzcińska *et al.* (2001)]. The CERN experiment using the LEAR facility detects more transitions, and, owing to the improved precision, for targets with Z≥48 in most cases allows to resolve the fine structure components. For targets from $^{16}$O up to $^{238}$U several shifts and with per atom are now available. Furthermore, in some deformed target nuclei, this information is enriched by providing additional evidence concerning quadrupole excitation (E2). So far, theoretical analyses have been concentrated on the data from Table 16.4 and only a preliminary analysis [Wycech (2001)] covers the full data base. It was quite early

Table 16.4 Compilation of $\bar{p}$ atomic data, references to the experimental papers may be traced down from [Batty *et al.* (1995)]

| Nucl | level | $\epsilon$ (keV) | $\Gamma$ (keV) | $\Gamma_u$ (eV) |
|---|---|---|---|---|
| C | 3d | -0.004 ± 0.01 | 0.042 ± 0.018 | 0.036 ± 0.013 |
| N | 3d | -0.04 ± 0.03 | 0.218 ± 0.075 | 0.136 ± 0.019 |
| $^{16}$O | 3d | -0.112 ± 0.02 | 0.495 ± 0.045 | 0.603 ± 0.022 |
| $^{18}$O | 3d | -0.195 ± 0.02 | 0.64 ± 0.04 | 0.795 ± 0.023 |
| F | 3d | -0.44 ± 0.07 | 1.365 ± 0.15 | 2.79 ± 0.16 |
| Na | 3d | -2.1 ± 0.3 | 2.9 ± 1.1 | 23.8 ± 7.4 |
| Si | 4f | -0.038 ± 0.039 | 0.11 ± 0.19 | – |
| P | 4f | -0.065 ± 0.023 | 0.446 ± 0.069 | 1.69 ± 0.23 |
| S | 4f | -0.06 ± 0.04 | 0.65 ± 0.1 | 2.95 ± 0.58 |
| Cl | 4f | – | – | 7.8 ± 1.7 |
| K | 4f | – | – | 26.2 ± 5.5 |
| Fe | 5g | -0.01 ± 0.31 | 0.54 ± 0.32 | 4. ± 2.4 |
| Y | 6h | -0.15 ± 0.16 | 0.8 ± 0.32 | 6.8 ± 1.7 |
| Zr | 6h | -0.45 ± 0.1 | 0.7 ± 0.21 | 6.4 ± 1.6 |
| $^{92}$Mo | 6h | -0.46 ± 0.08 | 1.4 ± 0.3 | 19.5 ± 1.2 |
| $^{98}$Mo | 6h | -0.55 ± 0.16 | 2.3 ± 0.7 | 47.9 ± 5.3 |
| Sn | 7i | – | – | 3. ± 1.5 |
| I | 7i | – | – | 9.7 ± 5.2 |
| Ba | 7i | -0.306 ± 0.101 | 1.864 ± 0.522 | – |
| Pr | 7i | – | – | 24. ± 21.4 |

established that an optical potential linear the nuclear density gave a good fit to the $\bar{p}$-atom data of the day and that its real part corresponded to an attractive potential. Since absorption plays a crucial role, the phenomenological potential contained also a very strong imaginary component. The most complete analysis of the data given in Table 16.4 has been presented by [Batty *et al.* (1995)]. To fit the $\bar{p}$-atomic data across the periodic table these authors introduced a phenomenological density-dependent $\bar{p}$-nucleus

Table 16.5 Compilation of $\bar{\text{p}}$ atomic data from LEAR as reported by [Trzcińska *et al.* (2001)] (since in this book we use Warsaw convention, our shift has opposite sign).

| Nucl | level | $\Gamma$(eV) | $\epsilon$(eV) | $\Gamma_u(eV)$ | $\epsilon_u$(eV) |
|---|---|---|---|---|---|
| $^{16}$O | 3d | 484±25 | -103±10 | | |
| $^{40}$Ca | 5g | | -5±12 | 0.059±0.018 | |
| $^{42}$Ca | 5g | | -17±14 | 0.080±0.028 | |
| $^{43}$Ca | 5g | | -62±30 | 0.073±0.042 | |
| $^{44}$Ca | 5g | | -31±10 | 0.077±0.023 | |
| $^{48}$Ca | 5g | | -33±12 | 0.116±0.017 | |
| $^{54}$Fe | 5g | 545±45 | -155±60 | 2.9±0.6 | |
| $^{56}$Fe | 5g | 545±54 | -167±22 | 3.3±0.5 | |
| $^{57}$Fe | 5g | 638±35 | -164±25 | 3.7±0.4 | |
| $^{58}$Fe | 5g | 2017±203 | 115±115 | 4.1±1.0 | |
| $^{59}$Co | 5g | 1370±370 | -254±111 | 5.2±0.7 | |
| $^{58}$Ni | 5g | 910±150 | -150±40 | 4.6±1.0 | |
| $^{60}$Ni | 5g | 1132±173 | -218±62 | 5.3±2.1 | |
| $^{62}$Ni | 5g | 1210±270 | -227±85 | 5.6±1.8 | |
| $^{64}$Ni | 5g | 1570±220 | -349±79 | 9.3±1.8 | |
| $^{90}$Zr | 6h | 1040±40 | -28±29 | 11.5±0.5 | |
| $^{96}$Zr | 6h | 1260±60 | -169±49 | 15.0±1.7 | |
| $^{106}$Cd | 7i | 199±60 | 26±20 | 3.8±0.5 | |
| $^{116}$Cd | 7i | 251±47 | 19±16 | 3.0±0.5 | |
| $^{112}$Sn | 7i | 387±17 | 5±11 | 4.2±0.6 | |
| $^{116}$Sn | 7i | 382±20 | -23±13 | 4.9±0.9 | |
| $^{120}$Sn | 7i | 474±21 | -31±13 | 5.6±0.6 | |
| $^{124}$Sn | 7i | 512±19 | -43±11 | 6.1±0.7 | |
| $^{122}$Te | 7i | 619±59 | -51±28 | 7.5±10 | 2±8 |
| $^{124}$Te | 7i | 548±51 | -51±17 | 7.5±12 | 4±5 |
| $^{126}$Te | 7i | 651±52 | -45±28 | 8.4±0.9 | -5±7 |
| $^{128}$Te | 7i | 627±60 | -64±18 | 11.1±1.7 | -15±4 |
| $^{130}$Te | 7i | 651±119 | -66±40 | 43.2±8.6 | 67±4 |
| $^{172}$Yb | 8j | 997±63 | 422±60 | 31±2 | 9±42 |
| $^{176}$Yb | 8j | 1121±60 | 342±36 | 37±2 | -10±31 |
| $^{208}$Pb | 9k | 312±26 | -95±35 | 5.9±0.8 | -38±18 |
| $^{209}$Bi | 9k | 506±50 | 8±53 | 6.9±1.3 | 14±20 |
| $^{232}$Th | 9k | 1534±206 | 1972±71 | 50.8±7.2 | 507±30 |
| $^{238}$U | 9k | 2362±221 | 2832±69 | 68.1±6.1 | 872±16 |

optical potential of the form

$$2\mu\, V(r) = -4\pi\,(1+\frac{\mu}{M})\left\{\left[b_0 + B_0\left(\frac{\rho(r)}{\rho(0)}\right)^{\alpha}\right]\rho(r) + b_1\,\delta\rho(r)\right\} \\ +4\pi(1+\frac{\mu}{M})^{-1}\,\boldsymbol{\nabla}[c_0\,\rho(r) + c_1\delta\rho(r)]\cdot\boldsymbol{\nabla}, \tag{16.71}$$

where $\mu$ is the p̄-nucleus reduced mass, M is the nucleon mass, and $\rho(r) = \rho_n(r) + \rho_p(r)$ and $\delta\rho(r) = \rho_n(r) - \rho_p(r)$ are the isoscalar and isovector density distributions, respectively. With $B_0$=0 the potential (16.71) reduces to the standard "t$\rho$" form involving the contribution from s- and p-waves. The latter generates the gradient terms in the potential (16.71). When the medium corrections are disregarded the coefficients $b_0(b_1)$, and $c_0(c_1)$ , may be tied up with the N̄N isocalar (isovector) scattering lengths and scattering volumes, respectively. To this end, the spin-averaged N̄N scattering amplitude in the center-of-mass frame for a given isospin I is written in the form of a partial wave expansion

$$f^{(I)}(k, \cos\theta) = \sum_{\ell}(2\ell + 1)\, f_{\ell}^{(I)}(k)\, P_{\ell}(\cos\theta), \tag{16.72}$$

where $\theta$ and k are scattering angle and momentum, respectively. Now, the b and c coefficients in (16.71) are obtained by taking zero-momentum limit of the isospin-averaged amplitudes

$$b_0 = \lim_{k\to 0}\left[\frac{3}{4} f_0^{(1)}(k) + \frac{1}{4} f^{(0)}(k)\right]; \quad c_0 = \lim_{k\to 0}\frac{3}{k^2}\left[\frac{3}{4} f_0^{(1)}(k) + \frac{1}{4} f^{(0)}(k)\right];$$

$$b_1 = \lim_{k\to 0}\left[\frac{1}{4} f_0^{(1)}(k) - \frac{1}{4} f^{(0)}(k)\right]; \quad c_1 = \lim_{k\to 0}\frac{3}{k^2}\left[\frac{1}{4} f_0^{(1)}(k) - \frac{1}{4} f^{(0)}(k)\right].$$

The partial waves $f_{\ell}^{(I)}(k)$ have been calculated using various N̄N potentials available in the literature, and [Batty *et al.* (1995)] took

$$b_0 = (-0.9 + \mathrm{i}0.9)\,\mathrm{fm}; \qquad c_0 = (-0.8 + \mathrm{i}1.9)\,\mathrm{fm}^3, \tag{16.73}$$

as representative values. The above value of $b_0$ may be compared with the p̄p scattering length (taken with opposite sign)

$$-A_c = \lim_{k\to 0}\left(\tfrac{1}{2} f_0^{(1)}(k) + \tfrac{1}{2} f_0^{(0)}(k)\right),$$

whose value has been extracted from the p̄p X-ray spectrum (14.31). Despite different weights resulting from isospin averaging occurring in $b_0$ and in -$A_c$, their numerical values are quite close which indicates weak isospin dependence of the s-wave amplitudes.

Taking the simplest variant of the potential (16.71) in which only the term linear in the density is retained (i.e. setting $B_0$=0 and $c_0$=$c_1$=0), [Batty *et al.* (1995)] obtained a good fit to the data. However, the best fit value of $b_0$

$$b_0 = [(2.5 \pm 0.2) + \mathrm{i}\,(3.5 \pm 0.3)]\,\mathrm{fm}$$

is very different from the free-space value (16.73) and there is a change of sign in the real part of $b_0$. Similarly as for kaonic atoms, the simple first order optical potential is inadequate and the medium corrections are very important. Including an isovector term $b_1$ brings only a marginal improvement and the $b_1$ parameter is not determined at all by the data. Leaving out the isovector correction, i.e. setting $b_1$=0, but including the p-wave contribution in (16.71) brings significant reduction of $\chi^2$ and in this case [Batty *et al.* (1995)] obtained the following values for their best-fit parameters

$$b_0 = [(4.7 \pm 0.4) + \mathrm{i}\,(5.0 \pm 0.6)]\,\mathrm{fm}; \tag{16.74a}$$

$$c_0 = -[(4.3 \pm 0.2) + \mathrm{i}\,(3.1 \pm 0.8)]\,\mathrm{fm}^3. \tag{16.74b}$$

We note here not only a change of sign in Re $b_0$ but also that the parameter $c_0$ acquires a negative imaginary part which for free-space scattering would have been totally inadmissible as it corresponds to emissive rather than absorptive interactions.

With the added *ad hoc* $B_0$ term in their optical potential (16.71), [Batty *et al.* (1995)] managed to find several families of density dependent potentials offering a good fit to the data. However, since this analysis had been effected before the new LEAR data were made available, there is not much point in discussing here all the details as the fits based on the full sample might change the situation. Nevertheless, it is interesting to note that [Batty *et al.* (1995)] succeeded in obtaining the so called constrained fits, where the parameters $b_0$ and $c_0$ were held fixed at the values (16.73) corresponding to free-scattering, and where the parameters $B_0$ and $\alpha$ were varied. Whilst these density dependent potentials do not have emissive components and are certainly more satisfactory, the interpretation of the $B_0$ term is not clear and we are still far from understanding $\bar{\mathrm{p}}$-atoms phenomena at a microscopic level.

## 16.5 $\Sigma^-$ Atoms

Because of a very short lifetime of the hyperons $\sim 10^{-10}$ sec stopping hyperon beams are not available. Nevertheless, $\Sigma^-$ atoms can be formed in result of secondary interactions involving $\mathrm{K}^-$ stopping in the target.

Through the K$^-$p→ $\Sigma^-\pi^+$ reaction channel a sufficient number of $\Sigma^-$ is produced and they have chances to be trapped into atomic orbits in the target material. Therefore, the X-ray spectra from both $\Sigma^-$ and K$^-$ atoms may observed in the same experiment. In practise, this coexistence limits the accuracy of the $\Sigma^-$ data since most of the X-ray lines are relatively weak and must be discerned from the much stronger K$^-$ atomic transitions. The data base for $\Sigma^-$ atoms is presented in Table 16.6, the appropriate measurements have been conducted at CERN, Rutherford-Appleton Laboratory and at Brookhaven. The $\Sigma^-$ atoms data together with $\Sigma$ hyper-

Table 16.6 Compilation of $\Sigma^-$ atom data ( CERN cf. [Backenstoss *et al.* (1975)], RAL cf. [Batty *et al.*(1978)], BNL cf. [Powers *et al.* (1993)]).

| Nucl | level | $\epsilon$ (eV) | $\Gamma$ (eV) | $\Gamma_u$ (eV) | ref |
|---|---|---|---|---|---|
| $^{12}$C | 3d | - | - | 0.031 ± 0.012 | CERN |
| $^{16}$O | 3d | 320.± 230. | - | 1.0 ± 0.7 | RAL |
| $^{24}$Mg | 4f | 25. ± 40. | - | 0.11 ± 0.09 | RAL |
| $^{27}$Al | 4f | 68. ± 28. | 43. ± 75. | 0.24 ±0.06 | RAL |
| $^{28}$Si | 4f | 169.± 36. | 220.±110. | 0.41 ±0.1 | RAL |
| $^{32}$S | 4f | 360. ± 220. | 870.±700. | 1.5 ± 0.8 | RAL |
| $^{40}$Ca | 5g | - | - | 0.41± 0.22 | CERN |
| $^{48}$Ti | 5g | - | - | 0.65± 0.42 | CERN |
| $^{138}$Ba | 8j | - | - | 2.9 ± 3.5 | CERN |
| $^{184}$W | 9k | 214.± 60. | 18. ± 149. | 2. ± 2. | BNL |
| $^{208}$Pb | 9k | 422.± 56. | 428.±158. | 17. ± 3. | BNL |

nuclear data are the only source of information on the low-energy nuclear interaction of the $\Sigma$ hyperon.

Our knowledge of the free-space $\Sigma$N and $\Lambda$N interactions is very limited but there is some cross-section data bearing on the s-wave YN interaction ($\Sigma$ and $\Lambda$ are denoted collectively by the symbol Y). The $\Sigma$N data are very scarce, consisting only of total cross sections for $\Sigma^-p \to \Sigma^-p, \Sigma^0n, \Lambda n$ and $\Sigma^+p \to \Sigma^+p$ reactions in the momentum range 110-180 MeV/c, and some differential cross sections. The momentum range is limited by the production momentum of the $\Sigma$ in the $K^-p \to \Sigma^\pm\pi^\mp$ capture reaction at rest and is equal 173.3 MeV/c and 181.5 MeV/c, respectively, for $\Sigma^-$ and $\Sigma^+$. Needless to say, the present scattering data give no means for discriminating between singlet and triplet YN spin states. In this situation the values of the zero-energy parameters have been estimated from an analysis of the data that contains a great deal of theoretical input. The most extensive analyses of that type have been performed by the Nijmegen group. Their basic philosophy takes advantage of the fact that the NN data can be described well within an one-boson-exchange picture and therefore with the assistance of SU(3) and SU(6) fixing the relative strengths of the coupling constants, the model can be extended to incorporate YN interaction. The advantageous feature of this approach is that it allows to reduce dramatically the number of free parameters which need to be adjusted from the YN data. Actually, several Nijmegen YN potentials have been developed so far and they are known as the D and F models, the Soft-Core (SC) model and finally the New-Soft-Core (NSC) model (for a review, cf. [Rijken (2001)]). Unfortunately, the ΣN sector is not fixed well by the data and different Nijmegen models give different interactions. As a result, the ΣN zero-energy scattering parameters vary from one model to another, although the fits to the ΣN data are in each case equally good. Since the YN scattering data are so scarce, new data on other aspects of the YN interaction are of particular value; they might well favor one particular model and disagree with another and this is precisely where the $\Sigma^-$ atoms may well fit in.

The quantity directly bearing on the $\Sigma^-$-atom optical potential is the spin and isospin averaged ΣN scattering length $b_0$

$$b_0 = \tfrac{2}{3}(\tfrac{1}{4}\,a_3^s + \tfrac{3}{4}\,a_3^t) + \tfrac{1}{3}(\tfrac{1}{4}\,a_1^s + \tfrac{3}{4}\,a_1^t),$$

where the superfix s (t) denotes spin singlet (triplet) and 1 and 3 are isospin indices corresponding to the values $\frac{1}{2}$ and $\frac{3}{2}$, respectively. We note that the isoscalar scattering length $b_0$ is a complex quantity, reflecting the strong absorptive reaction $\Sigma N \to \Lambda N$ taking place for I=$\frac{1}{2}$, but such a $\Sigma \to \Lambda$ de-excitation reaction is forbidden for the $\Sigma^-$n and $\Sigma^+$p systems, by charge conservation. The estimate from Model F gives $b_0 = (1.04 + \mathrm{i}\,0.45)$ fm but a recent one-boson-exchange calculation reported by the Jülich [Reuber *et al.* (1994)] group calls for a bigger attraction with $b_0 = (1.38 + \mathrm{i}\,0.45)$ fm and [Batty *et al.* (1997)] eventually suggested that a representative value of the free-space ΣN isoscalar scattering length would be

$$b_0 \approx (1.2 + \mathrm{i}\,0.45)\ \mathrm{fm}.$$

The earlier $\Sigma^-$ atom data could be rather well reproduced by a "t$\rho$"potential of the standard form

$$2\mu\, V(r) = -4\pi\, b_0\, (1 + \frac{m}{M})\, A\, \rho(r), \tag{16.75}$$

with m being the sigma mass, where $b_0$ was regarded as as an adjustable parameter. The best fit of [Batty *et al.*(1978)] was obtained with

$$b_0 = [(0.35 \pm 0.04) + \mathrm{i}\,(0.19 \pm 0.03)]\ \mathrm{fm},$$

which is a long way from the free-space scattering length value. The attempts to include in this analysis the BNL data, comprising of the heavy W and Pb species, were less successful which provided incentive for studying more sophisticated forms of the optical potential. A good fit to all $\Sigma^-$ atom data has been obtained by using a density dependent potential of the form (16.61), except that the imaginary part must be supplemented by an isovector component. The main source of $\Sigma$ absorption is the $\Sigma N \to \Lambda N$ reaction (releasing about 80 MeV) but $\Sigma^-$ cannot be absorbed on a single neutron and therefore the imaginary part of the optical potential must be proportional solely to the proton density $\rho_p$. Formally, this is stipulated by setting Im $b_1$=-Im $b_0$. The density dependent potential used by [Batty *et al.* (1997)] was taken in the form

$$\begin{aligned} 2\mu\, V(r) = & -4\pi\,(1+\frac{\mu}{M})\left\{b_0 + B_0\left[\frac{\rho(r)}{\rho(0)}\right]^{\alpha}\right\}\rho(r) \\ & -4\pi\,(1+\frac{\mu}{M})\left\{b_1 + B_1\left[\frac{\rho(r)}{\rho(0)}\right]^{\alpha}\right\}\delta\rho(r), \end{aligned} \tag{16.76}$$

where $\rho(r) = \rho_p(r) + \rho_n(r)$ and $\delta\rho(r) = \rho_n(r) + \rho_p(r)$ with the neutron (proton) density is normalized to N(Z). Comparing with (16.75), the potential (16.76) contains two extra density-dependent terms proportional to $B_0$ and $B_1$. Fits to the data were made with and without $B_1$ for various combinations of the parameters. The addition of the density dependent terms allows to reproduce all $\Sigma^-$ atomic data quite well. The real part of the best fit optical potential determined by [Batty *et al.* (1997)] is repulsive in the nuclear interior changing into attraction at the nuclear surface. Very recently, the some conclusion has been reached by [Dąbrowski *et al.* (2002)], who have used the Nijmegen YN potentials to construct the $\Sigma^-$ optical potential. According to their calculations, among the different Nijmegen models only model F yields a repulsive optical potential in the nuclear interior. At the same time, model F leads to the best agreement with the atomic $\Sigma^-$ data.

## 16.6 Deeply Bound Pionic Atoms

The conventional method of producing pionic atoms by stopping negatively charged pions in the target material and subsequently observing the de-excitation X-ray spectrum has natural limitations in that the atomic cascade terminates at the so called last orbital from which the pion gets absorbed by the nucleus. Since all the deeper lying levels have large overlap with the nuclear density the absorption is so strong that there is no chance to see another X-ray transition and these states are unaccessible for the X-ray spectroscopy. Such states are usually referred to as *deeply bound pionic atom states.* At first sight, it might have appeared that these deeply bound levels should be so broad that there is little motivation for trying to detect them by using a different experimental method. However, elaborate calculations employing standard optical potentials revealed [Friedman *et al.* (1985); Toki *et al.* (1988)] that even in a heavy nucleus the lowest 1s or 2p atomic states should be sufficiently narrow to make them well defined, i.e. their widths were predicted to be considerably smaller than the level separation. In the explanation for this phenomenon offered by [Friedman *et al.* (1985)] the key role plays the repulsive component present in the optical potential experienced by the pion. As a result, the pion is pushed out towards the nuclear periphery where the absorptive part of the optical potential is considerably reduced, as compared with that from the nuclear interior. Formally, this follows from (2.16), where the reduction of $\Gamma$ is a consequence of the fact that the pion density distribution peaks in the nuclear periphery and the overlap with the nuclear density is significantly reduced. Since the predicted deeply bound pionic atom levels could not have been reached by X-ray spectroscopy, to populate them several alternative methods have been proposed (cf. [Toki *et al.* (1988); Toki *et al.* (1989)]), in particular pion transfer reactions such as (n,p) and (d,$^2$He), or proton pick-up processes like (n,d) and (d,$^3$He) (cf. Fig. 16.7). In the latter reaction, pionic atom is formed by converting a neutron form a target nucleus into a pion while the emerging proton is picked-up by the projectile. There have been also other suggestions of producing deeply bound pionic states, e.g. by $(\gamma, \pi^+)$ and (e, e') reactions (cf.[Nieves *et al.* (1990); Nieves *et al.* (1991)]), radiative pion capture in-flight $(\pi^-, \gamma)$ [Nieves *et al.* (1992); Nieves *et al.* (1993)] and via the decay of hypernuclei [Bando *et al.* (1989)].

The pion transfer reaction experiments and the radiative process $(\pi^-, \gamma)$ did not bring a clear evidence for the production of deeply bound pio-

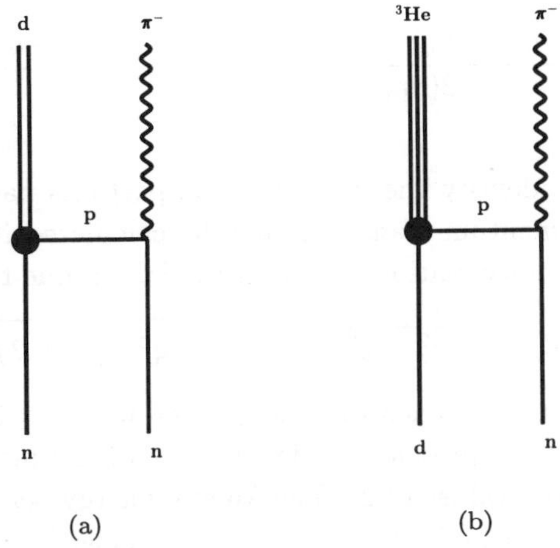


Fig. 16.7 The simplest proton pick-up diagrams

nic states. The experiments using (n,d) reaction at TRIUMF [Yamazaki (1993)] and the (p,$^2$He) experiment at RCNP [Matsuoka *et al.* (1995)] both failed to identify discrete levels but observed an increased yield below threshold for free $\pi^-$ production. In the first successful observation of deeply bound $\pi^- \otimes {}^{207}$Pb states, the reaction $^{208}$Pb(d,$^3$He) was used and the experiment was performed with the Fragment Separator using the deuteron beam from the synchrotron facility at GSI [Yamazaki *et al.* (1996); Gilg *et al.* (2000)]. Recently, this experiment was also repeated [Geissel *et al.* (2002)] for the $^{206}$Pb target and well separated 1s and 2p states of $\pi^- \otimes {}^{205}$Pb were observed. The extracted binding energies and the widths are:

$$B_{1s} = 6.762 \pm 0.061 \text{ MeV}; \qquad \Gamma_{1s} = 0.764 {}^{+0.171}_{-0.154} \text{ MeV};$$

and

$$B_{2p} = 5.110 \pm 0.045 \text{ MeV}; \qquad \Gamma_{2p} = 0.321 {}^{+0.060}_{-0.062} \text{ MeV}.$$

The reaction (d,$^3$He) has several advantageous features, but what is particularly important, with a special choice of kinematics it allows to make the pion production recoilless. Indeed, quite generally, when the ejectile is more massive than the projectile, at certain "magic" energy the momentum transfer can be brought to zero. In the concrete n(d,$^3$He)$\pi^-$ reaction the "magic" kinetic energy of the projectile $T_d$ is

$$\frac{(m_\pi + m_{He})^2 - (m_n + m_d)^2 + 2m_\pi(m_n + m_d - m_\pi - m_{He})}{2(m_n - m_\pi)} = 455\,\text{MeV},$$

where the indices identify the masses of the particles participating in the reaction. The momentum transfer q may be calculated from the equation expressing energy conservation, which in the lab. frame takes the form

$$m_d + T_d + m_n = \sqrt{m_\pi^2 + q^2} + \sqrt{m_{He}^2 + q^2 + p^2 - 2\,p\,q\,z}, \tag{16.77}$$

where p is the incident momentum and z=cos $\theta$ with $\theta$ being the scattering angle. In Fig. 16.8 we present the dependence of the momentum transfer q upon $T_d$ for three values of z. The kinetic energy selected for the GSI experiment was 604.3 MeV, not far from the "magic" energy so that the pion momentum was still small in comparison with the Fermi momentum.

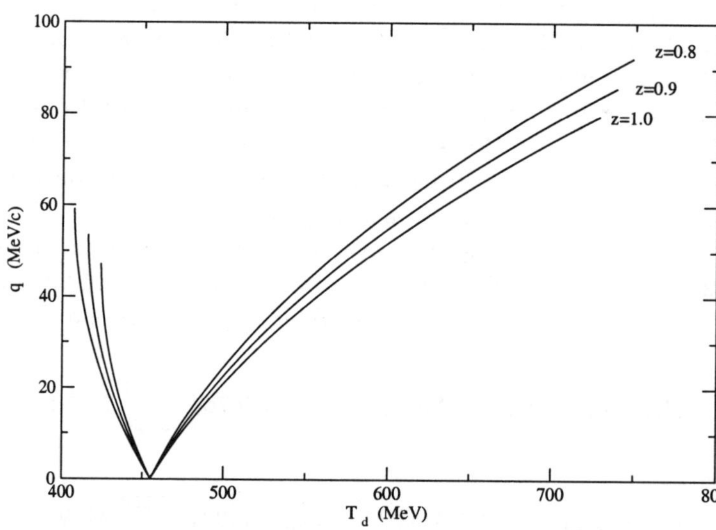


Fig. 16.8 Pion momentum transfer q versus deuteron kinetic energy $T_d$ for the elementary $d + n \to {}^3He + \pi^-$ process (z=cos $\theta$ where $\theta$ is the scattering angle).

The measured excitation spectrum, i.e. the double differential cross section as a function of the excitation energy, contains contributions which originate from quasi-free pion production, formation of bound pionic states and non-pionic nuclear excitation. In order to deduce the binding energies and the widths of the deeply bound pionic atom states it was necessary to confront the measured spectrum with the calculated spectrum obtained by using the existing pion optical potentials. The theoretically predicted spectra agree remarkably well with experiment which shows that the deeply bound $\pi^-$ states (1s and 2p) in heavy nuclei exist and that the mechanism of their formation has been well understood. Furthermore, the optical pion potential whose parameters have been adjusted to reproduce the pionic atom states established from X-ray spectroscopy, remains applicable also for the deeply bound levels.

# Bibliography


Abramowitz M. and I.A. Stegun (editors) *Handbook of Mathematical Functions* (Dover, New York, 1965).

Adeva, B. *et al.* (1995) CERN proposal, available from the Web page *http://dirac.web.cern.ch/DIRAC/*.

Adhikari, S. K. and Tomio, L. (1987) *Phys. Rev.* **C 36**, 1275.

Afanasyev, L. *et al.* (1993) *Phys. Lett.* **B 308**, 200.

Afnan, I. R. and Thomas, A. W. (1974) *Phys. Rev.* **C 10**, 109.

Anannthanarayan, B., Colangelo, G., Gasser, J. and Leutwyler, H. (2001) *Phys. Rep.* **353**, 207.

Babikov, V. V. *The Method of Phase Function in Quantum Mechanics*, 2nd ed., Nauka, Moscow 1976 (in Russian).

Backenstoss, G. (1970) *Ann. Rev. Nucl. Sci.* **20**, 467.

Backenstoss, G., Bunaciu, T., Egger, J. *et al.* (1975) *Z. Phys.* **A 273**, 137.

Badalyan, A. M., Kok, L. P., Polikarpov, M. I., and Simonov, Yu. A. (1982) *Phys. Rep.* **82**, 31.

Bahaoui, A., Fayard, C., Lamot, C.H. and Mizutani T. (1990) *Nucl. Phys.* **A505**, 335c.

Bahaoui, A., Fayard, C., Mizutani T. and Saghai, B. (2002) *Phys. Rev.* **C66**, 057001.

Baca, A., Garcia-Recio, C. and Nieves, J. (2000) *Nucl. Phys.* **A 673**, 335.

Bando, H., Yamazaki, Y. and Zofka, J. (1989) *Phys. Rev.***C40**, 875.

Bang, J., Ershov, S. N., Gareev, F. N. and G.S. Kazacha, G. S. (1980) *Nucl. Phys. A* **339**, 89.

Barnett, A. R., Feng, D. H., Steed, J. E., and Goldfarb, L. J. B. (1974) *Comput. Phys. Comm.* **8**, 377.

Barrett, R.C. and Deloff, A. (1999) *Phys. Rev.* **C60**, 025201.

Baru, V. V. and A. E. Kudryavtsev, A. E. (1997) *Phys. Atom. Nucl.* **60**, 1475.

Batty, C. J., Biagi, S. F., Blecher, M. *et al.* (1978) *Phys. Lett.* **B 74**, 27.

Batty, C. J. (1982) *Sov. J. Part. Nucl.* **13**, 71.

Batty, C. J. (1989) *Rep. Progr. Phys.* **52**, 1165.

Batty, C. J., Friedman, E and Gal, A. (1995) *Nucl. Phys.* **A 592**, 487.

Batty, C. J., Friedman, E. and Gal, A. (1997) *Phys. Rep.* **287**, 385.

Berggren, T. (1968) *Nucl. Phys. A* **109**, 265.

Berggren, T. (1971) *Nucl. Phys. A* **169**, 353.

Bertero, M. and Viano, G. A. (1967) *Nucl. Phys. B* **1**, 317.

Bertin, A. *et al.* (1996) *Phys. Lett.* **B 369**, 77.

Bianco, S., Bragadireanu, A. M., Fabbri, F.L, Guaraldo, C. *et al.* (1999) *Rivista N. Cim.* **22**, 1.

Bilenky, S. M., Nguen, V. K., Nemenov, L. L. *et al.* (1969) *Yad. Fiz.* **10**, 812.

Bird, P. M., Clough, A. S., Parker, K. R. *et al.* (1983) *Nucl. Phys.* **A404**, 482.

Blatt, J. M. (1967) *J. Comp. Phys.* **1**, 382.

Bohm, A. and Gadella, M. (1989) *Dirac kets, Gamow Vectors, and Gelfand Triplets*, Springer Lectures Notes in Physics Vol. 348, Springer, Berlin.

Brueckner, K. A. (1953) *Phys. Rev.* **89**, 834.

Bryan, R. A. and Phillips, R. J. N (1968) *Nucl. Phys.* **B 5**, 201; *ibid.* **B 7**, 481.

Burke, P. G. (1977) *Potential Scattering in Atomic Physics* Plenum, New York.

Calogero, F. (1967) *Variable Phase Approach to Potential Scattering*, Academic Press, New York.

Camac, M., McGuire, A. D., Platt, J.B. and Schulte H. J. (1952) *Phys. Rev.***88**, 134.

Chand, R. and Dalitz, R. H. (1962) *Ann. Phys. (N.Y.)* **20**, 1.

Carbonell, J., Protasov, K. V. and Zenoni, A. (1997) *Phys. Lett.* **B 397**, 345.

Caswell, W. E. and Lepage, G. P. (1986) *Phys. Lett.* **167B**, 437.

Chen, J. C. Y. and Chen, A. C. (1972) *Adv. At. Mol. Phys.* **8**, 71.

Cieplý, A., Fredman, E., Gal, A. and Mareš, J. (2001) *Nucl. Phys.* **A 696**, 173.

Conboy, J. E. Miller, D. J., Bedford, N. *et al.* (1986) *J. Phys. G* **12**, 1143.

Cornille, H. and Martin, A. (1962) *Nuovo Cimento* **26**, 298.

Ĉoté, J., Lacombe, M., Loiseau, B. *et al.* (1982) *Phys. Rev. Lett.* **48**, 1319.

Curtis, A. R. (1964) *Coulomb Wave Functions*, Vol II, Royal Soc. Math. Tables, Cambridge Univ. Press, London.

Dąbrowski, J. Rożynek, J and Anagnostatos, G. S. (2002) *Eur. Phys. J.* **A14**, 125.

Dalitz, R. H. and Deloff, A. (1986) Unpublished work.

Dalitz, R. H. and Tuan, S. F. (1960) *Ann. Phys. (N.Y.)* **10**, 307.

Dalitz, R. R., McGinley, J., Belyea, C., and Anthony, S. (1982) in *Proc. of Int. Conf. on Hypernuclear and Kaon Physics*, Heidelberg.

Davies, J. D., Pyle, G. J., Squier, G. T. A., Batty, C. J. *et al.* (1979) *Phys. Lett.* **83B**, 55.

de Alfaro, V. and Regge, T. (1965) *Potential Scattering* North-Holland, Amsterdam.

Deloff, A. (1976) *Phys. Rev. C* **13**, 730.

Deloff, A. (1978) *Nucl. Phys. A* **303**, 412.

Deloff, A. (1981) *Phys. Rev. C* **23**, 128.

Deloff, A. and Law, J (1980) *Phys. Rev. C* **21**, 2048.

de Maag, J.W., Kok, L.P. and Van Haeringen, H. (1984) *J. Math. Phys.* **25**, 684.

Deser, S., Goldberger, M. L., Baumann, K. and Thirring, W. (1954) *Phys. Rev.* **96**, 774.

Dover, C. and Richard, J. M. (1980) *Phys. Rev* **C 21**, 1466.

Eisenberg, J. M. Hüfner, J. and Moniz, E. J. (1973) *Phys. Lett.* **49B**, 381.
Eisenberg, J. M. and Koltun, D. S. (1980) *Theory of meson interactions with nuclei*, Wiley, New York.
Eisenstein, R. A., and Tabakin, F. (1976) *Comput. Phys. Comm.* **12**, 237.
Elster, Ch., Liu, L. C. and Thaler, R. M. (1993) *J. Phys. G* **19**, 2123
Erdelyi, A., Magnus, W., Oberhettinger, F. and Tricomi, F. (1953) *Higher Transcendental Functions*, Mc Graw-Hill, New York.
Ericson, T. E. O. and Weise, W. (1988) *Pions and Nuclei*, (Clarendon, Oxford).
Ericson, T.E.O and Hambro, L. (1977) *Ann. Phys (N.Y.)* **107**, 44.
Ericson, M. and Ericson, T. E. O. (1966) *Ann. Phys. (N.Y.)* **36**, 323.
Fäldt, G. (1977) *Physica Scripta* **16**, 81.
Ferreira, E. M. and Texeira, A. F. F. (1966) *J.Math. Phys.* **7**, 1201.
Feshbach, H. (1993) *Theoretcal nuclear physics II*, Wiley, New York.
Feynman, R. P. (1939) *Phys. Rev.* **56**, 340.
Foldy, L.L. and Walecka, J. D. (1969) *Ann. Phys. (N.Y.)* **54** 447.
Frazer, W. R. and Hendry, A. W. (1964) *Phys. Rev.* **134 B**, 1307.
Friedman, E. and Soff, G. (1985) *J. Phys. G Nucl. Phys.* **11**, L37.
Friedman, E., Gal, A. and Batty, C.J. (1994) *Nucl. Phys.* **A 579**, 518.
Fullerton, L. W. and Rinker, G. A. (1976) *Phys. Rev* **A13**, 1283
Gal, A., Toker, G. and Alexander, Y. (1981) *Ann. Phys. (N.Y.)* **137**, 341.
Gal, A. (2001) *Nucl. Phys.* **A691**, 268.
Gamow, G. (1928) *Z. Phys.* **51**, 204.
Garcia-Calderon, G. and Peierls, R. E. (1976) *Nucl. Phys. A* **265**, 443.
Gashi, A., Matsinos, E., Oades, G. C. *et al.*(2000) *Nucl. Phys A* (to be published), hep-ph/0009081.
Gashi, A., Oades, G.C., Rasche, G. and Woolcock, W. S. (1998) *Nucl. Phys.* **A628**, 101.
Gashi, A., Oades, G.C., Rasche, G. and Woolcock, W. S. (2002) *Nucl. Phys.* **A699**, 732.
Gasser, J. and Leutwyler, H. (1984) *Ann. Phys.* **158**, 142.
Gasser, J. and Leutwyler, H. (1985) *Nucl. Phys.* **B250**, 645.
Gasser, J., Lyubovitskij, V. E., Rusetsky, A. and Gall, A. (2001) *Phys. Rev.* **D 64**, 016008.
Geissel, H., Gilg, H., Gillitzer, A. *et al.* (2002) *Phys. Rev. Lett.***88**, 122301.
Gibbs, W. R., Ai Li and Kaufmann, W. B. (1998) *Phys. Rev.* C **57**, 784.
Gilg, H., Gillitzer, A., Knülle, M. *et al.* (2000) *Phys. Rev.* **C62**, 025201.
Goebel, C. (1958) *Phys. Rev. Lett.* **1**, 337.
Goldstone, J. (1961) *Nuovo Cim.* **19**, 154.
Goldstone, J., Salam, A. and Weinberg, S. (1962) *Phys. Rev.* **127**, 965.
Gotta, D., Anagostopoulos, D. F., Augsburger, M., *et al.* (1999) *Nucl. Phys.* **A 660**, 283.
Green, A. M. and Wycech, S. (1982) *Nucl Phys.* **377**, 441.
Hagiwara, K. et. al. (2002) *Phys. Rev.* **D66**, 010001.
Hamilton, J., Øverbö, I. and Tromborg, B. (1973) *Nucl. Phys. B* **60**, 443.
Harrington, D. (1965) *Phys. Rev.* **139**, B691.
Hauser, P., Kirch, K., Simons, L. M. *et al.* (1998) *Phys. Rev.* **C 58**, R1869.

Heitlinger, K., Backer, R., A. Badertscher, Blüm, P., *et al.* (1992) *Z. Phys. A* **342**, 359.
Heller, E. J. (1975) *Phys. Rev. A* **12**, 1222.
Hellmann, H. (1937) *Quantenchemie*, Deuticke and Co., Leipzig.
Hemingway, R. (1985) *Nucl. Phys.* **B253**, 742.
Hernandez, E. and Mondragon, A. (1984) *Phys. Rev. C* **29**, 722.
Hetherington, J. H. and Schick, L. H. (1965) *Phys. Rev.* **137**, B935.
Hetherington, J. H. and Schick, L. H. (1965) *Phys. Rev.* **165**, 1647.
Hokkyo, N. (1965) *Prog. Theor. Phys.* **33**, 1116.
Holstein, B. R. (1999) *Phys. Rev.* **D 60**, 114030.
Horvath, D. and Lambrecht, R. M. (1984) *Exotic Atoms - a bibliography 1939-1982*, Elsevier, Amsterdam.
Huang, K. (1963) *Statistical Mechanics*, (Wiley, New York), Chap. XIII.
Hull Jr., M. H. and Breit, G. (1959) in *Encyclopedia of Physics*, edited by S. Flügge, Springer, Berlin, Vol. XLI/1, p. 408.
Hulthen, L. (1948) *Ark. Mat. Astron. Fys. A* **35** No. 25.
Humblet, J. (1984) *Ann. Phys. (N.Y.)* **155**, 462.
Humblet, J. (1985) *J.Math. Phys.* **26**, 656.
Ito, T. M., Hayano, R. S., Nakamura, S. N., *et al.* (1998) *Phys. Rev.* **C 58**, 2366.
Iwasaki, M., Ito, T. M., Nakamura, S. N., Terada, T. P., *et al.* (1997) *Phys. Rev. Lett.* **78**, 3067.
Izycki, M., Backenstoss, G., Tauscher, L., Blüm P. *et al.* (1980) *Zs. Phys.* **A 297**, 11.
Jones, W. B. and Thorn, W. J. (1980) *Continued fractions Analytic Theory and Applications*, Addison-Wesley, Reading, Mass.
Kamalov, S.S., Oset, E. and Ramos, A. (2001) *Nucl. Phys.* **A690**, 494.
Karnakov, B. M., Kudryavtsev, A. E., Mur, V. D. and Popov, V. S. (1987) *Nuovo Cim. B* **98**, 63.
Karnakov, B. M., Mur, V. D. and Popov, V. S. (1985) *Yad. Fiz.* **41**, 1258.
Kaufmann, W. B. (1979) *Phys. Rev.* **C 19**, 440.
Kerman, A. K., McManus, H. and Thaler, R. M. (1951) *Ann. Phys. (N. Y.)* **8**, 551.
Kim, J. K. (1967) *Phys. Rev. Lett.* **19**, 1074.
Kisslinger, L. S. (1955) *Phys. Rev.* **98**, 761.
Kohn, W. (1948) *Phys. Rev.* **74**, 1763.
Kohono, M. and Weise, W. (1986) *Nucl. Phys.* **A454**, 429.
Kok, L. P. and van Haeringen, H. (1981) *Ann. Phys. (N.Y.)* **131**, 426.
Kok, L. P., de Maag, J. W., Brouwer, H. H. and van Haeringen, H. (1982) *Phys. Rev. C* **26**, 2381.
Kolybasov, V. M. and A. E. Kudryavtsev, A. E. (1973) *Sov. Phys. JETP* **36**, 18.
Konijn, J., De Laat, C., Taal, A. and Koch, J. H. (1990) *Nucl. Phys.* **A519**, 773.
Kónya, B., Lévai, G. and Papp, Z. (1997) *J. Math. Phys.* **38**, 4832.
Kónya, B., Lévai, G. and Papp, Z. (2000) *Phys. Rev. C* **61**, 034302.
Krell, M. (1971) *Phys. Rev. Lett.* **26**, 584.
Krell, M. and Ericson, T. E. O. (1968) *J. Comp. Phys.* **3**, 202.
Krell, M. and Ericson, T. E. O. (1969) *Nucl. Phys. B* **11**, 521.

Kwon, Y. R. and Tabakin, F. (1978) *Phys. Rev.* **C18**, 932.
Lacombe, M., Loiseau, B, Vinh Mau, R. *et al.* (1981) *Phys. Lett.* **B 101**, 139.
Lambert, E. (1969) *Helv. Phys. Acta* **42**, 677.
Landau, L. and Smorodinsky, J. (1944) *J. Phys. USSR* **8**, 154.
Landau, L. and Smorodinsky, J. (1944) *Zh. Eksp. Teor. Fiz.* **14**, 269.
Lippmann, B. A. and Schwinger, J. (1950) *Phys. Rev.* **79**, 469.
Lucchese, R. R., Takasuka, K. and McKoy, V. (1986) *Phys. Rep.* **131**, 147.
Lutz, M. F. M. and Korpa, C. L. (2002) *Nucl. Phys.* **A 700**, 309.
Machleidt, R., Holinde, K. and Elster, Ch. (1987) *Phys. Rep.* **149**, 1.
Mandelzweig, V. (1977) *Nucl. Phys. A* **292**, 333.
Martin, A. D. (1981) *Nucl. Phys.* **B179**, 33.
Martin, B. R., Morgan, D., and Shaw, G. (1976) *Pion–Pion Interactions in Particle Physics* Academic Press, London.
Matsuoka, N., Noro, T., Tamura, K. *et al.* (1995) *Phys. Lett.* **B359**, 39.
McVoy, K. W. (1967) in *Fundamentals in Nuclear Theory*, ed. by A. DeShalit and C. Villi IAEA, Vienna, p. 499.
Messiah, A. (1969) *Quantum Mechanics* North-Holland, Amsterdam.
Mizutani, T. and Koltun, D. (1977) *Ann. Phys. (NY)* **109**, 1.
Mukunda, N. (1978)*Am. J. Phys.* **46**, 910.
Muskhelishvili, N. I. (1953) *Singular Integral Equations* Nordhoff, Groningen.
Mutchler, G. S. *et al.* (1988) *Phys. Rev.* **D 38**, 742.
Nagarajan, M. A., Wang, W.L., Ernst, D.J. and Thaler, R.M. (1975) *Phys. Rev.* **C11**, 1167.
Newton, R. G. (1982) *Scattering Theory of Waves and Particles*, 2nd edition Springer, New York.
Nieves, J. and Oset, E. (1990) *Phys. Lett.* **B244**, 368.
Nieves, J. and Oset, E. (1991) *Phys. Rev.* **C43**, 1937.
Nieves, J. and Oset, E. (1992) *Phys. Lett.* **B282**, 24.
Nieves, J. and Oset, E. (1993) *Nucl. Phys.* **A553**, 595c.
Nikolayev, M. V. and Olkhovsky, V. S. (1973) *Lett. Nuovo Cimento* **8**, 703.
Nussenzveig, H. M. (1972) *Causality and Dispersion Relation* Academic Press, New York.
Oades, G. C. and Rasche, G. (1970) *Nucl. Phys.* **B20**, 333.
Ojha, P. C. (1986) *Phys. Rev. A* **34**, 969.
Oset, E. and Ramos, A. (2000) *Nucl. Phys.* **A 671**, 99.
Partensky, A. and Ericson, M. (1967) *Nucl. Phys. B* **1**, 382.
Peierls, R. E. (1948) *Proc. Cam. Phil. Soc.* **44**, 242.
Peresypkin, V. V. and Petrov, N. M. (1974) *Nucl. Phys.* **A 220**, 277.
Pislak, S., Appel, R., Atoyan, G. S. *et al.* (2001) *Phys. Rev. Lett.* **87**, 221801.
Powers, R. J., Eckhause, M., Guss, P. P. *et al.* (1993) *Phys. Rev.* **C 47**, 1263.
Press, W. H., Teukolsky, S. A, Vetterling, W. T. and Flannery, B. P. (1992) *Numerical Recipes the Art of Scientific Computing*, 2nd Edition, Cambridge Univ. Press, Cambridge.
Protasov, K. V., Bonomi, G., Lodi RIzzini, E. and Zenoni, A. (2000) *Eur. Phys. J.* **A 7**, 429.
Rijken, Th. J. (2001) *Nucl. Phys.* **A**

Reuber, A., Holinde, J. and Speth J. (1994) *Nucl. Phys.* **A 570**, 543.
Romo, W. (1968) *Nucl. Phys.* **A116**, 618.
Rosselet, L. *et al.* (1977) *Phys. Rev.* **D15**, 574.
Rotenberg, M. (1970) *Adv. At. Mol. Phys.* **6**, 233.
Roy, S. M. (1971) *Phys. Lett.* **B 36**, 353.
Sanatmarina, C. (2001) in *Workshop on Hadronic Atoms*, hep-ph/0112293.
Schröder, H. C., Badertscher, A., Goudsmit, P.F.A. *et al.* (1999) *Phys. Lett.* **B 469**, 25.
Schwinger, J. (1964) *J. Math. Phys.* **5**, 1606.
Seaton, M. J. (2002) *Comput. Phys. Comm.* **146**, 225.
Seki, R. (1971) *Phys. Rev. C* **5**, 1196.
Seki, R. and A.H. Cromer, A. H. (1987) *Phys. Rev.* **156**, 93.
Siegel, P. B. and Gibbs, W. R. *Phys. Rev. C* **33**, 1407.
Siegert, A. F. J. (1939) *Phys. Rev.* **56**, 750.
Sigg, D., Badertscher, A., Goudsmit, P. F. A., Leisi, H. J. L. and Oades, G. C. (1996) *Nucl. Phys. A* **609**, 310.
Smith, B. T., Boyle, J. M., Garbow, B. S. *et al.* (1976) *Matrix Eigensystem Routines — EISPACK Guide* 2nd Edition, vol. 6 of Lecture Notes in Computer Science, Springer-Verlag, New York.
Spuller, J. *et al.* (1977) *Phys. Lett. B* **67**, 479.
Toker, G. Gal, A. and Eisenberg, J.M. (1982) *Nucl. Phys.* **A362**, 405.
Toki, H. and Yamazaki, T. (1988) *Phys. Lett* **B123**, 129.
Toki, H., Hirenzaki, S., Yamazaki, T. and Hayano, R. S. (1989) *Nucl. Phys A* **501**, 563.
Thompson, I. J. and A.R. Barnett, A. R. (1986) *J.Comput. Phys.* **64**, 490.
Thompson, I. J. and Barnett, A. R. (1985) *Comput. Phys. Comm.* **36**, 363.
Torres, M, Dalitz, R. H. and Deloff, A. (1986) *Phys. Lett.* **174B**, 213.
Trueman, T. L. (1961) *Nucl. Phys.* **26**, 57.
Trzcińska, A., Jastrzębski, J., Czosnyka, T., *et al.* (2001) *Nucl. Phys.* **A 692**, 176c.
Uretsky, J. L. and Palfrey Jr., T. R. (1961) *Phys. Rev.* **121**, 1798.
van Haeringen, H. (1975) *Nucl. Phys. A* **253**, 355.
van Haeringen, H. (1977) *J.Math. Phys.* **18**, 927.
Watson, K. M. (1953) *Phys. Rev.* **89**, 575
Weinberg, S. (1963) *Phys. Rev.* **131**, 440.
Weinberg, S. (1967) *Phys. Rev. Lett.* **18**, 188.
Weinberg, S. (1968) *Phys. Rev.* **166**, 1568.
Weinberg, S. (1979) *Physica* **A96**, 327.
Wigner, E. P. (1948) *Phys. Rev.* **73**, 1002.
Wilkinson, J. H. and Reinsch, C. (1971) *Handbook of Automatic Computation* Springer-Verlag, New York.
Wycech, S. (2001) *Nucl. Phys.* **A 692**, 29c.
Yamani, H. A. and Fishman, L. (1975) *J. Math. Phys.* **16**, 410.
Yamazaki, T. (1993) *Nucl. Phys.* **A553**, 221c.
Yamazaki, T., Hayano, R. S., Itahashi, K. *et al.* (1996) *Z. Phys A* **355**, 219.
Yan, Y., Tegen, R., Gutsche, T. and Faessler, A. (1997) *Phys. Rev. C* **56**, 1596.

Zankel, H., Plessas, W. and Haidenbauer, J. (1983) *Phys. Rev.* **C 28**, 538.
Zel'dovich, Ya. (1959) *Fiz. Tverd. Tela (Leningrad)* **1**, 1637.
Zel'dovich, Ya. (1961) *Zh. Eksp. Teor. Fiz.* **39**, 776.
Zenoni, A., Bianconi, A., Bocci,. F *et al.* (1999) *Phys. Lett.* **B 461**, 405.

# Index